Artificial Intelligence and Algorithms with Their Applications

Artificial Intelligence and Algorithms with Their Applications

Guest Editors

Gang Hu
Lan Cheng
Guanqiu Qi

MDPI

Basel • Beijing • Wuhan • Barcelona • Belgrade • Novi Sad • Cluj • Manchester

Guest Editors

Gang Hu
Department of Computer Information Systems
SUNY Buffalo State University
Buffalo
USA

Lan Cheng
Department of Mathematical Science
State University of New York at Fredonia
Fredonia
USA

Guanqiu Qi
Department of Computer Information Systems
SUNY Buffalo State University
Buffalo
USA

Editorial Office
MDPI AG
Grosspeteranlage 5
4052 Basel, Switzerland

This is a reprint of the Special Issue, published open access by the journal *Mathematics* (ISSN 2227-7390), freely accessible at: https://www.mdpi.com/journal/mathematics/special_issues/TH1BV1E24B.

For citation purposes, cite each article independently as indicated on the article page online and as indicated below:

Lastname, A.A.; Lastname, B.B. Article Title. *Journal Name* **Year**, *Volume Number*, Page Range.

ISBN 978-3-7258-6874-2 (Hbk)
ISBN 978-3-7258-6875-9 (PDF)
https://doi.org/10.3390/books978-3-7258-6875-9

Contents

About the Editors

Gang Hu

Gang Hu is an Associate Professor in the Computer Information Systems Department at SUNY Buffalo State University and currently serves on the University AI Task Force. He received his Ph.D. in Computer Science from Dalhousie University. His research spans deep learning, machine learning, image and natural language processing, human–computer interaction, multimedia systems, and explainable AI (XAI). His current projects include research into AI-driven negotiation chatbots, autonomous vehicle navigation, human activity recognition, text-to-image generation, and symbolic edge token systems.

Lan Cheng

Lan Cheng is an Associate Professor of Applied Mathematics in the Department of Mathematical Sciences at SUNY Fredonia. She earned her Ph.D. in Applied Mathematics from the University of Pittsburgh. Her research interests include actuary sciences, stochastic processes, integral and partial differential equations, free boundary problems, asymptotic expansion, singular perturbation, variational inequalities, numerical computation and analysis, quantitative risk management, and econometrics. Her teaching focuses on calculus, mathematical modeling, interest theory, ordinary and partial differential equations, probability, and statistics.

Guanqiu Qi

Guanqiu Qi is currently an Associate Professor in the Computer Information Systems Department at SUNY Buffalo State University. He received his Ph.D. degree in Computer Science from Arizona State University in 2014. His primary research interests include deep learning, machine learning, and image processing, and also span many aspects of software engineering, such as SaaS (Software-as-a-Service), TaaS (Testing-as-a-Service), big data testing, combinatorial testing, and service-oriented computing. He has (co-)authored over 130 technical papers in these areas.

mathematics MDPI

Editorial

Preface for "Artificial Intelligence and Algorithms with Their Applications"

Gang Hu [1,*], Lan Cheng [2] and Guanqiu Qi [1]

1 Department of Computer Information Systems, SUNY Buffalo State University, Buffalo, NY 14222, USA
2 Department of Mathematical Science, State University of New York at Fredonia, Fredonia, NY 14063, USA
* Correspondence: hug@buffalostate.edu

Artificial Intelligence (AI) has become a driving force of advancements in science, engineering, and society. While recent progress has been fueled by large-scale data and massive computing resources, the long-term success of AI systems fundamentally depends on sound mathematical modeling, robust algorithms, and principled theoretical foundations. This Special Issue, *Artificial Intelligence and Algorithms with Their Applications*, brings together eleven peer-reviewed papers that collectively highlight the role of mathematical rigor and algorithmic innovation in modern AI research.

1. Perception and Computer Vision

An overarching theme of this collection is deep learning–based perception and computer vision. In Contribution 1, Turkmen and Akgun propose a hybrid UNet incorporating Transformer attention and a novel perceptual loss function for monocular depth estimation, achieving state-of-the-art accuracy on benchmark datasets. Complementing this, Hu (Contribution 2) provides a comprehensive mathematical survey of deep edge-detection algorithms, tracing the progression from early convolutional models to recent attention-based approaches and generative paradigms. This survey offers a valuable reference for researchers in symbolic and low-level vision.

Interpretability in vision models is addressed by Chavarro et al. (Contribution 3), who conduct a comparative evaluation of Class Activation Mapping (CAM) techniques. This work emphasizes the importance of eXplainable AI (XAI) in critical application domains, such as plant disease detection. Additionally, Yang et al. (Contribution 11) exploit local depth properties to enhance 3D feature matching.

2. Decision-Making and Control

Another salient focus of the Special Issue is learning-based decision-making and control. Autonomous decision-making under uncertainty has a wide application domain. Mei et al. (Contribution 4) have developed a deep reinforcement learning framework for autonomous air-combat maneuver generation, combining hierarchical policies, recurrent modeling, and self-play training to address challenges in complex, dynamic, and partially observable environments. From an industrial perspective, Contribution 5 introduces an AI-driven predictive maintenance framework supported by digital twin technology. This paper illustrates how incorporating machine learning, optimization, and system modeling can enhance reliability and efficiency in smart manufacturing.

3. Security and Trustworthy AI

Secure and trustworthy AI systems form a central pillar of modern research. Abdel-Basset et al. (Contribution 6) present a blockchain-integrated federated learning approach for pandemic diagnosis in smart cities, addressing data privacy and decentralization. Relatedly, Reddy et al. (Contribution 7) develop a hybrid fuzzy trust mechanism for IoT-embedded systems, while Kiran et al. (Contribution 10) enhance IoT security via multi-level blockchain and adaptive clustering.

4. Pattern Matching and Semantic Analysis

Finally, the Special Issue broadens its scope through pattern matching, text mining, and semantic analysis. Na et al. (Contribution 8) present an efficient approximate order-preserving pattern matching algorithm, which extends classical string-matching techniques to manage noisy and real-valued time-series data. In Contribution 9, a mathematically grounded framework is proposed for extracting spatial information from non-spatial text using conceptual spaces, entropy measures, and spatial statistics, highlighting the applicability of AI algorithms to natural language processing and semantic inference.

5. Conclusions

Collectively, the eleven papers in this Special Issue showcase the breadth, depth, and mathematical foundations of contemporary AI research. By addressing perception, learning, optimization, interpretability, and text analysis, this collection underscores the essential role of algorithms and applied mathematics in advancing reliable, explainable, and high-performance artificial intelligence systems.

Conflicts of Interest: The authors declare no conflict of interest.

List of Contributions:

1. Turkmen, H.; Akgun, D. A Hybrid UNet with Attention and a Perceptual Loss Function for Monocular Depth Estimation. *Mathematics* **2025**, *13*, 2567. https://doi.org/10.3390/math13162567.
2. Hu, G. A Mathematical Survey of Image Deep Edge Detection Algorithms: From Convolution to Attention. *Mathematics* **2025**, *13*, 2464. https://doi.org/10.3390/math13152464.
3. Chavarro, A.; Renza, D.; Moya-Albor, E. ConvNext as a Basis for Interpretability in Coffee Leaf Rust Classification. *Mathematics* **2024**, *12*, 2668. https://doi.org/10.3390/math12172668.
4. Mei, J.; Li, G.; Huang, H. Deep Reinforcement-Learning-Based Air-Combat-Maneuver Generation Framework. *Mathematics* **2024**, *12*, 3020. https://doi.org/10.3390/math12193020.
5. Nagy, M.; Figura, M.; Valaskova, K.; Lăzăroiu, G. Predictive Maintenance Algorithms, Artificial Intelligence Digital Twin Technologies, and Internet of Robotic Things in Big Data-Driven Industry 4.0 Manufacturing Systems. *Mathematics* **2025**, *13*, 981. https://doi.org/10.3390/math13060981.
6. Abdel-Basset, M.; Alrashdi, I.; Hawash, H.; Sallam, K.; Hameed, I.A. Towards Efficient and Trustworthy Pandemic Diagnosis in Smart Cities: A Blockchain-Based Federated Learning Approach. *Mathematics* **2023**, *11*, 3093. https://doi.org/10.3390/math11143093.
7. Reddy, N.V.R.; Padmaja, P.; Mahdal, M.; Seerangan, S.; Vimal, V.; Talasila, V.; Cepova, L. Hybrid Fuzzy Rule Algorithm and Trust Planning Mechanism for Robust Trust Management in IoT-Embedded Systems Integration. *Mathematics* **2023**, *11*, 2546. https://doi.org/10.3390/math11112546.
8. Na, J.C.; Kim, Y.; Kang, S.; Sim, J.S. Order-Preserving Pattern Matching with Partition. *Mathematics* **2024**, *12*, 3381. https://doi.org/10.3390/math12213381.
9. Abbasi, O.R.; Alesheikh, A.A.; Razavi-Termeh, S.V. Get Spatial from Non-Spatial Information: Inferring Spatial Information from Textual Descriptions by Conceptual Spaces. *Mathematics* **2023**, *11*, 4917. https://doi.org/10.3390/math11244917.

10. Kiran, A.; Mathivanan, P.; Mahdal, M.; Sairam, K.; Chauhan, D.; Talasila, V. Enhancing Data Security in IoT Networks with Blockchain-Based Management and Adaptive Clustering Techniques. *Mathematics* **2023**, *11*, 2073. https://doi.org/10.3390/math11092073.
11. Yang, E.; Chen, F.; Wang, M.; Cheng, H.; Liu, R. Local Property of Depth Information in 3D Images and Its Application in Feature Matching. *Mathematics* **2023**, *11*, 1154. https://doi.org/10.3390/math11051154.

mathematics

MDPI

Article

A Hybrid UNet with Attention and a Perceptual Loss Function for Monocular Depth Estimation

Hamidullah Turkmen [1] and Devrim Akgun [2,*]

1 Computer and Informatics Engineering Department, Institute of Natural Science and Technology, Sakarya University, Esentepe Campus, 54050 Serdivan, Sakarya, Türkiye; hamidullah.nazari1@ogr.sakarya.edu.tr
2 Software Engineering Department, Faculty of Computer and Information Sciences, Sakarya University, 54050 Serdivan, Sakarya, Türkiye
* Correspondence: dakgun@sakarya.edu.tr

Abstract

Monocular depth estimation is a crucial technique in computer vision that determines the depth or distance of objects in a scene using a single 2D image captured by a camera. UNet-based models are a fundamental architecture for monocular depth estimation, due to their effective encoder–decoder structure. This study presents an effective depth estimation model based on a hybrid UNet architecture that incorporates ensemble features. The new model integrates Transformer-based attention blocks to capture global context and an encoder built on ResNet18 to extract spatial features. Additionally, a novel Boundary-Aware Depth Consistency Loss (BADCL) function has been introduced to enhance accuracy. This function features dynamic scaling, smoothness regularization, and boundary-aware weighting, which provides sharper edges, smoother depth transitions, and scale-consistent predictions. The proposed model has been evaluated on the NYU Depth V2 dataset, achieving a Structural Similarity Index Measure (SSIM) of 99.8%. The performance of the proposed model indicates increased depth accuracy compared to state-of-the-art methods.

Keywords: monocular depth estimation; autonomous driving; transformer attention; hybrid UNet model; boundary-aware depth consistency loss

MSC: 68T07

1. Introduction

Depth estimation is an essential component of autonomous vehicle systems, allowing them to perceive their environment and make informed decisions accurately. Currently, various methods are used for depth estimation. For instance, point cloud LiDAR data [1] is a highly effective method to precisely and accurately estimate the depth of a scene. However, obtaining such detailed and highly accurate depth maps is often costly. High hardware costs, processing power requirements, and complexity limit active sensing methods such as LiDAR. This has directed researchers to search for more cost-effective, flexible, and efficient solutions. In this context, passive sensing methods like monocular depth estimation have emerged. These techniques, which use a single camera to estimate depth, are more affordable and versatile, making them suitable for many applications. Image-based depth estimation has evolved significantly from basic principles to today's advanced technologies. Early depth estimation techniques relied on the exploitation of depth cues like focus [2], vanishing points [3], and shadow information [4]. However, the applicability

 https://doi.org/10.3390/math13162567

of these methods has often been restricted to constrained scenes. The evolution of computer vision led to the adoption of handcrafted engineered features (SIFT [5], SURF [6], PHOG [7]) and probabilistic graphical models (CRF [8], MRF [9]) for monocular depth estimation. These techniques were often integrated with machine learning approaches to learning model parameters or non-parametric representations [8,9]. The emergence of deep learning algorithms has brought great advantages to image processing [10–12] and especially depth estimation methods [13]. The advent of neural networks, particularly Convolutional Neural Networks (CNNs) [14], has enhanced image depth detection capabilities. Different models, such as ResNet [15], VGG [16], MobileNet [17], encoder–decoder [18], and UNet [19] have been developed using CNN methods and components. Later, with the development of Transformer and Attention layers, monocular depth estimation studies began to be used in various applications such as autonomous driving, robotic navigation, and augmented reality.

Duong et al. [20] proposed URNet, an improved UNet architecture with a residual backbone for monocular depth estimation. This proposed model effectively handles depth estimation in autonomous vehicles, showing good results when tested on the KITTI dataset. A UNet++-like architecture integrating characteristics for lightweight and effective depth estimation was presented by Wang et al. [21]. In this work, they tried to increase the accuracy rate of depth estimation results by further reducing the parameters using light channel attention based on convolution in the decoder stage. Xue et al. [22], in their study, proposed Hierarchical Self-Supervised Monocular Absolute Depth Estimation for Autonomous Driving Applications (DNet), which enhances object-level depth estimation using hierarchical constraints and densely connected layers, ideal for self-supervised monocular depth estimation. Tadepalli et al. [23] proposed an EfficientNet-B0 backbone-based hybrid UNet architecture for monocular dense-depth map estimation. Their approach uses EfficientNet-B0 for efficient feature extraction and bilinear upsampling for depth map generation; they tried to achieve good results with fewer parameters. Li et al. [24] proposed an Adaptive Semantic Fusion Framework for unsupervised monocular depth estimation, integrating segmentation data to enhance depth consistency. This approach is effective for dealing with difficult scenarios. The paper [25] proposed a lightweight model called RCCNet that uses Reducing Channel Convolution (RCConv) to extract local features efficiently. RCCNet has achieved comparable results with existing methods while reasonably reducing the computational cost. In this study, Shen et al. [26] proposed a new model for autonomy drives called DNA-Depth. This method was developed for monocular depth estimation in night scenarios. It uses Fourier transform for domain alignment in order to handle the domain shift between day and night images. To deal with moving light sources, DNA-Depth uses an unsupervised joint learning architecture to jointly estimate depth, optical flow, and ego-motion. In the study [27], the Stereoscopic Pyramid Transformer-Depth (SPT-Depth) model is proposed for monocular depth estimation, which utilizes pyramid Transformers and multi-scale feature fusion to integrate shallow and deep semantic information. The model employs a unique training strategy with shift and scale-invariant loss functions and edge smoothing techniques to reduce noise and enhance robustness. On the NYU Depth V2 dataset, SPT-Depth has achieved a 10% reduction in Absolute Relative Error and a 17% reduction in RMSE. Sharma et al. [28] proposed the 2T-UNet model for robust stereo depth estimation, which uses twin convolution towers with different weights to avoid explicit stereo correspondence matching. To improve scene geometry prediction, 2T-UNet, in contrast to conventional stereo algorithms, inputs both the left and right stereo images along with monocular depth clue information. When tested on the difficult Scene Flow dataset, it outperformed the reasonable outcomes. In their study, Tang et al. [29] proposed CATNet for monocular depth estimation, handling the assignment as though it were an ordinal regression issue. The model uses an encoder–decoder architecture that minimizes complexity and parameters while trying to preserve depth estimation accuracy. CATNet incorporates a Multi-dimensional Convolutional Attention (MCA) module and a Dual Attention Transformer (DAT) module to enhance multi-scale feature refinement and global feature extraction. Experimental results on the KITTI and NYU datasets demonstrate that CATNet achieves performance comparable to state-of-the-art Transformer-based models. Robin et al. [30] proposed a monocular end-to-end depth estimation model called UCorr. This method has overcome challenges such as thin-wire obstacles in autonomous drone navigation. UCorr uses a temporal correlation layer trained on artificial data. The UCorr depth estimate shows how autonomous drone safety and accuracy might be enhanced in practical situations. A monocular depth estimation network called AMENet was suggested by Wu et al. [31] to increase autostereoscopic accuracy. The AMENet model is constructed by combining a CNN for local feature extraction and a Vision Transformer (ViT) for global semantic feature extraction. In contrast to other datasets, they have highlighted that the improved model demonstrated a discernible improvement in accuracy when evaluated on the KITTI dataset. In this study, Xi et al. [32] proposed a new approach for monocular depth estimation, called LapUNet, which considers the difficulties in preserving structural details and sharpness of edges in depth maps. In this hybrid model, they used a ResNeXt101 encoder and a decoder with a Dynamic Laplacian Residual U-shape module to improve the preservation of high-frequency features during upsampling. They tried to show that the experimental results on the NYU Depth v2 dataset are much better.

Depth estimation is a fundamental technique in computer vision that enables the extraction of three-dimensional scenes from two-dimensional images. Monocular depth estimation finds applications in various areas such as autonomous systems, robotics, augmented reality, and medical imaging, where accurately determining the distance of objects plays an important role. Increasing the accuracy of the models allows applications to make safe decisions when interacting with their environment. Traditional UNets are good at finding fine-grained local details, but they need to be improved to understand the more general scene context. While large UNet models can achieve high accuracy rates, they often demand considerable computational resources, reducing their practicality for real-time applications on resource-constrained hardware. This paper improves the limitations of current hybrid models in effectively integrating local and global information for depth estimation by designing a hybrid architecture. The new model uses a ResNet18 encoder for rich local feature extraction and integrates Transformer-based attention blocks, allowing the model to learn long-range dependencies and understand relationships between distant objects, improving scene understanding. Another improvement by the present work is the Boundary-Aware Depth Consistency Loss (BADCL) function, which combines a boundary-aware function to sharpen edges, a smoothness regularizer for uniform surfaces, and a dynamic scaling factor to ensure scale-consistent predictions.

The contributions of this paper can be summarized as follows:

- A new hybrid model combines a ResNet18-based encoder for extracting local spatial features with Transformer attention blocks that are designed to capture global context and long-range dependencies.
- A Boundary-Aware Depth Consistency Loss (BADCL) function has been introduced to improve training accuracy.
- Comparative results show that the proposed approach achieves similar results with fewer trainable parameters.

The remainder of this paper is structured as follows: Section 2 describes the materials and methods, providing an overview of the UNet model, the practical application of monocular depth estimation, and the NYU-Depth V2 dataset used for evaluation. Section 3 introduces the proposed method, describing the architecture of the Transformer-based hybrid UNet, the innovative Boundary-Aware Depth Consistency Loss (BADCL) function, and the metrics used for performance evaluation. In Section 4, we present the experimental results, demonstrating the efficiency of the model on the NYU Depth Dataset V2 using various test metrics. Finally, Section 5 concludes the paper by summarizing the key contributions and suggesting potential directions for future work.

2. Materials and Methods

2.1. UNet Model

The UNet architecture is a well-known and popular deep learning model. Its flexible design and ability to capture multi-scale spatial information make it an ideal choice for a variety of computer vision tasks, including depth estimation. The UNet model is made up of two main parts: the encoder and the decoder. The encoder uses convolutional layers to extract features and reduce spatial resolution. Conversely, the decoder uses transposed convolutions (up-convolutions) to gradually upsample the feature map and recover the original spatial dimensions. Skip connections between the encoder and decoder help preserve fine-grained information, which is crucial for accurate depth prediction. UNet is crucial for effectively combining local and global feature information, essential for accurate depth estimation. This multi-scale feature fusion capability has made the UNet suitable for monocular depth estimation. In the encoder block of the UNet, novel model architectures can be constructed by employing various backbone networks such as ResNet, VGG, MobileNet, etc. These approaches have become increasingly prevalent in contemporary deep learning applications, particularly in tasks like depth estimation.

2.2. Monocular Depth Estimation in Practice

Monocular depth estimation (single-camera depth estimation) extracts depth information from a single image and offers advantages over stereo vision, such as simplicity, cost-effectiveness, and versatility. Although it may be less accurate due to intrinsic uncertainty, advances in deep learning techniques are improving its accuracy and extending its applications in areas such as autonomous driving, robotics, and augmented reality. These applications require the depth information to perceive their environment and make accurate decisions. Monocular depth estimation is used in such systems because they offer lower-cost and lighter alternatives. However, obtaining depth information from a single camera can be difficult compared to sensors like stereo cameras or LiDAR, as it requires extracting 3D information from 2D images.

Monocular depth estimation uses deep learning methods to analyze 2D images and attempt to predict the depth value of each pixel in the scene, as shown in Figure 1. Such systems provide significant advantages, especially for low-cost applications. This technique is typically used in autonomous vehicles for applications such as environmental perception, road lane and obstacle detection, tracking, and navigation. The accuracy of monocular depth estimation is generally lower than that of stereo or LiDAR-based systems and may encounter some limitations. Therefore, many autonomous vehicles combine monocular depth estimation with other sensors (e.g., stereo cameras, LiDAR, or radar) to obtain better results [33]. However, monocular depth estimation enables depth information acquisition with low-cost camera systems and can be used instead of expensive sensors such as stereo cameras or LiDAR. However, accurate depth estimation from a single image is difficult due to scene complexity, lighting conditions, and visibility constraints. Therefore, researchers are working on deep learning-based methods to increase the accuracy of algorithms and improve real-time performance.

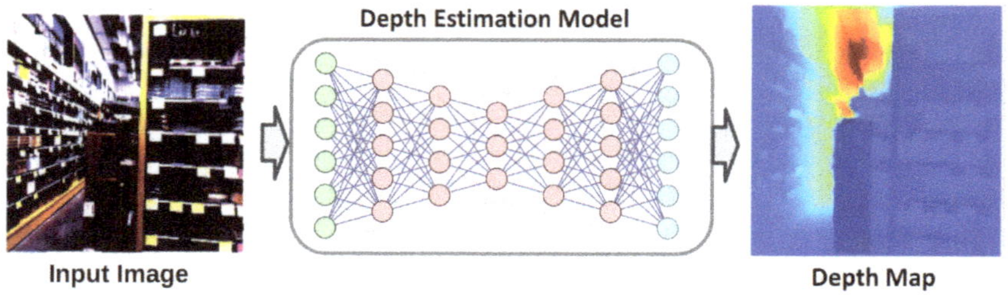

Figure 1. Monoculardepth estimation from a single camera image [34].

2.3. NYU-Depth V2

The NYU-Depth V2 [34] is a widely used dataset for depth estimation in autonomous and indoor scene understanding. This dataset consists of color image RGB and depth data from 464 different indoor scenes in three cities, captured using a Microsoft Kinect sensor. The dataset consists of 1449 RGB-D image pairs with dense pixel-level semantic annotations that enable multi-class segmentation. Missing depth values in the labeled data are preprocessed to ensure completeness, increasing its utility for supervised learning tasks. Additionally, 407,024 unlabeled frames from continuous video sequences are included to support unsupervised and self-supervised learning approaches. The dataset is organized by dividing the NYU Depth V2 training set into three distinct subsets. A random split is used to separate the data using the training set, with 70% allocated for the training set, 10% for the validation set, and the remaining 20% for the test set. The total number of samples for the dataset is split into 35,481 for train, 5068 for validation, and 10,139 for testing. Figure 2 shows sample images from the NYU-Depth V2 dataset.

Figure 2. Examples of RGB images and corresponding depth maps from NYUv2 dataset [34].

Before training the model, each image in the dataset is resized to a fixed resolution of 224 × 224 pixels, which is then applied to the ResNet backbone. After resizing, the pixel values are normalized using the mean and standard deviation derived from the ImageNet dataset. This step is crucial since the model uses weights pretrained on ImageNet. Finally, the processed image is converted into a PyTorch 2.5.1 tensor, preparing it for input into the neural network. For the depth maps, the preprocessing step involves a multi-step normalization process. First, the raw depth values, provided initially in millimeters, are converted to meters by dividing them by 1000. Next, these values are clipped to enforce a maximum depth of 10 meters, thereby addressing outliers or invalid measurements. The final step normalizes the depth map to a range of [0, 1] by dividing it by 10. No data augmentation is applied.

3. Proposed Method

3.1. Transformer-Based Hybrid UNet

The Transformer-enhanced UNet model improves the local feature extraction capabilities of the traditional UNet while including the global context understanding extracted by the Transformer mechanisms. This integration results in a more effective model for monocular depth estimation. The hybrid model is based on the encoder–decoder structure of UNet. In the encoder section, the ResNet18 model is used to extract multi-scale feature maps from the input image, providing local edge, texture, and structural information. In addition to local features, transfer blocks are used to incorporate global context, allowing the model to understand the relationships between distant regions. In particular, the Multi-Head Self-Attention (MHSA) [35] mechanism learns long-range relationships in the input data. MHSA uses the attention mechanism to focus on critical locations to comprehend the scene structure in detail.

In the decoder section, low-resolution feature maps are integrated with the global features processed by the Transformer. This combination allows for merging high-resolution depth maps, effectively preserving both local details and global context. Moreover, skip connections in the UNet enable the transfer of detailed spatial information from the encoder to the decoder. Therefore, skip connections enhance depth estimation accuracy, especially in critical areas such as object boundaries. The improved Boundary-Aware Depth Consistency Loss (BADCL) function of the proposed model further enhances accuracy at object boundaries while ensuring depth consistency in homogeneous regions. The Transformer-based hybrid lightweight architecture combines both local features and global context, contributing to monocular depth estimation applications in autonomous vehicles and robotic systems.

The proposed hybrid depth estimation model consists of three main components, the encoder, Transformer layers, and decoder, with fourteen layers. Based on the pre-trained ResNet18 backbone, the encoder includes five layers: ResNet Block 1 through ResNet Block 5, which progressively downsample the input image from 224×224 to 7×7 while extracting hierarchical features. Following the encoder, four Transformer attention layers are applied to the deepest features from the encoder to capture global context and long-range dependencies. Each Transformer layer uses MHSA and feedforward networks to improve the feature representations. The decoder then obtains the high-resolution depth map using four decoder blocks. These blocks perform upsampling operations and use skip connections from the encoder layers. This process ends with the output depth map layer, which produces a single-channel depth map at the original resolution of $224 \times 224 \times 1$. The model details are shown in Table A1 in Appendix A.

Figure 3 illustrates the overall structure of the proposed hybrid UNet model for depth estimation. The structure of the model is organized as follows: First, the encoder section includes ResNet blocks (from 1 to 4). These are followed by a FlattenBlock. Subsequently, Transformer blocks are employed to capture global context and long-range dependencies.

Finally, the decoder blocks complete the process by generating a depth map. Figure 4 explains the details of the blocks.

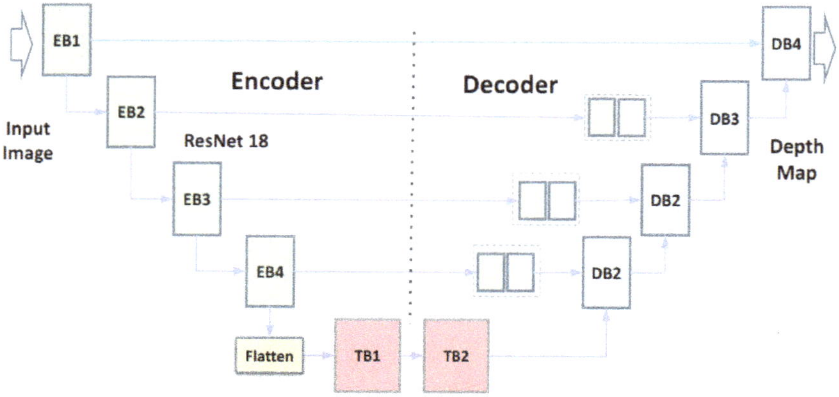

Figure 3. Proposed hybrid Unet depth estimation model.

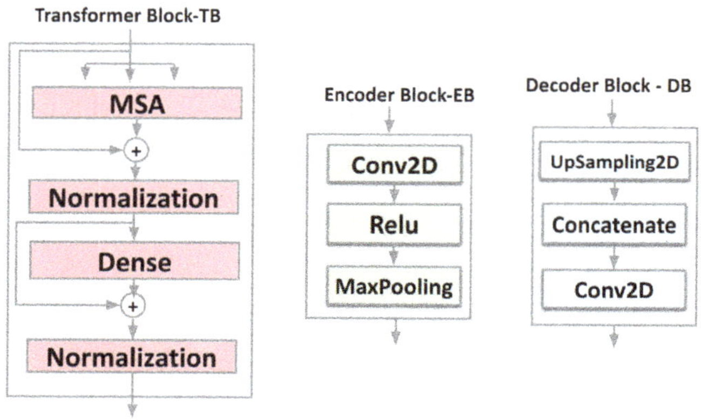

Figure 4. The content of the layers inside each block.

The encoder blocks consist of a Conv2D, ReLU, and a MaxPooling layer. The Flatten-Block takes the output of the previous layer as input, flattens it using an internal flatten layer, reshapes it, and then sends its output to the next block. The operation of the Transformer blocks involves the following steps: The features from the previous layer are first passed to the Multi-Head Self-Attention (MSA) layer, and the output is combined with the previous attention mechanism before being fed into the Normalization layer. The output from the Normalization layer is directed to the Dense layer. The output of the Dense layer is also combined with the previous attention mechanism and sent back to the Normalization layer. This process continues iteratively, providing input similar to that of the next layer. The decoder blocks contain an UpSampling2D, Concatenate, and Conv2D layer.

3.2. Hybrid Loss Function

In deep learning, the loss function is important for efficiently training the model parameters. Selecting a suitable loss function has the potential to improve the performance of the model, optimize its efficiency, and enhance its ability to focus on the intended target. It might not always be possible to provide pixel-level accuracy with a single loss function for intricate applications like depth estimation.

In this study, new loss functions were proposed. Within this scope, the Boundary-Aware Depth Consistency Loss (BADCL) is introduced by improving the boundary-aware [36] loss. BADCL is a loss function designed to enhance the accuracy of monocular depth estimation by highlighting the essential features of depth maps. It consists of 3 components: boundary awareness, smoothness regularization, and dynamic scaling to optimize monocular depth estimation.

3.2.1. Boundary-Aware Loss

Boundary-aware loss focuses on improving depth accuracy near high-gradient regions (e.g., object boundaries) where sharp changes in depth are crucial [37]. The mathematical equation is presented below:

$$L_{\text{boundary}} = \frac{1}{N}\sum_{i=1}^{N} B(i) \cdot |\hat{D}(i) - D(i)| \tag{1}$$

In this formula, $\hat{D}(i)$ represents the predicted depth value at pixel i, while $D(i)$ represents the corresponding ground truth depth value. The term $|\hat{D}(i) - D(i)|$ calculates the absolute error in depth prediction for pixel i. The boundary map $B(i)$ is a weighting factor in the loss function and indicates the importance of pixels near high-gradient areas of the image. The overall loss is computed by summing these weighted errors for all N pixels and normalizing by N.

$$B(i) = \exp\left(-\sqrt{\left(\frac{\partial I}{\partial x}\right)^2 + \left(\frac{\partial I}{\partial y}\right)^2}\right) \tag{2}$$

3.2.2. Smoothness Regularization Loss

Smoothness is only implemented in low-gradient regions, which are weighted by the image gradients [36].

$$L_{\text{smoothness}} = \frac{1}{N}\sum_{i=1}^{N}\left(\left|\frac{\partial \hat{D}}{\partial x}(i)\right| \cdot \exp\left(-\left|\frac{\partial I}{\partial x}(i)\right|\right) + \left|\frac{\partial \hat{D}}{\partial y}(i)\right| \cdot \exp\left(-\left|\frac{\partial I}{\partial y}(i)\right|\right)\right) \tag{3}$$

The predicted depth map is defined as $\hat{D}$, with its gradients along the horizontal and vertical directions represented as $\frac{\partial \hat{D}}{\partial x}$ and $\frac{\partial \hat{D}}{\partial y}$, respectively. These gradients show changes in the depth prediction. The image gradients, defined as $\frac{\partial I}{\partial x}$ and $\frac{\partial I}{\partial y}$, represent the intensity changes in the input image for the horizontal and vertical directions. The exponential terms $\exp\left(-|\frac{\partial I}{\partial x}|\right)$ and $\exp\left(-|\frac{\partial I}{\partial y}|\right)$ serve as weights that emphasize the smoothness loss in regions where the image gradients are low, indicating less texture or detail. The weighted depth gradients are summed for all N pixels and then normalized by N.

3.2.3. Dynamic Scaling Loss

Scales the predicted depth to align it with the ground truth [38]. The equation is given below.

$$L_{\text{scaling}} = \frac{1}{N}\sum_{i=1}^{N} s \cdot |\hat{D}(i) - D(i)| \tag{4}$$

For each pixel i, the term $|\hat{D}(i) - D(i)|$ calculates the absolute error between the predicted and the ground truth depth values. This error is scaled by a dynamic factor s, which aligns the overall scale of the predicted depth map with that of the ground truth. The dynamic scaling factor is calculated as follows:

$$s = \frac{\sum_{i=1}^{N} D(i)}{\sum_{i=1}^{N} \hat{D}(i) + \epsilon} \tag{5}$$

The scaling factor s is determined by the ratio of the sum of the ground truth depth values $\sum_{i=1}^{N} D(i)$ to the sum of the predicted depth values $\sum_{i=1}^{N} \hat{D}(i)$. A small constant ϵ is added to the denominator to prevent division by zero.

3.3. Combined Loss Functions

The total loss function combines the three components with weighting factors:

$$\begin{aligned} L_{\text{BADCL}} = \lambda_{\text{boundary}} \cdot L_{\text{boundary}} \\ + \lambda_{\text{smoothness}} \cdot L_{\text{smoothness}} \\ + \lambda_{\text{scaling}} \cdot L_{\text{scaling}} \end{aligned} \tag{6}$$

3.4. Hyperparameters

- $\lambda_{\text{boundary}}, \lambda_{\text{smoothness}}, \lambda_{\text{scaling}}$: Control the contribution of each term to the final loss.

The three components are combined with weighting factors to form the BADCL (Boundary-Aware Depth Consistency Loss) function. BADCL is designed to enhance the accuracy of monocular depth maps while ensuring the overall scale of the depth map is correct. As a result, monocular depth estimates provide more reliable and accurate predictions.

3.5. Performance Metrics

3.5.1. SSIM (Structural Similarity Index Measure)

SSIM is a metric used to evaluate the structural similarity between two images. It considers pixel-level differences and factors like luminance, contrast, and structural similarity. The Structural Similarity Index (SSIM) is defined as

$$\text{SSIM}(x, y) = \frac{(2\mu_x \mu_y + C_1)(2\sigma_{xy} + C_2)}{(\mu_x^2 + \mu_y^2 + C_1)(\sigma_x^2 + \sigma_y^2 + C_2)} \tag{7}$$

The luminance, represented by the means μ_x and μ_y of the two image patches x and y, measures the similarity in brightness levels. The contrast, determined by the standard deviations σ_x and σ_y, measures the difference in intensity variations between the images. The structural similarity, captured by the covariance σ_{xy}, reflects the correlation of pixel patterns and textures in the two images. Two stabilizing constants, $C_1 = (K_1 L)^2$ and $C_2 = (K_2 L)^2$, where L is the dynamic range of pixel values and K_1, K_2 are small constants, ensure numerical stability by preventing division by zero.

3.5.2. MSE (Mean Squared Error)

The MSE between the predicted depth map (x) and the ground truth depth map (y) is calculated as

$$\text{MSE} = \frac{1}{N} \sum_{i=1}^{N} (x_i - y_i)^2 \tag{8}$$

MSE is computed over all N pixels in the image, where N represents the total number of pixels. For each pixel i, the predicted depth value is defined as x_i, and the ground truth depth value is defined as y_i. The squared difference $(x_i - y_i)^2$ measures the error for pixel i, and the summation operator collects these errors over all pixels. Finally, dividing by N gives the average squared error to measure the quality of the result.

3.5.3. ARE (Absolute Relative Error)

The ARE metric for a predicted depth map (x) and a ground truth depth map (y) is calculated as

$$\text{ARE} = \frac{1}{N}\sum_{i=1}^{N}\frac{|x_i - y_i|}{y_i} \tag{9}$$

Here, N represents the total number of pixels in the depth map. For each pixel i, the predicted depth value is denoted as x_i, while the corresponding ground truth value is y_i. The absolute difference, $|x_i - y_i|$, measures the magnitude of the error at pixel i. Additionally, the term $\frac{|x_i - y_i|}{y_i}$ expresses this error relative to the true depth value. Summing these relative errors for all pixels and dividing by N gives the average relative error for the computed depth map. This measure is useful for evaluating depth estimation performance, where the magnitude of depth values varies throughout the map.

3.5.4. δ-Metrics

Delta Metrics (δ-Metrics) are crucial accuracy measures in depth estimation. They evaluate how closely predicted depth values align with ground truth values by measuring their adherence to specific error thresholds ($< \delta$). $\delta < 1.25$, $\delta < 1.25^2$, and $\delta < 1.25^3$ are widely used standards for evaluating depth estimation performance in the literature. Delta Metrics (δ) are defined as

$$\begin{aligned} \delta_k = {} & \text{Percentage of predictions } d_i \\ & \text{such that: } \max\left(\frac{d_i}{g_i}, \frac{g_i}{d_i}\right) < \delta^k \end{aligned} \tag{10}$$

For each prediction d_i and its corresponding ground truth depth value g_i, the term $\max\left(\frac{d_i}{g_i}, \frac{g_i}{d_i}\right)$ represents the relative error. This ensures that predictions close to the ground truth are considered accurate. The metric uses a threshold δ^k, where δ is a base threshold value (e.g., 1.25), and k defines the tolerance for larger errors. Typically, k can take values of 1, 2, or 3, representing increasingly lenient levels of error acceptance. The final value of δ_k indicates the percentage of predictions that meet condition $\max\left(\frac{d_i}{g_i}, \frac{g_i}{d_i}\right) < \delta^k$.

3.5.5. Log10 Error (Mean Logarithmic Error)

The Log10 Error provides a scale-invariant and relative accuracy metric for depth estimation. The Log10 Error is calculated as

$$\text{Log10} = \frac{1}{N}\sum_{i=1}^{N}|\log_{10}(d_i) - \log_{10}(g_i)| \tag{11}$$

The Log10 Error metric is applied to all N pixels in the depth map, where N represents the total number of pixels. The logarithmic transformation, represented as $\log_{10}(d_i)$ for the predicted values and $\log_{10}(g_i)$ for the ground truth values, ensures that the metric is scale-invariant. To calculate the Log10 Error, the sum operator accumulates the absolute differences $|\log_{10}(d_i) - \log_{10}(g_i)|$ for all pixels and divides by N.

4. Experimental Results

The experiments were carried out on a computer equipped with a 12-core CPU, 64 GB of RAM, and an NVIDIA® GeForce RTX™ 4090 24 GB memory. The operating system used was Ubuntu 22.04. The model implementation utilized Python 3.12.5 and the PyTorch library, specifically designed for deep learning applications. For training, the proposed model was trained for 30 epochs in the NYU Depth V2 dataset, which is a widely recognized dataset for in-depth estimation research. The loss function weights were empirically determined and set as follows: $\lambda_{\text{boundary}} = 3.0$, $\lambda_{\text{smoothness}} = 1.2$, and $\lambda_{\text{scaling}} = 1.0$. The constant ϵ in Equation (5), used for numerical stability, is set to 1×10^{-8}. The other training details like backbone, optimizer, and learning rate are summarized in Table 1.

Table 1. Training details.

Parameter	Value / Specification
Backbone	ResNet-18 (Pretrained on ImageNet)
Optimizer	AdamW (using PyTorch defaults)
Learning Rate	1×10^{-4} (constant, no schedule)
Batch Size	8
Number of Epochs	100
Random Seed	42 (for dataset splits)

The proposed hybrid model presented promising performance for various widely used monocular depth estimation metrics, including Root Mean Squared Error (RMS), Mean Relative Error (MRE), Mean Log10 Error, and thresholded accuracy measures ($\delta < 1.25$, $\delta < 1.25^2, \delta < 1.25^3$). In addition, the SSIM value, which highlights the model's ability to preserve spatial structure during depth prediction, has also increased to 0.998. Table 2 compares our model against other existing methods, while Figure 5 illustrates that after training the model for 100 epochs, the training loss approaches 0.0005, and the validation loss reaches approximately 0.0006. This consistent reduction in loss demonstrates the training convergence of the proposed model.

Table 2. Performance comparison of various existing methods for monocular depth estimation using error metrics and accuracy at different thresholds ($\delta < 1.25, \delta < 1.25^2, \delta < 1.25^3$). The proposed model outperforms the others.

Method	Error Metrics (Lower is Better)			Accuracy (Higher is Better)		
	ARE	**RMSE**	**Log10**	$\delta < 1.25$	$\delta < 1.25^2$	$\delta < 1.25^3$
Mancini et al. [39]	0.312	0.565	0.336	0.809	0.786	0.911
Xu et al. [40]	0.125	0.593	0.057	0.806	0.952	0.986
Alhashim et al. [41]	0.123	0.465	0.053	0.846	0.974	0.994
Li et al. (VGG16) [42]	0.152	0.611	0.064	0.789	0.955	0.988
Li et al. (VGG19) [42]	0.146	0.617	0.063	0.795	0.958	0.991
Li et al. (ResNet50) [42]	0.143	0.635	0.063	0.788	0.958	0.991
Rudolph et al. [43]	0.138	0.501	0.058	0.823	0.961	0.990
Lee et al. [44]	0.131	0.538	–	0.837	0.971	0.994
Guizilini et al. [45]	0.072	2.727	0.120	0.932	0.984	0.994
Basak et al. [46]	0.103	0.388	–	0.892	0.978	**0.995**
Das et al. (Enc-Dec-IRv2) [47]	**0.064**	**0.228**	0.032	0.893	0.967	0.985
Ignatov et al. [48]	0.090	0.322	0.039	0.929	0.991	**0.998**
Hybrid Ensemble Unet (Proposed)	**0.063**	0.237	**0.026**	**0.982**	**0.996**	**0.998**

The best result in each column is highlighted in bold.

The performance of the monocular depth estimation model can be improved by changing its architecture and loss function. In this study, the backbone of the proposed model was constructed using Resnet18, which has residual connections to help avoid the problem of vanishing gradients. In the experiments, the light model was observed to have the best results compared to the models in the literature. Furthermore, loss functions are critical in solving monocular depth estimation problems. The proposed hybrid loss function, a combination of boundary-aware loss, smoothness regularization loss, and dynamic scaling loss, considerably improved depth estimation accuracy.

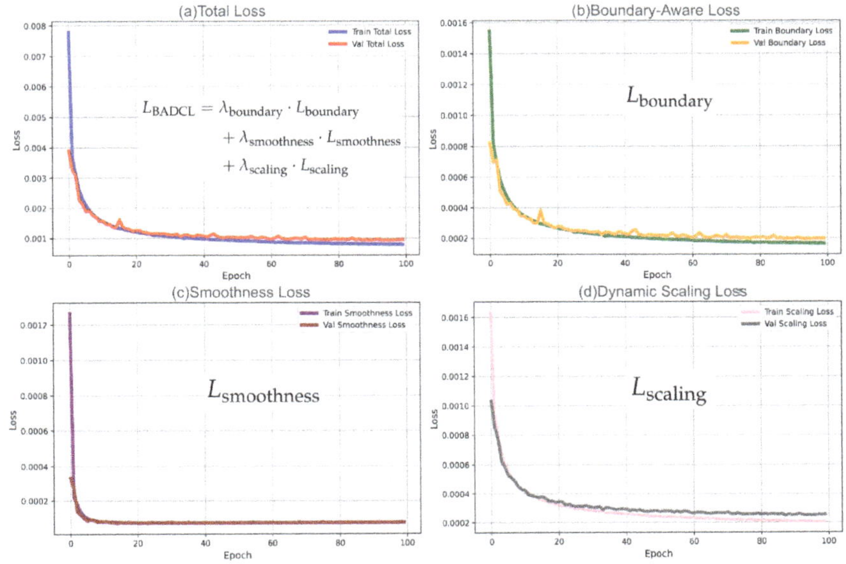

Figure 5. Training and validation loss plots for (**a**) total loss, (**b**) boundary-aware loss, (**c**) smoothness loss, and (**d**) dynamic scaling loss.

Although our method achieved good performance results as visually shown by Figure 6, a closer inspection of the predicted maps in complex scenes shows some areas for improvement. Specifically, our model tends to smooth over fine, high-frequency details in visually dense scenes such as the store aisle depicted in our results. This is the inherent difficulty in resolving ambiguous, repetitive patterns from a single 2D image, where the model may average out features to maintain overall structural consistency. A second limitation is observed with thin objects like chair legs, where predictions can show blurring at the boundaries. This is likely caused by the low pixel count of such structures, where the convolutional encoder fails to extract robust features without some spatial uncertainty.

The accuracy and efficiency of the model make it suitable for a range of applications. For instance, in augmented reality, accurate depth maps are essential for properly placing virtual objects within real-world scenes, ensuring realistic occlusion by actual objects. Since the model was evaluated on the NYU-Depth V2 indoor dataset, it is particularly well suited for indoor robotics, enabling tasks such as obstacle avoidance, mapping, and safe navigation in cluttered environments. Additionally, the model holds significant potential for autonomous driving systems, where accurate depth information is critical for environmental perception. However, because the model was trained and tested exclusively on the NYU-Depth V2 dataset, its performance in dynamic outdoor lighting and complex weather conditions remains untested. Furthermore, as a monocular system, it is inherently more susceptible to challenges due to scene complexity and visibility constraints compared to multi-sensor systems that utilize LiDAR or stereo cameras.

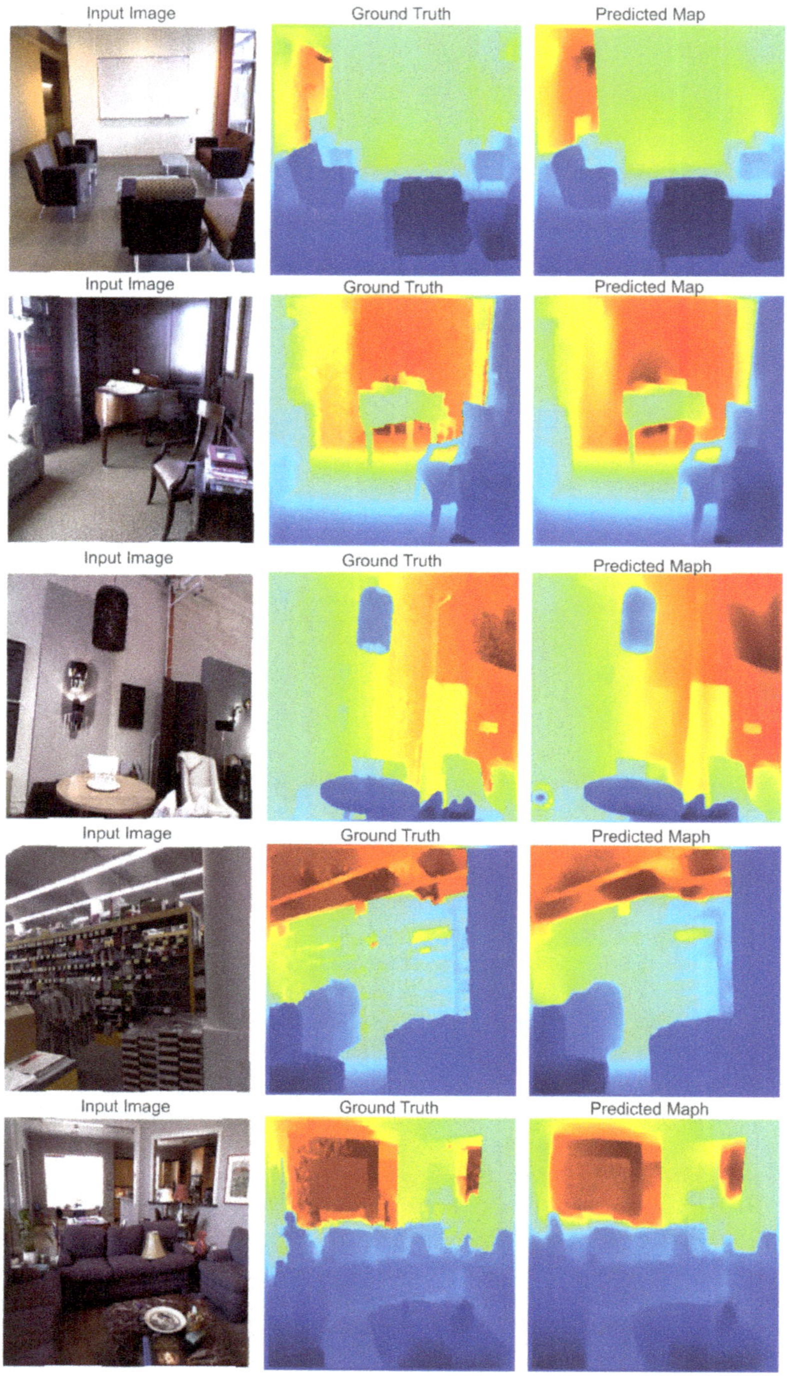

Figure 6. The figure presents input RGB images, ground truth depth maps, and predicted maps.

Computational Complexity

This study developed an efficient model suitable for real-time applications on resource-constrained hardware, utilizing a lightweight ResNet18 backbone. The resulting model comprises approximately 4.6 million parameters and achieves promising performance with fewer parameters than larger architectures. The model processes a fixed input resolution of 224×224 pixels. The operational load for a single forward pass is 43.02 Giga-Mult-Adds. Since each multiply-accumulate (mult-add) operation accounts for two floating-point operations (FLOPs), the model's overall complexity is approximately 86.04 GFLOPS. Therefore, the computational cost to process a single frame of 224×224 is equivalent to approximately 86.04 GFLOPS. The architecture itself contains 4.6 million trainable parameters, which provide a computationally efficient solution for monocular depth estimation.

Table 3 shows the experiments conducted for the FPS tests and provides a performance analysis of the model in real-time applications. While a single-image batch achieves a high throughput of nearly 250 images per second, the performance scales significantly with larger batch sizes, increasing to 1200 images per second with a batch size of 32. Furthermore, the latency, or the average time to process a single image, decreases considerably as the batch size increases. The processing time drops from 4 ms for a single image to under 1 ms per image for all batch sizes of 4 or greater. The performance gains begin to plateau at larger batch sizes (16 and 32), which suggests that the GPU is approaching its maximum utilization for this particular model. Overall, the results confirm that the model can operate beyond typical real-time thresholds such as 30–60 FPS.

Table 3. Model inference performance by batch size.

Batch Size	Throughput (Images/s)	Batches/s	Latency (ms/Image)
1	249.84	249.84	4.003
4	1011.47	252.87	0.989
8	1087.35	135.92	0.920
16	1176.84	73.55	0.850
32	1212.19	37.88	0.825

5. Conclusions

This study presents a novel hybrid depth estimation model aimed at improving the accuracy of monocular depth estimation for applications in areas such as autonomous driving and robotics. The model builds on existing methods by combining a ResNet18-based encoder for local feature extraction with Transformer attention layers to capture global contexts. Another improvement of the study is the BADCL, a new loss function that integrates boundary awareness, smoothness regularization, and dynamic scaling. This approach enhances depth estimation accuracy, especially around critical areas like object edges, while ensuring smoother predictions in uniform regions. Experimental evaluations on the NYU Depth V2 dataset showed quantitatively that the proposed model achieved an Absolute Relative Error (ARE) of 0.063, a Log10 error of 0.026, and a thresholded accuracy ($\delta < 1.25$) of 0.982. With a Structural Similarity Index (SSIM) of 99.8%, these results have significant implications for practical applications. The increased accuracy enables autonomous systems to make safer decisions when interacting with their environment. Additionally, the proposed method accomplishes these results with fewer trainable parameters, addressing the demand for efficient models suitable for real-time applications on resource-constrained hardware. Future work will explore more lightweight architectures and effective loss functions to enhance performance for different datasets.

Author Contributions: Conceptualization, H.T. and D.A.; methodology, H.T. and D.A.; software, H.T.; validation and data curation, H.T.; original draft preparation, H.T. and D.A.; review and editing, H.T. and D.A.; visualization, H.T. and D.A.; supervision, D.A. All authors have read and agreed to the published version of the manuscript.

Funding: This research received no external funding.

Data Availability Statement: The NYU2 dataset used in the study are openly available at https://cs.nyu.edu/~fergus/datasets/nyu_depth_v2.html [34] (accessed on 1 February 2025).

Conflicts of Interest: The authors declare no conflicts of interest.

Abbreviations

The following abbreviations are used in this manuscript:

AMENet	A Monocular Depth Estimation Network
ARE	Absolute Relative Error
BADCL	Boundary-Aware Depth Consistency Loss
CATNet	Convolutional Attention and Transformer
CNN	Convolutional Neural Network
CRF	Conditional Random Field
DAT	Dual Attention Transformer
DB	Decoder Block
DNet	Hierarchical Self-Supervised Monocular Absolute Depth Estimation
EB	Encoder Block
FLOPs	Floating-Point Operations
LapUNet	Laplacian Residual U-shape Network
LiDAR	Light Detection and Ranging
MCA	Multi-dimensional Convolutional Attention
MHSA	Multi-Head Self-Attention
MRF	Markov Random Field
MSE	Mean Squared Error
PHOG	Pyramid Histogram of Oriented Gradients
RCConv	Reducing Channel Convolution
REL	Mean Relative Error
ResNet	Residual Network
RMS	Root Mean Squared Error
SIFT	Scale-Invariant Feature Transform
SPT-Depth	Stereoscopic Pyramid Transformer-Depth
SSIM	Structural Similarity Index Measure
SURF	Speeded Up Robust Features
TB	Transformer Block
UNet	A U-shaped encoder–decoder network architecture
ViT	Vision Transformer

Appendix A

Table A1. Hybrid depth estimation model summary.

Layer (Type)	Output Shape	Param #
HybridDepthEstimationModel	[8, 1, 224, 224]	-
CNNEncoder	[8, 64, 56, 56]	-
Sequential	[8, 64, 56, 56]	-
Conv2d	[8, 64, 112, 112]	9408
BatchNorm2d	[8, 64, 112, 112]	128
ReLU	[8, 64, 112, 112]	-
MaxPool2d	[8, 64, 56, 56]	-
Sequential	[8, 64, 56, 56]	-
BasicBlock	[8, 64, 56, 56]	73,984
BasicBlock	[8, 64, 56, 56]	73,984
Sequential	[8, 128, 28, 28]	-
BasicBlock	[8, 128, 28, 28]	230,144
BasicBlock	[8, 128, 28, 28]	295,424
Sequential	[8, 256, 14, 14]	-
BasicBlock	[8, 256, 14, 14]	919,040
BasicBlock	[8, 256, 14, 14]	1,180,672
TransformerAttention	[8, 128, 28, 28]	-
MultiheadAttention	[8, 784, 128]	66,048
LayerNorm	[8, 784, 128]	256
Sequential	[8, 784, 128]	131,712
LayerNorm	[8, 784, 128]	256
TransformerAttention	[8, 256, 14, 14]	-
MultiheadAttention	[8, 196, 256]	263,168
LayerNorm	[8, 196, 256]	512
Sequential	[8, 196, 256]	525,568
LayerNorm	[8, 196, 256]	512
UpsampleBlock	[8, 128, 28, 28]	-
Conv2d	[8, 128, 28, 28]	442,496
BatchNorm2d	[8, 128, 28, 28]	256
Conv2d	[8, 128, 28, 28]	147,584
BatchNorm2d	[8, 128, 28, 28]	256
UpsampleBlock	[8, 64, 56, 56]	-
Conv2d	[8, 64, 56, 56]	110,656
BatchNorm2d	[8, 64, 56, 56]	128
Conv2d	[8, 64, 56, 56]	36,928
BatchNorm2d	[8, 64, 56, 56]	128
UpsampleBlock	[8, 64, 112, 112]	-
Conv2d	[8, 64, 112, 112]	73,792
BatchNorm2d	[8, 64, 112, 112]	128
Conv2d	[8, 64, 112, 112]	36,928
BatchNorm2d	[8, 64, 112, 112]	128
ConvTranspose2d	[8, 32, 224, 224]	32,800
Conv2d	[8, 1, 224, 224]	289
Total params: 4,653,313		
Trainable params: 4,653,313		
Non-trainable params: 0		
Total mult-adds (G): 43.02		

References

1. Li, Y.; Ma, L.; Zhong, Z.; Liu, F.; Chapman, M.A.; Cao, D.; Li, J. Deep learning for lidar point clouds in autonomous driving: A review. *IEEE Trans. Neural Netw. Learn. Syst.* **2020**, *32*, 3412–3432. [CrossRef]
2. Tang, C.; Hou, C.; Song, Z. Depth recovery and refinement from a single image using defocus cues. *J. Mod. Opt.* **2015**, *62*, 441–448. [CrossRef]
3. Tsai, Y.M.; Chang, Y.L.; Chen, L.G. Block-based vanishing line and vanishing point detection for 3D scene reconstruction. In Proceedings of the 2006 IEEE International Symposium on Intelligent Signal Processing and Communications, Yonago, Japan, 12–15 December 2006; pp. 586–589. [CrossRef]
4. Zhang, R.; Tsai, P.S.; Cryer, J.E.; Shah, M. Shape-from-shading: A survey. *IEEE Trans. Pattern Anal. Mach. Intell.* **1999**, *21*, 690–706. [CrossRef]
5. Lowe, D.G. Object recognition from local scale-invariant features. In Proceedings of the Seventh IEEE International Conference on Computer Vision, Kerkyra, Greece, 20–27 September 1999; Volume 2, pp. 1150–1157. [CrossRef]
6. Bay, H. Surf: Speeded up robust features. In *Computer Vision—ECCV*; Springer: Berlin/Heidelberg, Germany, 2006. [CrossRef]
7. Bosch, A.; Zisserman, A.; Munoz, X. Image classification using random forests and ferns. In Proceedings of the 2007 IEEE 11th International Conference on Computer Vision, Rio De Janeiro, Brazil, 14–21 October 2007; pp. 1–8. [CrossRef]
8. Lafferty, J.; McCallum, A.; Pereira, F. Conditional random fields: Probabilistic models for segmenting and labeling sequence data. In Proceedings of the ICML, Williamstown, MA, USA, 28 June–1 July 2001; Volume 1, p. 3.
9. Cross, G.R.; Jain, A.K. Markov random field texture models. *IEEE Trans. Pattern Anal. Mach. Intell.* **1983**, *PAMI-5*, 25–39. [CrossRef]
10. Güney, E.; Bayılmış, C.; Çakar, S.; Erol, E.; Atmaca, Ö. Autonomous control of shore robotic charging systems based on computer vision. *Expert Syst. Appl.* **2024**, *238*, 122116. [CrossRef]
11. Yolcu, G.; Oztel, I.; Kazan, S.; Oz, C.; Palaniappan, K.; Lever, T.E.; Bunyak, F. Facial expression recognition for monitoring neurological disorders based on convolutional neural network. *Multimed. Tools Appl.* **2019**, *78*, 31581–31603. [CrossRef] [PubMed]
12. Sazak, H.; Kotan, M. Automated Blood Cell Detection and Classification in Microscopic Images Using YOLOv11 and Optimized Weights. *Diagnostics* **2024**, *15*, 22. [CrossRef] [PubMed]
13. Yang, W.J.; Wu, C.C.; Yang, J.F. Residual Vision Transformer and Adaptive Fusion Autoencoders for Monocular Depth Estimation. *Sensors* **2024**, *25*, 80. [CrossRef]
14. O'Shea, K. An introduction to convolutional neural networks. *arXiv* **2015**, arXiv:1511.08458. [CrossRef]
15. Targ, S.; Almeida, D.; Lyman, K. Resnet in resnet: Generalizing residual architectures. *arXiv* **2016**, arXiv:1603.08029. [CrossRef]
16. Simonyan, K.; Zisserman, A. Very deep convolutional networks for large-scale image recognition. *arXiv* **2014**, arXiv:1409.1556. [CrossRef]
17. Sinha, D.; El-Sharkawy, M. Thin mobilenet: An enhanced mobilenet architecture. In Proceedings of the 2019 IEEE 10th Annual Ubiquitous Computing, Electronics & Mobile Communication Conference (UEMCON), New York, NY, USA, 10–12 October 2019; pp. 280–285. [CrossRef]
18. Badrinarayanan, V.; Kendall, A.; Cipolla, R. Segnet: A deep convolutional encoder-decoder architecture for image segmentation. *IEEE Trans. Pattern Anal. Mach. Intell.* **2017**, *39*, 2481–2495. [CrossRef] [PubMed]
19. Huang, H.; Lin, L.; Tong, R.; Hu, H.; Zhang, Q.; Iwamoto, Y.; Han, X.; Chen, Y.W.; Wu, J. Unet 3+: A full-scale connected unet for medical image segmentation. In Proceedings of the ICASSP 2020 IEEE International Conference on Acoustics, Speech and Signal Processing (ICASSP), Barcelona, Spain, 4–8 May 2020; pp. 1055–1059. [CrossRef]
20. Duong, H.T.; Chen, H.M.; Chang, C.C. URNet: An UNet-based model with residual mechanism for monocular depth estimation. *Electronics* **2023**, *12*, 1450. [CrossRef]
21. Wang, B.; Wang, S.; Dou, Z.; Ye, D. Deep Neighbor Layer Aggregation for Lightweight Self-Supervised Monocular Depth Estimation. *arXiv* **2024**, arXiv:2309.09272. [CrossRef]
22. Xue, F.; Zhuo, G.; Huang, Z.; Fu, W.; Wu, Z.; Ang, M. Toward Hierarchical Self-Supervised Monocular Absolute Depth Estimation for Autonomous Driving Applications. In Proceedings of the 2020 IEEE/RSJ International Conference on Intelligent Robots and Systems (IROS), Las Vegas, NV, USA, 25–29 October 2020; pp. 2330–2337. [CrossRef]
23. Tadepalli, Y.; Kollati, M.; Kuraparthi, S.; Kora, P. EfficientNet-B0 Based Monocular Dense-Depth Map Estimation. *Trait. Signal* **2021**, *38*, 1485–1493. [CrossRef]
24. Li, R.; Yu, H.; Du, K.; Xiao, Z.; Yan, B.; Yuan, Z. Adaptive Semantic Fusion Framework for Unsupervised Monocular Depth Estimation. In Proceedings of the ICASSP 2023 IEEE International Conference on Acoustics, Speech and Signal Processing (ICASSP), Rhodes Island, Greece, 4–10 June 2023; pp. 1–5. [CrossRef]

25. Dang, Y.; Li, C.; Zhang, L.; Gao, Y. RCCNet: Reducing Channel Convolution Network for Monocular Depth Estimation. In Proceedings of the 2023 4th IEEE International Conference on Computer Vision, Image and Deep Learning (CVIDL), Zhuhai, China, 12–14 May 2023; pp. 1–4. [CrossRef]
26. Shen, M.; Wang, Z.; Su, S.; Liu, C.; Chen, Q. DNA-Depth: A Frequency-Based Day-Night Adaptation for Monocular Depth Estimation. *IEEE Trans. Instrum. Meas.* **2023**, *72*, 2530112. [CrossRef]
27. Xia, Z.; Wu, T.; Wang, Z.; Zhou, M.; Wu, B.; Chan, C.; Kong, L.B. Dense monocular depth estimation for stereoscopic vision based on pyramid transformer and multi-scale feature fusion. *Sci. Rep.* **2024**, *14*, 7037. [CrossRef]
28. Sharma, M.; Choudhary, R.; Anil, R. 2T-UNET: A Two-Tower UNet with Depth Clues for Robust Stereo Depth Estimation. In Proceedings of the IEEE/CVF Conference on Computer Vision and Pattern Recognition, Seattle, WA, USA, 16–22 June 2024; pp. 757–764.
29. Tang, S.; Lu, T.; Liu, X.; Zhou, H.; Zhang, Y. CATNet: Convolutional attention and transformer for monocular depth estimation. *Pattern Recognit.* **2024**, *145*, 109982. [CrossRef]
30. Kolbeinsson, B.; Mikolajczyk, K. UCorr: Wire Detection and Depth Estimation for Autonomous Drones. In Proceedings of the International Conference on Robotics, Computer Vision and Intelligent Systems, Rome, Italy, 25–27 February 2024; Springer: Berlin/Heidelberg, Germany, 2024; pp. 179–192. [CrossRef]
31. Wu, T.; Xia, Z.; Zhou, M.; Kong, L.B.; Chen, Z. AMENet is a monocular depth estimation network designed for automatic stereoscopic display. *Sci. Rep.* **2024**, *14*, 5868. [CrossRef]
32. Xi, Y.; Li, S.; Xu, Z.; Zhou, F.; Tian, J. LapUNet: A novel approach to monocular depth estimation using dynamic laplacian residual U-shape networks. *Sci. Rep.* **2024**, *14*, 23544. [CrossRef]
33. Zou, Y.; Luo, Z.; Huang, J.B. Df-net: Unsupervised joint learning of depth and flow using cross-task consistency. In Proceedings of the European Conference on Computer Vision (ECCV), Munich, Germany, 8–14 September 2018; pp. 36–53.
34. Nathan Silberman, Derek Hoiem, Pushmeet Kohli ; Rob Fergus Indoor Segmentation and Support Inference from RGBD Images. In Proceedings of the ECCV, Florence, Italy, 7–13 October 2012. [CrossRef]
35. Wu, C.; Wu, F.; Ge, S.; Qi, T.; Huang, Y.; Xie, X. Neural news recommendation with multi-head self-attention. In Proceedings of the 2019 Conference on Empirical Methods in Natural Language Processing and the 9th International Joint Conference on Natural Language Processing (EMNLP-IJCNLP), Hong Kong, China, 3–7 November 2019; pp. 6389–6394. [CrossRef]
36. Borse, S.; Wang, Y.; Zhang, Y.; Porikli, F. Inverseform: A loss function for structured boundary-aware segmentation. In Proceedings of the IEEE/CVF Conference on Computer Vision and Pattern Recognition, Nashville, TN, USA, 20–25 June 2021; pp. 5901–5911.
37. Ngoc, M.Ô.V.; Chen, Y.; Boutry, N.; Chazalon, J.; Carlinet, E.; Fabrizio, J.; Mallet, C.; Géraud, T. Introducing the Boundary-Aware loss for deep image segmentation. In Proceedings of the British Machine Vision Conference (BMVC) 2021, Online, 22–25 November 2021.
38. Abdusalomov, A.; Umirzakova, S.; Shukhratovich, M.B.; Kakhorov, A.; Cho, Y.I. Breaking New Ground in Monocular Depth Estimation with Dynamic Iterative Refinement and Scale Consistency. *Appl. Sci.* **2025**, *15*, 674. [CrossRef]
39. Mancini, M.; Costante, G.; Valigi, P.; Ciarfuglia, T.A.; Delmerico, J.; Scaramuzza, D. Toward domain independence for learning-based monocular depth estimation. *IEEE Robot. Autom. Lett.* **2017**, *2*, 1778–1785. [CrossRef]
40. Xu, D.; Wang, W.; Tang, H.; Liu, H.; Sebe, N.; Ricci, E. Structured attention guided convolutional neural fields for monocular depth estimation. In Proceedings of the IEEE Conference on Computer Vision and Pattern Recognition, Salt Lake City, UT, USA, 18–22 June 2018; pp. 3917–3925. [CrossRef]
41. Alhashim, Ibraheem; Wonka, Peter High quality monocular depth estimation via transfer learning. *arXiv* **2018**, arXiv:1812.11941. [CrossRef]
42. Li, J.; Klein, R.; Yao, A. A two-streamed network for estimating fine-scaled depth maps from single rgb images. In Proceedings of the IEEE International Conference on Computer Vision, Venice, Italy, 22–29 October 2017; pp. 3372–3380.
43. Rudolph, M.; Dawoud, Y.; Güldenring, R.; Nalpantidis, L.; Belagiannis, V. Lightweight monocular depth estimation through guided decoding. In Proceedings of the 2022 IEEE International Conference on Robotics and Automation (ICRA), Philadelphia, PA, USA, 23–27 May 2022; pp. 2344–2350. [CrossRef]
44. Lee, J.H.; Kim, C.S. Monocular depth estimation using relative depth maps. In Proceedings of the IEEE/CVF Conference on Computer Vision and Pattern Recognition, Long Beach, CA, USA, 16–17 June 2019; pp. 9729–9738.
45. Guizilini, V.; Ambrus, R.; Pillai, S.; Raventos, A.; Gaidon, A. 3d packing for self-supervised monocular depth estimation. In Proceedings of the IEEE/CVF Conference on Computer Vision and Pattern Recognition, Seattle, WA, USA, 14–19 June 2020; pp. 2485–2494. [CrossRef]
46. Basak, H.; Ghosal, S.; Sarkar, M.; Das, M.; Chattopadhyay, S. Monocular depth estimation using encoder-decoder architecture and transfer learning from single RGB image. In Proceedings of the 2020 IEEE 7th Uttar Pradesh Section International Conference on Electrical, Electronics and Computer Engineering (UPCON), Prayagraj, India, 27–29 November 2020; pp. 1–6. [CrossRef]

47. Das, D.; Das, A.D.; Sadaf, F. Depth Estimation From Monocular Images with Enhanced Encoder-Decoder Architecture. *arXiv* **2024**, arXiv:2410.11610. [CrossRef]
48. Ignatov, D.; Ignatov, A.; Timofte, R. Virtually Enriched NYU Depth V2 Dataset for Monocular Depth Estimation: Do We Need Artificial Augmentation? In Proceedings of the Synthetic Data for Computer Vision Workshop@ CVPR 2024, Seattle, WA, USA, 17–21 June 2024.

mathematics

MDPI

Review

A Mathematical Survey of Image Deep Edge Detection Algorithms: From Convolution to Attention

Gang Hu

Department of Computer Information Systems, SUNY Buffalo State University, Buffalo, NY 14222, USA; hug@buffalostate.edu; Tel.: +1-716-8785112

Abstract

Edge detection, a cornerstone of computer vision, identifies intensity discontinuities in images, enabling applications from object recognition to autonomous navigation. This survey presents a mathematically grounded analysis of edge detection's evolution, spanning traditional gradient-based methods, convolutional neural networks (CNNs), attention-driven architectures, transformer-backbone models, and generative paradigms. Beginning with Sobel and Canny's kernel-based approaches, we trace the shift to data-driven CNNs like Holistically Nested Edge Detection (HED) and Bidirectional Cascade Network (BDCN), which leverage multi-scale supervision and achieve ODS (Optimal Dataset Scale) scores 0.788 and 0.806, respectively. Attention mechanisms, as in EdgeNAT (ODS 0.860) and RankED (ODS 0.824), enhance global context, while generative models like GED (ODS 0.870) achieve state-of-the-art precision via diffusion and GAN frameworks. Evaluated on BSDS500 and NYUDv2, these methods highlight a trajectory toward accuracy and robustness, yet challenges in efficiency, generalization, and multi-modal integration persist. By synthesizing mathematical formulations, performance metrics, and future directions, this survey equips researchers with a comprehensive understanding of edge detection's past, present, and potential, bridging theoretical insights with practical advancements.

Keywords: edge detection; computer vision; convolutional neural networks (CNNs); attention mechanisms; transformer; generative models; BSDS500; NYUDv2

MSC: 68T07

1. Introduction

Edge detection, a cornerstone of computer vision, involves identifying boundaries within images where intensity changes abruptly. It serves as a fundamental preprocessing step for numerous downstream tasks, including object recognition, image segmentation, and autonomous navigation [1–7]. These edges, conveying essential structural and perceptual information, have been one of the central focuses of research in computer vision and image processing for decades, evolving from hand-crafted filters to sophisticated deep learning models. Early methods like Sobel [8], Canny [9] and others [10–12] relied on gradient-based operations, offering simplicity but struggling with noise and complex scenes. The advent of convolutional neural networks (CNNs) marked a paradigm shift, introducing trainable feature extraction that significantly improved accuracy on benchmarks like BSDS500 [13]. More recently, attention mechanisms and generative approaches have pushed the frontier, achieving state-of-the-art performance by capturing global context and refining edge quality. Figure 1 shows the timeline of edge detection paradiams, from gadient-based methods to attention-based methods.

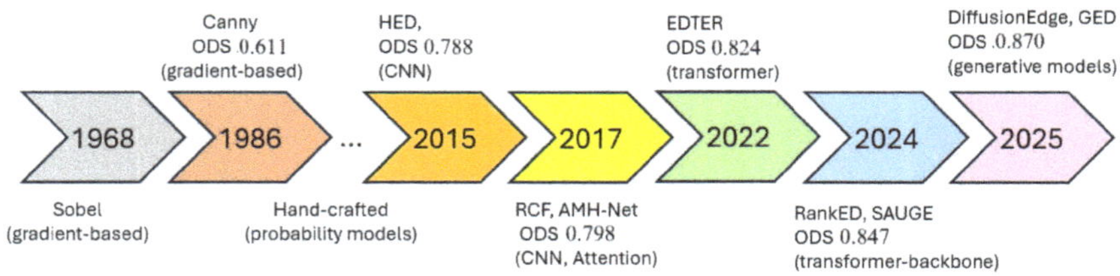

Figure 1. Timeline of edge detection paradigms, from gradient-based methods (Sobel, 1968) [8] to generative and transformer-backbone models (GED, 2025) [14], with key ODS milestones on BSDS500.

These advancements are driven by mathematical innovations such as transformer mechanisms (Equation (25)) and diffusion processes (Equation (51)). However, several key challenges remain unresolved, including computational inefficiency (e.g., the high complexity of transformers), limited generalization across datasets, insufficient multi-modal robustness, lack of theoretical clarity in parameter interpretation, and unstable training dynamics in generative models (Section 7).

While numerous surveys have reviewed edge detection methods, few provide a mathematically grounded perspective that unifies traditional, CNN-based, and emerging paradigms under a consistent analytical framework. Therefore, a survey that addresses these critical challenges through mathematical analysis is both timely and essential for advancing the real-world applicability of edge detection techniques.

This survey traces the evolution of edge detection. Besides the traditional gradient-based methods, it focuses on mainstream supervised deep learning methods for general edge detection, including CNN-based approaches, such as Holistically Nested Edge Detection (HED) [15], encoder–decoder (Bidirectional Cascade Network (BDCN) [16] and NHP [17]), attention-transformer-driven methods, including VST [18], EDTER [19], EdgeNAT [20], hybrid transformer-backbone (e.g., RankED [21] and SAUGE [22]), and generative algorithms (e.g., GED [14]), while excluding unsupervised, graph-based, reinforcement learning-based, and salient object detection methods due to their limited mathematical development or specific application focus.

Our objective is to provide a comprehensive, mathematically grounded analysis of these developments, synthesizing their strengths and limitations (Section 6) and identifying future directions (Section 7). Section 2 reviews Sobel and Canny, establishing the baseline. Section 3 examines early CNNs (HED [15], RCF [23]) and encoder–decoder innovations (BDCN [16], DexiNed [24], and NHP [17]), highlighting loss function designs and time efficiency. Section 4 explores attention-based and generative methods, from AMH-Net [25] to GED, emphasizing hybrid paradigms. Section 5 details the evaluation frameworks, while Section 6 compares all methods across accuracy and efficiency. Section 7 presents the current challenges and possible directions, and Section 8 concludes with reflections on the field's trajectory.

2. Traditional Gradient-Based Edge Detection Methods

Before the advent of deep learning, edge detection relied on traditional methods rooted in mathematical operations on image gradients. These approaches, developed decades ago, remain foundational due to their simplicity and interpretability, though they lack the adaptability of modern techniques. This section examines two classic methods: the Sobel operator [8], introduced in 1968, and the Canny edge detector [9], proposed in 1986. Both are based on detecting intensity changes in an image I and have significantly influenced the development of subsequent techniques, laying the groundwork for the data-driven paradigms explored in Section 3.

2.1. Sobel Operator

The Sobel operator [8], developed by Irwin Sobel in 1968 as part of early computer vision research, is a lightweight method for approximating image gradients using convolution kernels. For a grayscale image $I(x,y)$, it computes the gradient components in the horizontal (G_x) and vertical (G_y) directions:

$$G_x = I * K_x, \quad G_y = I * K_y, \tag{1}$$

where $*$ denotes convolution, and the kernels are

$$K_x = \begin{bmatrix} -1 & 0 & 1 \\ -2 & 0 & 2 \\ -1 & 0 & 1 \end{bmatrix}, \quad K_y = \begin{bmatrix} -1 & -2 & -1 \\ 0 & 0 & 0 \\ 1 & 2 & 1 \end{bmatrix}. \tag{2}$$

The gradient magnitude, representing edge strength, is

$$|\nabla I| = \sqrt{G_x^2 + G_y^2}, \tag{3}$$

often approximated as $|G_x| + |G_y|$ for computational efficiency. A threshold T is applied to $|\nabla I|$ to produce a binary edge map $\hat{y}$.

Sobel's simplicity—requiring only small 3×3 kernels, has linear complexity, $O(hw)$, for image dimensions $h \times w$ but struggles with noise and complex scenes. Also, it lacks robustness to complex scenes. Its fixed kernels cannot adapt to varying edge characteristics, limiting its performance on datasets like BSDS500 [13], where no standard ODS F-score (Section 5.2) is typically reported due to its heuristic nature.

2.2. Canny Edge Detector

John Canny's edge detector [9], introduced in 1986, builds on gradient-based methods like Sobel with a multi-stage pipeline to improve edge quality. It begins with noise reduction via Gaussian smoothing:

$$I_\sigma = I * G_\sigma, \quad G_\sigma(x,y) = \frac{1}{2\pi\sigma^2} e^{-\frac{x^2+y^2}{2\sigma^2}}, \tag{4}$$

where G_σ is a Gaussian kernel with standard deviation σ. Gradient magnitude $|\nabla I_\sigma|$ and direction $\theta = \arctan(G_y/G_x)$ are then computed, typically using Sobel kernels.

Canny introduces two key post-processing steps: non-maximum suppression (NMS) and hysteresis thresholding. NMS refines edge maps by thinning edges to a single pixel width, retaining only the local maxima in $|\nabla I_\sigma|$ along θ, reducing false positives. Hysteresis uses two thresholds, T_{low} and T_{high}, to classify strong edges ($|\nabla I_\sigma| > T_{\text{high}}$) and connect weak edges ($T_{\text{low}} < |\nabla I_\sigma| \leq T_{\text{high}}$) if adjacent to strong ones, yielding a binary edge map $\hat{y}$.

Canny's sophisticated approach offers improved noise robustness and edge continuity compared to the Sobel operator, but its reliance on hand-tuned parameters (σ, T_{low}, T_{high}) limits its adaptability to varying image conditions (see Figure 2). Table 1 summerizes their difference.

Table 1. Comparison of traditional edge detection methods in Section 2. Both rely on gradient operations, with Canny enhancing Sobel via multi-stage processing, yet both lack the adaptability of deep learning approaches.

Method	Year	Key Operation	Limitations
Sobel [8]	1968	Gradient magnitude ($\|\nabla I\|$)	Noise-sensitive, non-adaptive
Canny [9]	1986	Gaussian + NMS + Hysteresis	Parameter-dependent, heuristic

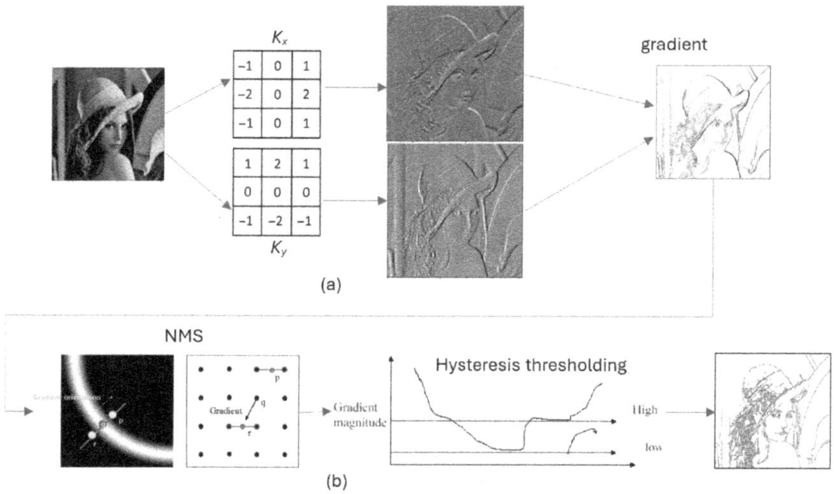

Figure 2. (**a**) Sobel operator applying K_x and K_y kernels to compute the gradient magnitude; (**b**) NMS (non-maximum suppression) and hysteresis thresholding in Canny pipeline after gradient computation on a sample image.

3. Convolutional Neural Networks for Edge Detection

The advent of deep learning marked a transformative shift in edge detection, replacing hand-crafted filters like Sobel [8] and Canny [9] with data-driven convolutional neural networks (CNNs). We examine the progression of CNN-based methods, which leverage hierarchical feature extraction to outperform traditional approaches on benchmarks like BSDS500 [13]. We divide this evolution into two phases: early CNN approaches, exemplified by Holistically Nested Edge Detection (HED) [15] and Richer Convolutional Features (RCFs) [23], and encoder–decoder innovations, represented by the Bidirectional Cascade Network (BDCN) [16] and Dense Extreme Inception Network (DexiNed) [24]. These methods laid the groundwork for the attention-based paradigms explored in Section 4, with their loss functions critical to optimizing edge predictions.

3.1. Early CNN Approaches

Convolutional operations are particularly effective for local feature extraction such as edges due to their inherent spatial locality and ability to capture intensity variations. A convolution layer applies a kernel $K \in \mathbb{R}^{k\times k}$ across an image $I \in \mathbb{R}^{h\times w}$, computing the local weighted sums at each location:

$$(I * K)(x,y) = \sum_{i=-r}^{r}\sum_{j=-r}^{r} K(i,j)\cdot I(x+i,y+j), \quad (5)$$

where $r = \lfloor k/2 \rfloor$. This operation aggregates pixel intensities in a local neighborhood, making it sensitive to spatial discontinuities in intensity, which are characteristic of edges. Classical edge detectors such as Sobel, Prewitt, and Roberts are special cases of fixed convolutional kernels designed to approximate gradients. In modern convolutional neural networks (CNNs), these filters are learned from data, and empirical evidence consistently shows that the first-layer filters resemble oriented edge detectors. This occurs because minimizing loss on visual tasks naturally encourages the network to detect strong local contrast patterns. Furthermore, convolution enforces translation invariance and parameter sharing, making it both efficient and robust for detecting edges regardless of their position in the image. Therefore, convolution serves as a mathematically grounded and empirically validated foundation for edge extraction in both classical and deep learning-based frameworks.

The initial application of CNNs to edge detection adapted pretrained networks like VGG-16 [26] to produce dense edge maps. Two representative approaches, HED and RCF, are described here, with HED introducing multi-scale supervision and RCF emphasizing feature richness, both employing loss functions that prioritize hierarchical learning.

3.1.1. Holistically Nested Edge Detection (HED)

Proposed by Xie and Tu [15] in 2015, HED was among the first to apply CNNs holistically to edge detection, achieving an ODS F-score of 0.788 on BSDS500. HED modifies VGG16 by removing fully connected layers and extracting multi-scale feature maps $F_1, F_2, ..., F_5$ from its five blocks. Each block produces a side output via a 1×1 convolution followed by a sigmoid activation:

$$\hat{y}_i = \text{sigmoid}(\text{Conv}_{1\times1}(F_i)), \tag{6}$$

where $i = 1, ..., 5$, representing edge features at different scales—from local to global views. The side outputs are upsampled to the original image resolution, then fused by a weighted combination to produce the final edge map ($\hat{y}$):

$$\hat{y} = \text{sigmoid}(\sum_{i=1}^{5} w_i \hat{y}_i), \tag{7}$$

with w_i as learned weights.

HED formulates edge detection as a per-pixel binary classification problem and employs a multi-scale, class-balanced binary cross-entropy (BCE) loss to supervise learning at different stages of the network. The total loss combines the losses from multiple side outputs and the final fused output, defined as

$$L_{\text{HED}} = \sum_{m=1}^{M} \alpha_m \cdot \mathcal{L}_{\text{side}}^{(m)} + \alpha_f \cdot \mathcal{L}_{\text{fuse}}, \tag{8}$$

where M is the number of side outputs, $\mathcal{L}_{\text{side}}^{(m)}$ denotes the class-balanced BCE loss for the m-th side output, and $\mathcal{L}_{\text{fuse}}$ is the loss for the final fused prediction. Each loss term is computed as

$$L_{\text{BCE}} = -\beta \sum_{i \in Y_+} \log(\hat{y}_i) - (1-\beta) \sum_{i \in Y_-} \log(1-\hat{y}_i), \tag{9}$$

where Y_+ and Y_- represent the sets of edge and non-edge pixels, respectively, $\hat{y}_i$ is the predicted edge probability at pixel i, and $\beta = \frac{|Y_-|}{|Y_+|+|Y_-|}$ balances the contributions of positive and negative samples. This loss formulation encourages the model to accurately predict edge pixels while addressing the severe class imbalance typical in edge detection tasks. This loss encourages fine local edges from lower layers (e.g., F_1) and object context from higher layers (e.g., F_5), outperforming traditional methods (see Figure 3).

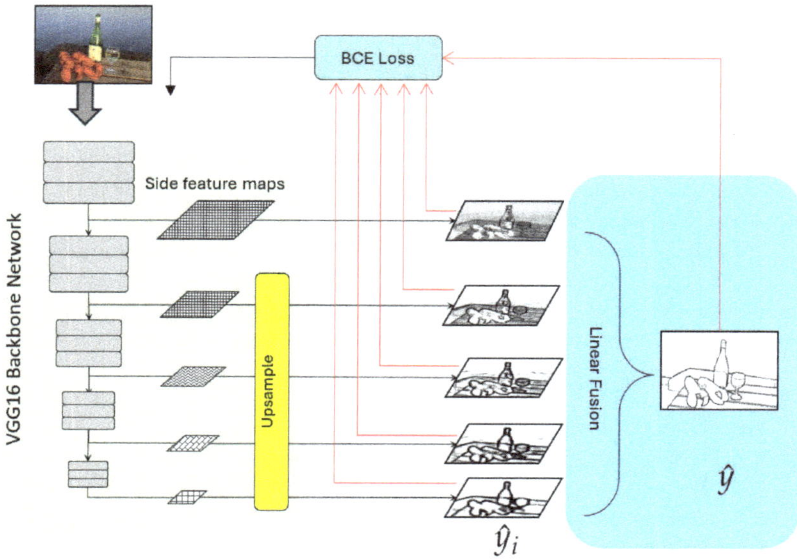

Figure 3. HED architecture: VGG16 backbone with side outputs ($\hat{y}_i$) from five blocks, fused into $\hat{y}$, supervised by BCE loss at multiple scales.

3.1.2. Richer Convolutional Features (RCFs)

Liu et al. [23] extended HED at CVPR 2017 with RCF, improving the ODS F-score to 0.798 on BSDS500. RCF also uses VGG16 but aggregates intermediate features within each block $k = 1, ..., 5$:

$$F_k = \sum_j \text{Conv}_{1\times1}(F_{k,j}), \tag{10}$$

producing side outputs $\hat{y}_k = \text{sigmoid}(\text{Conv}_{1\times1}(F_k))$ and a fused output $\hat{y} = \text{sigmoid}(\sum_{k=1}^{5} w_k \hat{y}_k)$.

The RCF loss function mirrors that of HED, using the same BCE form (Equation (9)):

$$L_{\text{RCF}} = L_{\text{BCE}}(y, \hat{y}) + \sum_{k=1}^{5} \alpha_k L_{\text{BCE}}(y, \hat{y}_k), \tag{11}$$

differing only in the source of side outputs (richer intra-block features vs. the HED block endpoints). This similarity enhances details while maintaining multi-scale supervision (see Figure 3), achieving higher accuracy at ODS 0.798 on BSDS500 dataset.

Given an input image of size $h \times w \times c$, the time complexity per convolutional layer in early CNN-based edge detection methods, such as HED and RCF, is generally proportional to

$$O(h \cdot w \cdot k^2 \cdot c \cdot c_{\text{out}}) \tag{12}$$

where h and w are the height and width of the input, c is the number of input channels, k is the kernel size, and c_{out} is the number of output channels. The total time complexity across all layers depends on the network's depth and structure:

$$O\left(\sum_{\ell=1}^{L} h_\ell \cdot w_\ell \cdot k_\ell^2 \cdot c_\ell \cdot c_{\ell,\text{out}}\right) \tag{13}$$

where L is the number of layers, and h_ℓ, w_ℓ, c_ℓ, $c_{\ell,\text{out}}$, and k_ℓ represent the parameters of the ℓ-th layer. Since later layers often operate on downsampled feature maps, the actual complexity is lower than a linear multiple of input size. In practice, for architectures like HED or RCF using VGG-style backbones (e.g., $L = 16$, $k = 3$), the overall complexity can be approximated by

$$O(L \cdot h \cdot w \cdot c) \tag{14}$$

assuming fixed small kernels and proportional input/output channels per layer.

3.2. Encoder–Decoder Innovations

Encoder–decoder architectures have become pivotal for edge detection, addressing the resolution and continuity limitations of early CNNs. The encoder downsamples an input image I into compact feature maps $F_1, \ldots, F_n$, capturing multi-scale features from local textures to global semantics. The decoder upsamples these, often with skip connections, to produce high-resolution edge maps $\hat{y}$, ensuring precise boundary localization. This paradigm excels in edge detection by balancing fine details with contextual understanding. Notably, early CNNs like HED and RCF (Section 3.1) exhibit an encoder–decoder-like structure: their VGG16 backbones downsample features, and upsampling occurs in fusion layers. However, they lack explicit decoder modules, relying on side outputs, upsampling, and simple linear fusion, unlike formal encoder–decoder designs. This survey is, to our knowledge, the first to frame HED and RCF in this light, revealing a precursor to the innovations in the Bidirectional Cascade Network (BDCN) [16], Dense Extreme Inception Network (DexiNed) [24], and Normalized Hadamard Product (NHP) [17] fusion, which introduce specialized modules for a robust encoder–decoder flavor.

3.2.1. Bidirectional Cascade Network (BDCN)

He et al. [16] introduced BDCN at CVPR 2019, achieving an ODS F-score of 0.828 on BSDS500. BDCN formalizes the encoder–decoder paradigm, using a VGG16 encoder with a Scale Enhancement Module (SEM) to refine features $F_1, \ldots, F_5$. The SEM processes each feature map:

$$F_l^{\text{SEM}} = \text{ReLU}(\text{BN}(\text{Conv}_{3\times3}(F_l))), \tag{15}$$

enhancing scale-specific edge details. These are upsampled by a decoder except F_1^{SEM}, where a bidirectional cascade combines outputs across scales:

$$C_l = \text{Conv}_{1\times1}(\text{Concat}(U_l, C_{l+1}^{\text{down}}, C_{l-1}^{\text{up}})), \tag{16}$$

where $U_l = \text{Upsample}(F_l^{\text{SEM}})$, and C_{l+1}^{down}, C_{l-1}^{up} are downward and upward cascade outputs. The edge map is $\hat{y}_l = \text{sigmoid}(\text{Conv}_{1\times1}(C_l))$, refining localization (see Figure 4).

The BDCN loss uses scale-specific BCE:

$$L_l = -\sum [y_l \log(\hat{y}_l) + (1 - y_l)\log(1 - \hat{y}_l)], \tag{17}$$

with total loss

$$L_{\text{BDCN}} = \sum_{l=1}^{L} L_l, \tag{18}$$

optimizing each layer's ground truth y_l. This leverages the cascade's decoding for precise edges.

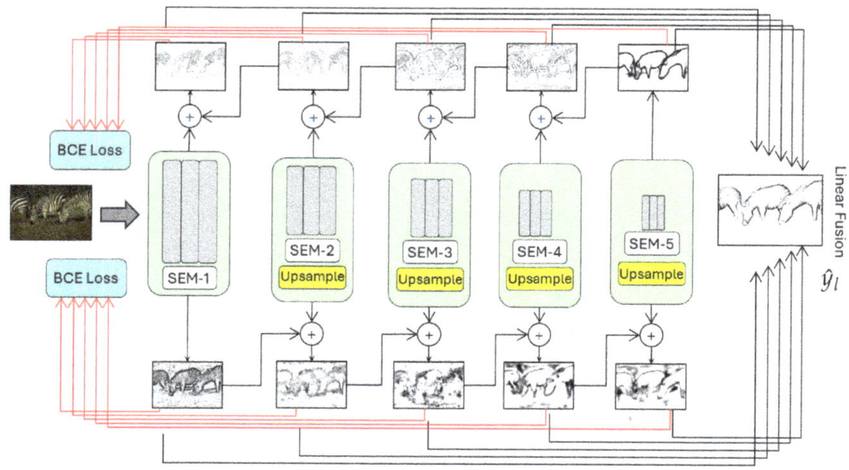

Figure 4. The BDCN bidirectional cascade: VGG16 encoder with SEM refines features ($F_1, ..., F_5$), upsampled and combined by a decoder's cascade into outputs ($\hat{y}_l$).

3.2.2. Dense Extreme Inception Network (DexiNed)

Xavier et al. [24] presented DexiNed in 2020, achieving an ODS F-score of 0.729 on BSDS500. DexiNed uses an Xception-inspired encoder for features $F_1, ..., F_6$, upsampled in a dense decoder:

$$F_{l,\text{up}} = \text{interp}(F_l), \tag{19}$$

fused as $F_{\text{fused}} = \text{Concat}(F_{1,\text{up}}, ..., F_{6,\text{up}})$, with outputs $\hat{y}_l$ and $\hat{y}$. Its BCE loss mirrors HED/RCF, enhancing edge continuity via dense fusion.

Recent work by Hu et al. [17] introduces normalized Hadamard product (NHP) fusion, achieving ODS 0.818 on BSDS500. NHP multiplies normalized VGG16 side outputs:

$$\text{MASE} = \prod_i \text{normalize}(\hat{y}_i), \tag{20}$$

producing Mutually Agreed Salient Edge (MASE) maps, refining edges. Like the BDCN cascade, the NHP operator formalizes the encoder–decoder flavor, surpassing HED/RCF.

3.3. *Multi-Scale Feature Fusion in Edge Detection*

The fusion process plays a vital role in CNN-based edge detection. Features extracted at multiple scales can effectively capture edges of varying thicknesses and complexities. Formally, let $F^{(m)}(x, y)$ denote the feature map or edge prediction at scale m for pixel location (x, y), where $m = 1, 2, \ldots, M$ indexes the scales from fine to coarse. A common approach to fuse these multi-scale features is a weighted summation:

$$F_{\text{fused}}(x, y) = \sum_{m=1}^{M} \alpha_m \cdot F^{(m)}(x, y), \tag{21}$$

where α_m are learnable or fixed weights that control the contribution of each scale.

Alternatively, concatenation followed by a convolutional layer can be used:

$$F_{\text{fused}} = \sigma\Big(W * \text{concat}(F^{(1)}, F^{(2)}, \ldots, F^{(M)}) + b\Big), \tag{22}$$

where W and b are the convolutional weights and bias, and σ is a nonlinear activation function (e.g., ReLU or sigmoid).

The fusion of multi-scale features offers several theoretical benefits for edge detection. Fine-scale features extracted from early layers, excel at capturing thin, localized edges. In contrast, coarse-scale features from deeper layers have larger receptive fields, able to capture thicker boundaries and broader contextual patterns. By combining these complementary representations, the model becomes capable of detecting edges across a wide range of spatial scales and complexities. Additionally, this fusion improves robustness by mitigating the effects of noise in fine scales and compensating for missing details in coarser scales. As a result, multi-scale fusion enhances both the precision and completeness of edge predictions.

For example, consider three scales with predicted edge probabilities at pixel (x, y): $F^{(1)} = 0.9, F^{(2)} = 0.6, F^{(3)} = 0.3$, and corresponding fusion weights: $\alpha_1 = 0.5$, $\alpha_2 = 0.3, \alpha_3 = 0.2$.

The fused prediction at this pixel is

$$F_{\text{fused}} = 0.5 \times 0.9 + 0.3 \times 0.6 + 0.2 \times 0.3 = 0.45 + 0.18 + 0.06 = 0.69.$$

This fusion combines strong fine-scale evidence with supporting coarser-scale information, yielding a more reliable edge prediction.

3.4. *Effectiveness and Efficiency of CNN-Based Approaches*

Table 2 summarizes the CNN-based edge detection methods. Despite advances, CNN-based approaches face limitations. Their convolutional layers, constrained by fixed receptive fields, struggle to capture long-range dependencies essential for complex boundaries in datasets like BSDS500 [13]. Simple fusion in HED and RCF risks diluting fine edges, while encoder–decoder-based approaches introduce sophisticated fusion components (BDCN cascade and NHP fusion) for higher ODS.

Theoretically, the time complexity of encoder–decoder-based approaches remains in the same order as HED and RCF, $O(L \cdot h \cdot w \cdot c)$, which is primarily based on convolutional backbones. However, due to its deeper supervision and multi-path design, these methods incur higher practical computational cost and memory usage compared to HED and RCF, limiting real-time use.

Moreover, CNN-based methods lack dynamic feature weighting, reducing adaptability to diverse edge patterns. Attention mechanisms address these challenges by capturing global context and prioritizing salient features, marking a significant evolution in edge detection.

Table 2. Comparison of CNN-based edge detection methods. ODS reflects BSDS500 performance; key loss features highlight similarities and differences.

Method	Year	Venue	ODS (BSDS500)	Key Loss Feature
HED [15]	2015	ICCV	0.788	BCE, 5 side outputs + fusion
RCF [23]	2017	CVPR	0.798	BCE, 5 richer side outputs + fusion
BDCN [16]	2019	CVPR	0.806	BCE, scale-specific layers
DexiNed [24]	2020	arXiv	0.729	BCE, 6 side outputs + dense fusion
NHP [17]	2024	J. Imaging	0.818	BCE, NHP fusion + MASE maps

4. Attention Mechanisms and Emerging Trends

The transition from CNN-based methods (Section 3) to attention-driven and generative paradigms marks a pivotal evolution in edge detection, emphasizing global context modeling and computational efficiency. While CNN-based and encoder–decoder approaches excel in hierarchical feature extraction and resolution recovery, they often struggle with capturing long-range dependencies due to their limited receptive fields. In contrast, attention mechanisms dynamically prioritize salient features across the entire image, overcoming the spatial limitations of conventional convolutional architectures (Section 3.2). These mechanisms enhance edge detection by being integrated into—or fully replacing—decoder components. This section highlights this natural progression, examining representative models such as AMH-Net [25], VST [18], EDTER [19], and EdgeNAT [20].

4.1. Attention-Based Contour Detection (AMH-Net)

AMH-Net [25] introduces a hybrid CNN–attention approach with an encoder–decoder flavor, using a ResNet-50 [27] encoder to extract multi-scale features $F_1, ..., F_n$ and an attention-gated CRF (AG-CRF) decoder to weight contour-relevant regions.

Unlike HED uniform fusion, AMH-Net introduces a two-stage fusion strategy. First, it uses AG-CRFs to refine and align features across scales by explicitly modeling contextual dependencies between different levels of abstraction. These AG-CRFs are enhanced with attention gates that dynamically weigh feature importance based on the current input, allowing the model to emphasize more informative contours and suppress irrelevant textures. In the second stage, the fused features are passed through additional convolutional layers to generate the final edge map. Equations (23) and (24) show its summarized formula:

$$M_{ij} = \text{softmax}(W \cdot [F_i, F_j]), \tag{23}$$

where W is a learned weight matrix, and $[F_i, F_j]$ denotes the concatenated features at positions i and j. The attention-gated output refines each feature map:

$$F_{i,\text{refined}} = F_i + \sum_j M_{ij} F_j, \tag{24}$$

which is then integrated into a CRF framework for contour prediction. The CRF optimizes the edge map $\hat{y}$ by minimizing an energy function combining unary and pairwise potentials, trained end-to-end with a cross-entropy loss akin to Equation (9).

This hierarchical and attention-driven fusion enables AMH-Net to produce more accurate and coherent edges, especially in cluttered or low-contrast regions. AMH-Net's AG-CRFs module with attention mechanism improves localization over CNN-only methods like HED [15] and RCF [23], demonstrates strong performance on standard edge detection benchmarks such as BSDS500 and NYUDv2 over earlier methods like HED and RCF in both precision and robustness, foreshadowing the shift to global attention in later models. Its reliance on a CNN backbone limits its scope compared to pure transformer approaches, but it marks a critical step in the evolution from convolution to attention.

4.2. Transformer-Based Edge Detection

The transformer-based attention mechanism [28] was originally developed for natural language processing (NLP) but was later introduced to computer vision with the Vision Transformer (ViT) architecture [29]. Its core component, the scaled dot-product attention, allows models to dynamically prioritize salient regions across the input feature map $X \in \mathbb{R}^{N \times C}$, and is defined as

$$\text{Attention}(Q, K, V) = \text{Softmax}\left(\frac{QK^T}{\sqrt{d_k}}\right)V, \tag{25}$$

where $Q = XW_Q$, $K = XW_K$, and $V = XW_V$ are query, key, and value projections, and d_k scales the dot product. This mechanism overcomes the fixed receptive field of convolutional operations, enabling the more effective modeling of long-range dependencies—a key advantage in vision tasks such as edge detection.

Among transformer-based vision models, the Vision Saliency Transformer (VST) [18], introduced at ICCV 2021, is notable for its ability to produce plausible object contours. VST is a transformer-based framework for salient object detection that explicitly incorporates boundary prediction as a complementary task. Built on a ViT-style encoder–decoder architecture, VST adopts a multi-task design in which saliency and boundary-specific tokens are decoded in parallel through a shared transformer decoder. This allows the model to simultaneously generate both a saliency map and a boundary map. The boundary prediction head is supervised using object contour annotations, enabling the network to learn spatially precise and semantically meaningful edges around salient regions. While VST is not intended as a general-purpose edge detector, its boundary outputs effectively highlight prominent object contours, particularly in cluttered or low-contrast scenes—achieving an S-measure of 0.91 on DUTS [30]. These saliency-aligned edge maps can be regarded as edge detection byproducts, especially valuable in applications focused on foreground object structure. However, they are not optimized for capturing background textures, fine-grained detail, or non-salient edges. Nonetheless, VST serves as a compelling example of how edge information can be jointly learned with saliency within a unified transformer-based framework.

Edge Detection with Transformer (EDTER) [19] (ECCV 2022) focuses on lightweight edge detection, achieving ODS 0.824 at 10 FPS on a Titan X. By integrating a CNN encoder with lightweight transformer layers, EDTER leverages a streamlined version of self-attention (Equation (25)) to refine edges efficiently. Its dual-task design, simultaneously optimizing edge detection and boundary refinement, enhances precision over DexiNed [24], though it struggles with severe class imbalance, addressed by later methods like RankED [21].

EdgeNAT [20] integrates a VGG16 encoder with a Dilated Attention Module, and introduces Scale-Adaptive Feature Multi-Level Aggregation (SCAF-MLA) as a decoder (see Figure 5). Its attention module follows Equation (25), while decoder SCAF-MLA is formalized as

$$\text{SCAF-MLA} = \sum_{i=1}^{N} \beta_i \cdot \text{Attention}(Q_i, K_i, V_i), \tag{26}$$

where β_i weights multi-level features. Achieving ODS 0.860 and 12.5 FPS on BSDS500, EdgeNAT combines CNN and transformer features to enhance efficiency, surpassing the BDCN cascade (Equation (16)). Its multi-level focus improves edge continuity but struggles with severe class imbalance.

Transformer-based methods have collectively advanced the field of edge detection by demonstrating the power and effectiveness of attention mechanisms in capturing the global context. Given an input image of size $h \times w$ and a patch size of $p \times p$, the image is divided into non-overlapping patches, resulting in

$$n = \frac{h \cdot w}{p^2} \tag{27}$$

where n is the number of input tokens (patches), h and w are the height and width of the input image, and p is the height and width of each square patch. These approaches utilize the scaled dot-product attention mechanism, defined over the space $\mathbb{R}^{n \times d}$, where d is the feature dimension of the key vectors and n is the number of image patches. The computational complexity of the attention operation is $O(n^2d)$, stemming from the pairwise interactions among all elements in the sequence. Substituting $n = hw/p^2$, the complexity becomes

$$O\left(\frac{h^2w^2d}{p^4}\right) \tag{28}$$

which reveals the quadratic dependence on image resolution. When the patch size p is fixed (as is common in practice), this can be approximated as

$$O(h^2w^2d) \tag{29}$$

where $d = 768{\sim}1280$, highlighting the substantial computational cost associated with applying full attention to high-resolution inputs as analyzed in Section 6.

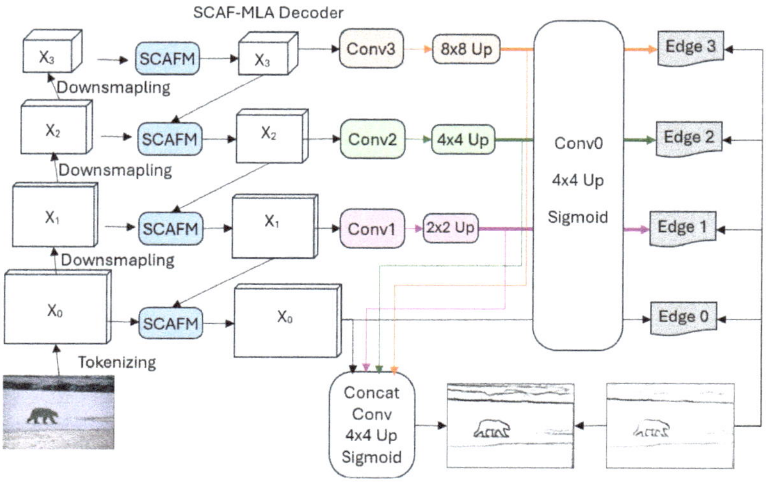

Figure 5. EdgeNAT's transformer architecture [20]: the VGG16 encoder feeds into a Dilated Attention Module, followed by SCAF-MLA decoder, with the attention weights visualized.

Compared to non-transformer-based methods, transformer-based edge detection models tend to be more computationally intensive due to the self-attention mechanism, which scales quadratically with the number of input tokens.

4.3. Transformer-Backbone Edge Detection

Recently, with the rising power of transformer models, the trend in edge detection has shifted away from traditional backbones like VGG or ResNet. From 2023 to 2025, transformer-based backbones have dominated the field, with methods such as RankED [21] and SAUGE [22] leveraging pretraining to achieve ODS scores above 0.824. Unlike architecture-focused approaches like EDTER [19] or EdgeNAT [20], these methods emphasize non-transformer innovations—such as ranking-based and sorting losses or uncertainty modeling—highlighting how transformer backbones can amplify algorithmic advancements.

RankED (CVPR 2024) employs a Swin Transformer backbone [31] (Swin-B, ~88 M parameters) to achieve ODS 0.824 on BSDS500, excelling in average precision (AP). Swin-B, pretrained on ImageNet-22K, processes 320 × 320 inputs through four hierarchical stages (patch size 4 × 4, feature dimensions 128 to 1024) using shifted window self-attention (Equation (25)). A lightweight convolutional decoder fuses multi-scale features, guided by a combined rank and sort loss:

$$L_{\text{rank}} = \frac{1}{|P|} \sum_{i \in P} \sum_{j \in N} \frac{H(p_j - p_i)}{\sum_{k \in P \cup N} H(p_k - p_i)}, \tag{30}$$

$$H(x) = \begin{cases} 0 & x < -\epsilon \\ \frac{x}{2\epsilon} + 0.5 & -\epsilon \leq x \leq \epsilon, \\ 1 & x > \epsilon \end{cases} \tag{31}$$

$$L_{\text{sort}} = \frac{1}{|P|} \sum_{i \in P} (\Delta_{\text{sort}}(i) - \Delta^*_{\text{sort}}(i)), \tag{32}$$

$$\Delta_{\text{sort}}(i) = \frac{1}{\sum_{j \in P} H(p_j - p_i)} \sum_{j \in P} H(p_j - p_i)(1 - c_j), \tag{33}$$

$$\Delta^*_{\text{sort}}(i) = \frac{\sum_{j \in P} H(p_j - p_i)[c_j > c_i](1 - c_j)}{\sum_{j \in P} H(p_j - p_i)[c_j > c_i]}, \tag{34}$$

$$L = L_{\text{rank}} + \alpha L_{\text{sort}}, \tag{35}$$

where P, N are edge and non-edge pixel sets, p_i are predicted scores, and $c_i \in [0,1]$ are ground-truth certainty scores, computed as

$$c_i = \frac{1}{K} \sum_{\ell=1}^{K} 1[\exists p \in \text{vicinity of } i : y_{\ell,p} = 1],$$

with K annotators and $y_{\ell,p}$ indicating edge labels in pixel i's neighborhood. $H(x)$ is a smooth step function with threshold ϵ, $[c_j > c_i]$ is 1 if $c_j > c_i$ else 0, and α is a weighting factor (e.g., 0.5, or 0 if only one annotation exists)[4] Certainty computation inferred from BSDS500; $\alpha = 1$ assumed based on typical loss weighting; ϵ and sampling strategy unavailable [21]. The rank loss optimizes AP, ranking edge pixels above non-edges to address class imbalance, while the sort loss aligns edge predictions with annotator certainty, tackling label uncertainty. Pairs are randomly sampled to reduce computational cost. RankED achieves robust boundaries in Multicue and NYUDv2, though high GPU memory demands limit efficiency compared to the EDTER transformer layers.

SAUGE (AAAI 2025) adapts the Segment Anything Model's Vision Transformer [32] (SAM-ViT, likely ViT-H, ∼632 M parameters) for edge detection, achieving ODS 0.847 per AAAI 2025 [22]. SAM-ViT, pretrained on SA-1B (1B masks, 11 M images), processes 320 × 320 inputs as 16 × 16 patches, producing 64 × 64 × 1280 embeddings via 32 transformer layers. Unlike the SAM prompt-driven segmentation, SAUGE uses only the image encoder, bypassing the prompt encoder. A lightweight decoder (∼1.5% parameters, ∼9.5 M) fuses multi-granularity features, regressing edges aligned with human label uncertainty (see Figure 6). The loss comprises

$$L^k_{\text{side}} = -\frac{1}{N} \sum_i \left[w_i \left(y^k_i \log p^k_i + (1 - y^k_i) \log(1 - p^k_i) \right) \right], \tag{36}$$

$$L_{\text{diversity}} = \frac{1}{M} \sum_{(m,n) \in C} |\text{XOR}(p^m, p^n)|, \tag{37}$$

$$L_{\text{guide}} = -\frac{1}{N} \sum_i \left[c_i \left(y^{\text{sobel}}_i \log p_i + (1 - y^{\text{sobel}}_i) \log(1 - p_i) \right) \right], \tag{38}$$

$$L = \sum_k L^k_{\text{side}} + \lambda_{\text{diversity}} L_{\text{diversity}} + \lambda_{\text{guide}} L_{\text{guide}}, \tag{39}$$

where L^k_{side} supervises the side outputs (coarse, medium, fine) with pseudo-labels y^k_i and weights w_i; $L_{\text{diversity}}$ enforces distinct side predictions; L_{guide} aligns the final edges with the SAM Sobel maps using confidence weights c_i; and $\lambda_{\text{diversity}}, \lambda_{\text{guide}}$ are fixed constants. This loss ensures uncertainty alignment, generalizing across Multicue and NYUDv2, outperforming HED. The SAUGE SAM-driven approach contrasts RankED's rank and sort innovation, marking a novel 2025 paradigm.

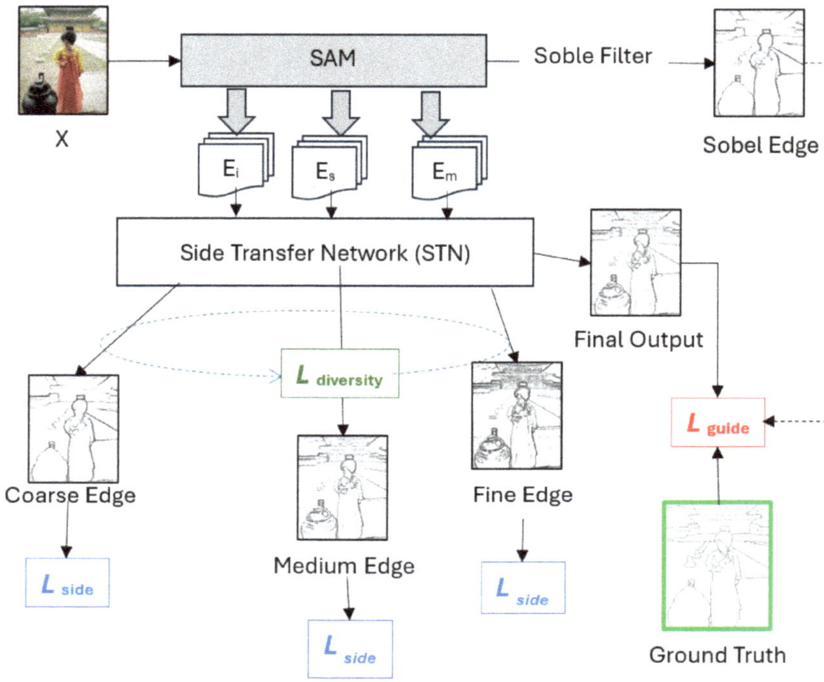

Figure 6. SAUGE architecture [22]: SAM is used as the backbone to extract features, and constructs edges at multiple granularity levels to align uncertainty with granularity. The final output is obtained by merging the features of side outputs.

By leveraging transformer backbones, both RankED and SAUGE incur a minimum computational complexity of $O(h^2w^2d)$ (Equation (29)) due to the self-attention mechanism. As hybrid architectures, they combine convolutional feature extractors with transformer-based attention modules to balance local inductive bias and global context modeling. However, this integration significantly increases the actual computational cost, which limits their efficiency in practice (Section 6).

4.4. Generative Models for Edge Detection

In recent years, generative models—especially diffusion-based methods and adversarial frameworks—have emerged as powerful tools for edge detection. Unlike traditional discriminative models that directly predict binary edge maps from input images, generative approaches learn to synthesize edge structures, often modeling uncertainty and fine-grained detail more effectively. Two prominent examples are DiffusionEdge and GED (Generative Edge Detection).

DiffusionEdge (AAAI 2024) [33] applies a diffusion probabilistic model (DPM) [34] in latent space, achieving ODS 0.834 on BSDS500. The forward process adds noise over T steps:

$$q(x_t|x_{t-1}) = \mathcal{N}(x_t; \sqrt{1-\beta_t}x_{t-1}, \beta_t I), \tag{40}$$

enabling denoising via

$$p_\theta(x_{t-1}|x_t) = \mathcal{N}(x_{t-1}; \mu_\theta(x_t, t), \Sigma_\theta(x_t, t)). \tag{41}$$

It uses a CNN-based encoder–decoder (∼200 M parameters) trained with uncertainty-aware cross-entropy:

$$L_{\text{DiffusionEdge}} = -\sum_i [c_i \log p_i + (1-c_i)\log(1-p_i)]. \tag{42}$$

Trained on BSDS500's 200 images with augmentation for <30 epochs, it generalizes to BIPED, leveraging uncertainty distillation for ambiguous edges.

In contrast, Generative Edge Detection (GED) [14] leverages the pretrained Stable Diffusion backbone (∼116 M parameters), repurposing its U-Net-like encoder–decoder architecture within a conditional GAN (cGAN) [35,36] framework for latent-to-latent edge prediction. Instead of generating full-resolution RGB images, GED operates directly in the latent space, predicting edge maps from noisy latent representations of the input image. This design enables more efficient and semantically meaningful edge generation. A key contribution of GED is its ability to generate edge maps at multiple levels of granularity, controlled through a conditioning mechanism:

$$L_{\text{GED}} = \frac{1}{N}\sum_i |p_i - y_i| + \lambda_g \sum_i |g_i - g_i^*|, \tag{43}$$

where $g_i \in [0,1]$ is a side output controlling edge thickness (e.g., 0 for thin, 1 for thick)[5] $\lambda_g \approx 0.1$ assumed for granularity balance; exact value unavailable [14]. Its single-step inference enhances efficiency, doubling DiffusionEdge's throughput, and supports applications like medical imaging. GED achieves ODS 0.870, OIS 0.880, and AP 0.934 on BSDS500. It demonstrates how high-level generative modeling can be repurposed for fine-grained, structure-aware prediction tasks.

Diffusion-based edge detection models typically operate over multiple iterative steps to refine predictions. The overall computational complexity can be approximated as $O(T \cdot L \cdot h \cdot w \cdot c)$, where T is the number of diffusion steps (e.g., 1000). Each step generally involves a forward pass through a deep neural network, making diffusion models more computationally intensive than feedforward approaches. Table 3 summarizes the differences between the two generative models.

Table 3. Comparison of Generative Edge Detection models.

Model	Type	Backbone	Notable Features	Output Characteristics
DiffusionEdge [33]	Latent diffusion	Custom denoising network (∼200 M)	Frequency-aware filtering, uncertainty distillation	Crisp, high-confidence edges
GED [14]	Finetuned Stable Diffusion	Stable Diffusion U-Net (∼116 M)	Latent-to-latent mapping, granularity control	Flexible, multi-scale edge maps

5. Datasets and Evaluation Metrics

The evaluation of edge detection algorithms, from traditional methods (Section 2) to deep learning approaches (Sections 3 and 4), relies on standardized datasets and metrics. This section reviews two pivotal datasets—BSDS500 [13] and NYUDv2 [37]—and key evaluation metrics, including the Optimal Dataset Scale (ODS) F-score and Optimal Image Scale (OIS) F-score. These tools provide a quantitative basis for comparing methods like HED (ODS 0.782) to SAUGE (ODS 0.860), highlighting advancements in accuracy and efficiency.

5.1. Datasets

The Berkeley Segmentation Dataset and Benchmark (BSDS500), introduced by Arbelaez et al. [13] in 2011, is the de facto standard for edge detection evaluation. It comprises 500 natural images (200 train, 100 validation, 100 test), each of size 481 × 321 pixels, with human-annotated ground truth edge maps y. Each image has multiple annotations (typically 5–7), reflecting subjective edge interpretations, averaged into a consensus map:

$$y(x,y) = \frac{1}{N} \sum_{n=1}^{N} y_n(x,y), \tag{44}$$

where y_n is the n-th annotator's binary edge map, and N is the number of annotators. BSDS500's diversity—spanning objects, textures, and scenes—challenges algorithms to detect perceptually significant edges, making it ideal for benchmarking methods.

The NYU Depth Dataset V2 (NYUDv2), proposed by Silberman et al. [37] in 2012, extends edge detection to multi-modal settings. It includes 1449 RGB-D images (381 train, 414 validation, and 654 test), captured indoors at 640 × 480 resolution, with depth (D) from Kinect and hand-labeled edge maps y. Annotations cover structural boundaries (e.g., walls and furniture), available in RGB, depth (HHA format: height, horizontal disparity, and angle), or combined modalities. NYUDv2 tests the algorithms' ability to leverage depth cues contrasting with BSDS500's RGB-only focus (see Figure 7).

Both BSDS500 and NYUDv2 datasets are widely used benchmarks in edge detection research due to their high-quality annotations and their ability to support meaningful comparisons across a wide range of algorithms. BSDS500 is favored for its rich human-labeled edge annotations, which reflect perceptual boundaries rather than simple gradients. This makes it particularly useful for evaluating how well edge detection algorithms capture semantically meaningful structures. Its standing use in literature enables reproducible and comparable performance assessment across methods [13]. NYUDv2, on the other hand, offers indoor scenes. Its depth images help explore multi-modal edge detection. This is especially valuable for developing and evaluating methods that fuse visual and geometric cues, which are critical for robotics, AR, and indoor navigation applications [37].

Even though both datasets are relatively small in size, which can pose challenges for training large network models and hinder generalization and contribute to overfitting, it is still suitable for benchmarking and remains crucial for evaluating edge detection performance under controlled and interpretable conditions. For model development, there is a growing need to complement them with larger, more diverse datasets to train and validate models in real-world scenarios.

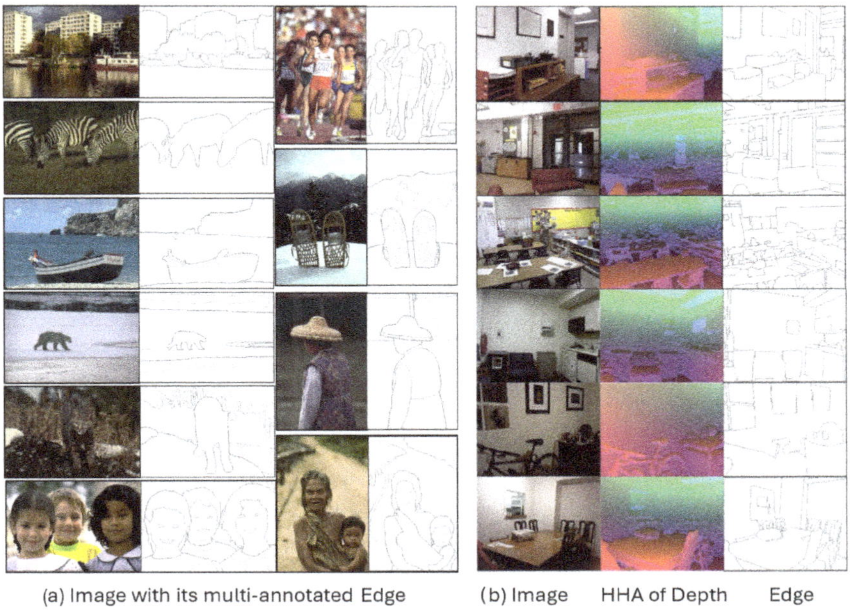

Figure 7. (**a**) BSDS500 RGB image with consensus edge map from multiple annotations; (**b**) NYUDv2 RGB-D pair (RGB + HHA) with structural edge map.

5.2. Evaluation Metrics

Edge detection methods typically produce edge maps, which must be thresholded to generate binary edge predictions. This is a highly imbalanced task, as edge pixels constitute a small fraction of all pixels compared to non-edge ones. Therefore, traditional performance metrics such as accuracy or Intersection over Union (IoU) are not appropriate, as they fail to capture the fine-grained nature of edge prediction.

To address this, the F-score metric—balancing precision and recall between a predicted edge map $\hat{y}$ and the ground truth y—was introduced by Martin et al. [38] as a standard evaluation criterion for edge detection. For a given threshold T, $\hat{y}_T = \{(x, y) \mid \hat{y}(x, y) > T\}$ is binarized, and precision (P) and recall (R) are

$$P = \frac{|\hat{y}_T \cap y|}{|\hat{y}_T|}, \quad R = \frac{|\hat{y}_T \cap y|}{|y|}, \tag{45}$$

where $|\cdot|$ denotes pixel count, accounting for a small matching tolerance (e.g., 2 pixels) due to annotation variability. The F-score is

$$F = \frac{2PR}{P + R}. \tag{46}$$

However, the performance of an edge detection model can vary significantly depending on the threshold used to binarize the soft edge map. A threshold that works well for one image might perform poorly on another. Since selecting a single optimal threshold across an entire dataset is challenging, two evaluation metrics are commonly adopted: ODS (Optimal Dataset Scale) and OIS (Optimal Image Scale).

- ODS (Optimal Dataset Scale): measures the best F-score across the entire dataset using a single fixed threshold that is applied to all images, e.g., 0.824 for EDTER on BSDS500;

- OIS (Optimal Image Scale): measures the best F-score achieved for each individual image using its own optimal threshold, and then averages these F-scores across all images in the dataset, e.g., 0.876 for EdgeNAT on BSDS500.

ODS = 1 indicates perfect edge detection performance across the dataset at a single, fixed threshold. This means the predicted edge maps align exactly with the ground truth edges for all images: every predicted edge pixel is correct (precision = 1), and every true edge pixel is detected (recall = 1). In this case, the number of predicted features is exactly what is needed—neither too many nor too few. The model extracts just the right edges, at the correct locations and scales, and does so consistently across the entire dataset without needing per-image threshold adjustment. This ideal scenario reflects not only a powerful model but also a highly effective balance in the edge extraction process.

In contrast, ODS = 0 reflects complete failure of the edge detection model. This could happen if the model predicts no edges at all (leading to zero recall), or if it predicts many edges that do not match the ground truth (leading to zero precision), or both. In terms of feature extraction, this implies that the model either extracts too few edges (missing important structures) or far too many irrelevant or noisy edges (over-segmentation). Both the under- and over-prediction of features severely reduce the F1-score. This situation highlights the importance of both correct feature quantity and quality—extracting either too few or too many edges leads to poor precision–recall balance and thus a low ODS score.

OIS typically exceeds ODS, reflecting flexibility to image-specific thresholds. Both metrics, reported extensively in Sections 3 and 4, quantify edge quality, with higher scores indicating better alignment with human perception.

Table 4 summarizes BSDS500 and NYUDv2, underscoring their roles in benchmarking edge detection. Together with ODS and OIS, they enable a comprehensive assessment of algorithmic progress, bridging traditional and deep learning eras.

Table 4. Comparison of key edge detection datasets. BSDS500 offers diverse RGB images with consensus edge maps, while NYUDv2 provides RGB-D pairs for multi-modal evaluation.

Dataset	Images	Modality	Ground Truth
BSDS500	500 (200/100/100)	RGB	Multiple human annotations
NYUDv2	1449 (381/414/654)	RGB-D	Structural boundaries

6. Comparative Analysis and Insights

This section synthesizes edge detection's evolution, from gradient-based methods (Section 2) to deep learning paradigms (Sections 3 and 4), balancing high-level architectural descriptions with detailed mathematical analysis to cater to both interdisciplinary readers and experts seeking rigor (Sections 2–4). It integrates the mathematical formulations of loss functions, attention mechanisms, diffusion processes, and computational complexity with performance metrics (ODS, OIS) from Tables 5 and 6, visualized in Figures 8 and 9.

Traditional methods like Sobel and Canny (ODS 0.611) use gradient convolutions:

$$G = \sqrt{(I * K_x)^2 + (I * K_y)^2}, \tag{47}$$

where K_x, K_y are Sobel kernels but struggle with noise due to fixed kernels (Section 2). CNN-based methods introduced trainable architectures. HED (ODS 0.788) employs multi-scale supervision:

$$L_{\text{HED}} = -\sum_{i=1}^{M}\sum_{j}\left[y_j \log p_j^{(i)} + (1-y_j)\log(1-p_j^{(i)})\right], \tag{48}$$

where $p_j^{(i)}$ is the prediction at scale i. BDCN (ODS 0.806) refines edges with bidirectional supervision:

$$L_{\text{BDCN}} = \sum_{i=1}^{M} \left[\ell_{\text{BCE}}(p_i^{\text{fwd}}, y) + \ell_{\text{BCE}}(p_i^{\text{bwd}}, y) \right], \tag{49}$$

enhancing continuity (Section 3.2). NHP (ODS 0.818) leverages Hadamard product fusion (Equation (20)) for robustness.

Table 5. Overall comparison of edge detection methods. ODS and OIS reflect BSDS500 and NYUDv2 accuracy (S-measure for VST on DUTS). Dashes denote unreported values.

Method	ODS (BSDS500)	OIS (BSDS500)	ODS (NYUD)	OIS (NYUD)	Paradigm
Sobel	-	-	-	-	Gradient
Canny	0.611	0.676	N/A	-	Gradient
HED	0.788	0.808	0.720	0.734	CNN
RCF	0.798	0.815	0.729	0.742	CNN
BDCN	0.806	0.826	0.748	0.763	CNN/Enc-Dec
DexiNed	0.729	0.745	0.658	0.674	CNN/Enc-Dec
NHP	0.818	0.837	0.779	0.792	CNN/Enc-Dec
AMH-Net	0.798	0.829	0.744	0.759	CNN+Attention
VST	N/A (S: 0.91)	-	-	-	Transformer
EDTER	0.824	0.841	0.774	0.789	Transformer
EdgeNAT	0.860	0.876	0.789	0.803	Transformer
RankED	0.824	0.840	0.780	0.793	Transformer Backbone
SAUGE	0.847	0.868	0.794	0.803	Transformer Backbone
DiffusionEdge	0.834	0.848	0.761	0.766	Generative
GED	0.870	0.880	-	-	Generative

Table 6. Computational complexity vs. ODS on BSDS500.

Method	Complexity	ODS
Sobel	$O(hw)$	0.611
HED	$O(Lhwc)$	0.788
BDCN	$O(Lhwc)$	0.806
EdgeNAT	$O(h^2w^2d)$	0.860
EDTER	$O(h^2w^2d)$	0.824
SAUGE	$O(h^2w^2d)$	0.847
GED	$O(TLhwc)$	0.870

Attention-driven methods, such as EDTER (ODS 0.824) and EdgeNAT (ODS 0.860), use transformer attention described in Equation (25). The EDTER attention-based feature aggregation balances local and global cues, while the EdgeNAT SCAF-MLA decoder optimizes edge localization. Transformer backbone-based methods, RankED (ODS 0.824) and SAUGE (ODS 0.847), further advance performance. The RankED ranking loss prioritizes edge significance:

$$L_{\text{RankED}} = \sum_{i,j} \max(0, \alpha - (s_i - s_j) \cdot \text{sign}(r_i - r_j)), \tag{50}$$

with $\alpha = 1$. The SAUGE diversity loss ($\lambda_{\text{diversity}} \approx 0.1$) enhances multi-scale edges. Generative models like DiffusionEdge (ODS 0.834) and GED (ODS 0.870) model edge distributions via diffusion:

$$q(x_t|x_{t-1}) = \mathcal{N}(x_t; \sqrt{1-\beta_t} x_{t-1}, \beta_t I), \tag{51}$$

where β_t controls noise. The GED loss (Equation (43)) with $\lambda_g \approx 0.1$ achieves state-of-the-art precision.

Figure 8 visualizes ODS on BSDS500, underscoring the progression from CNNs to attention-driven and generative paradigms. The GED ODS 0.870 surpasses that of EdgeNAT, at 0.860, reflecting the generative models' ability to capture multi-scale edges as detailed in Section 4.4. However, NYUDv2 performance varies, with SAUGE (ODS 0.794) and EdgeNAT (ODS 0.789) leading, while GED lacks reported metrics, indicating generalization challenges. Traditional methods like Canny (ODS 0.611) lag significantly, limited by handcrafted kernels (Section 2).

Figure 9 compares the loss convergence of HED, BDCN, and GED, representing the CNN-based, encoder–decoder, and generative paradigms, respectively. The GED diffusion-based loss shows noticeable instability, likely due to the complexity of its optimization process and the limited size of the training dataset (BSDS500: 200 training images). In contrast, HED demonstrates stable convergence with its binary cross-entropy (BCE) loss. The dashed loss curve for EDTER is based on simulated data, as no official training loss was reported. The fluctuations observed in this attention-based method (EDTER) suggest that complex architectures may require larger training datasets to achieve stable optimization.

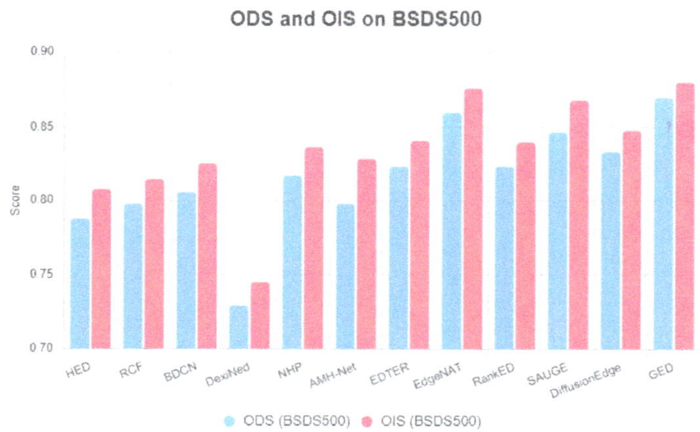

Figure 8. ODS on BSDS500, highlighting the progression from CNN-based methods (HED, RCF, and BDCN) to attention-driven (EdgeNAT, EDTER), transformer-backbone (RankED, SAUGE), and generative (DiffusionEdge, GED) paradigms.

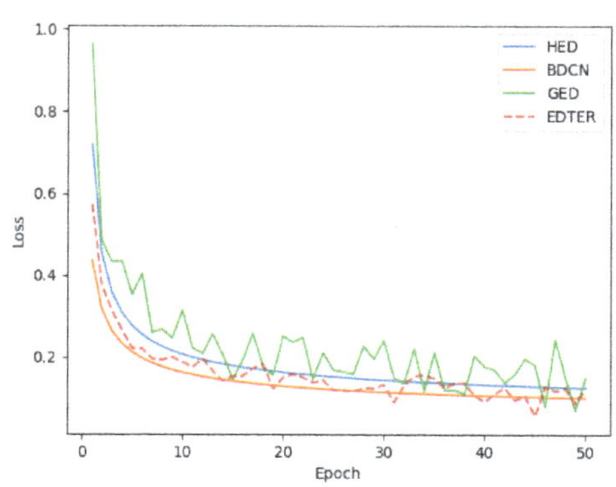

Figure 9. Loss convergence for HED (Equation (48)), BDCN (Equation (49)), and GED (Equation (51)) on BSDS500, representing CNN, encoder–decoder, and generative paradigms. The GED instability reflects complex diffusion losses.

Mathematically, the HED local supervision (Equation (48)) limits global context, while the BDCN bidirectional loss (Equation (49)) improves continuity. The EDTER and EdgeNAT attention (Equation (25)) captures long-range dependencies, boosting ODS. The RankED ranking loss (Equation (50)) prioritizes edges, but transformer complexity poses efficiency challenges. The GED diffusion (Equation (51)) excels in edge modeling but requires stabilization, highlighting gaps in optimization and multi-modal robustness (Section 7).

Computational complexity further explains performance-efficiency trade-offs. Sobel and Canny have linear complexity, $O(hw)$, for image dimensions $h \times w$, but lack adaptability. CNN-based methods like HED and BDCN scale with $O(Lhwc)$, where c is the number of channels, and L is the network depth, balancing efficiency and accuracy. Attention-driven and transformer-based methods (EdgeNAT, EDTER, RankED, and SAUGE) incur quadratic complexity, approximately $O(h^2w^2d)$, where d is the feature dimension, due to attention mechanisms (Equation (25)). For a typical 512 $\times$ 512 image, the dimensions are $d \approx 768{\sim}1280$, making transformers computationally heavy. Generative models like GED and DiffusionEdge require iterative sampling, with complexity $O(TLhwc)$, where T is the number of diffusion steps (e.g., $T = 1000$). This high cost limits real-time applications, underscoring the need for sparse attention or efficient diffusion strategies (Section 7).

Figure 10 shows qualitative comparison of samples in the BSDS500 testset from representative methods. They are RCF, BDCN, EDTER, DiffusionEdge and GED, representing different architectural paradigms (CNN based, encoder–decoder, attention based, and generative, respectively). The examples illustrate differences in edge localization, completeness, and thickness. While CNN-based RCF produces fine-grained edges, the transformer–attention approach EDTER captures global structure but may introduce noise. Generative diffusion-based GED shows strong boundary coherence, highlighting the strengths and trade-offs across paradigms.

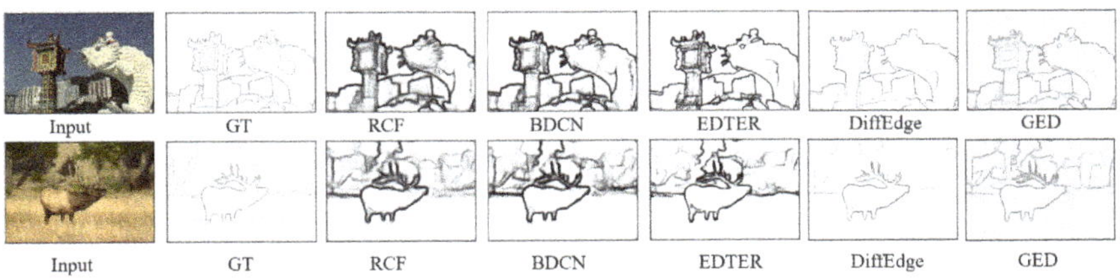

Figure 10. Qualitative comparison of edge detection outputs from representative methods. From left to right: input image, ground truth edge map, and predictions from RCF, BDCN, EDTER, DiffusionEdge and GED. These methods represent different architectural paradigms (CNN based, encoder–decoder, attention based, and generative, respectively).

7. Challenges

Edge detection has evolved from gradient-based methods (Section 2) to generative paradigms (Section 4.4), achieving high precision (e.g., GED: ODS 0.870, EdgeNAT: ODS 0.860) on BSDS500 [13]. However, challenges in efficiency, generalization, and robustness limit real-world deployment. Attention-based models like EdgeNAT and SAUGE (ODS 0.847) excel in accuracy but demand high computational resources, unlike lightweight CNNs such as HED (Section 3). Generalization remains elusive, with VST's S-measure 0.91 on DUTS [30] contrasting its BSDS500 limitations, and NYUDv2 results varying (e.g., SAUGE: ODS 0.794 vs. DiffusionEdge: 0.761). Robustness to diverse inputs, particularly in multi-modal settings like NYUDv2's RGB-D, is underutilized beyond EdgeNAT's HHA integration. Furthermore, Transformer-based methods (EdgeNAT, EDTER, and SAUGE) have high complexity ($O(n^2d)$), limiting real-time use (Table 6). Generative models (GED) require $O(Thw)$ due to iterative sampling. Future directions include sparse attention

($O(n \log nd)$) and optimized diffusion schedules to reduce costs while maintaining ODS (Section 6).

8. Conclusions

This survey traces the evolution of edge detection from Sobel's gradient-based filters to the GED generative framework, highlighting a field shaped by both mathematical and architectural innovation. Early CNN-based models such as HED (ODS 0.788) and BDCN (ODS 0.806) established the encoder–decoder paradigm (Section 3.2), which was later surpassed by attention-based approaches like EdgeNAT (ODS 0.860) and generative models such as GED (ODS 0.870) on the BSDS500 benchmark [13]. The mathematical foundations underpinning these advances—convolutions, multi-scale loss functions (Equation (48)), attention mechanisms (Equation (25)), and diffusion processes (Equation (51))—are analyzed in Section 6.

Despite significant progress, challenges remain in generalization and robustness. Transformer-based methods such as EDTER and EdgeNAT achieve strong performance (ODS 0.824–0.860) but incur the high computational complexity of $O(n^2d)$, while generative models like GED (ODS 0.870) face issues with training stability (Figure 9).

Future research should focus on designing more efficient architectures, such as sparse attention mechanisms [39,40] with $O(n \log n)$ complexity or hybrid CNN–transformer frameworks [41,42]. Enhancing generalization through domain-agnostic learning [43]—using self-supervised pretraining or leveraging GED diffusion priors [44]—could improve cross-dataset performance. To boost robustness, techniques like cross-modal attention [45,46] and graph-based fusion [47] hold promise for multi-modal edge detection. Furthermore, hybrid generative models that combine GED latent representations with EdgeNAT attention mechanisms may strike a balance between precision and efficiency, advancing the applicability of edge detection in domains such as robotics, medical imaging, and autonomous navigation.

This survey benefits researchers by providing (1) a clear overview of paradigms, aiding algorithm selection; (2) mathematical formulations for theoretical extensions; (3) quantitative comparisons for benchmarking; and (4) identified gaps like transformer complexity and generative instability, guiding future work. By consolidating these insights, the survey equips researchers to develop novel algorithms, optimize computational efficiency, and address challenges in generalization and robustness.

Funding: This research received no external funding.

Conflicts of Interest: The author declares no conflicts of interest.

References

1. Chen, L.C.; Papandreou, G.; Kokkinos, I.; Murphy, K.; Yuille, A.L. Deeplab: Semantic image segmentation with deep convolutional nets, atrous convolution, and fully connected crfs. *IEEE Trans. Pattern Anal. Mach. Intell.* **2017**, *40*, 834–848. [CrossRef]
2. Cordts, M.; Omran, M.; Ramos, S.; Rehfeld, T.; Enzweiler, M.; Benenson, R.; Franke, U.; Roth, S.; Schiele, B. The cityscapes dataset for semantic urban scene understanding. In Proceedings of the 2016 IEEE Conference on Computer Vision and Pattern Recognition (CVPR), Las Vegas, NV, USA, 27–30 June 2016; pp. 3213–3223.
3. Hu, G.; Reilly, D.; Alnusayri, M.; Swinden, B.; Gao, Q. DT-DT: Top-down human activity analysis for interactive surface applications. In Proceedings of the Ninth ACM International Conference on Interactive Tabletops and Surfaces, Dresden, Germany, 16–19 November 2014; pp. 167–176.
4. He, C.; Li, K.; Zhang, Y.; Tang, L.; Zhang, Y.; Guo, Z.; Li, X. Camouflaged object detection with feature decomposition and edge reconstruction. In Proceedings of the 2023 IEEE/CVF Conference on Computer Vision and Pattern Recognition (CVPR), Vancouver, BC, Canada, 17–24 June 2023; pp. 22046–22055.
5. Chen, C.; Wang, C.; Liu, B.; He, C.; Cong, L.; Wan, S. Edge intelligence empowered vehicle detection and image segmentation for autonomous vehicles. *IEEE Trans. Intell. Transp. Syst.* **2023**, *24*, 13023–13034. [CrossRef]

6. Ghandorh, H.; Boulila, W.; Masood, S.; Koubaa, A.; Ahmed, F.; Ahmad, J. Semantic segmentation and edge detection—Approach to road detection in very high resolution satellite images. *Remote Sens.* **2022**, *14*, 613. [CrossRef]
7. Hu, G.; Dixit, C.; Qi, G. Discriminative Shape Feature Pooling in Deep Neural Networks. *J. Imaging* **2022**, *8*, 118. [CrossRef]
8. Sobel, I.; Feldman, G. A 3 × 3 isotropic gradient operator for image processing. *Pattern Classif. Scene Anal.* **1968**, *1968*, 271–272.
9. Canny, J. A computational approach to edge detection. *IEEE Trans. Pattern Anal. Mach. Intell.* **1986**, *PAMI-8*, 679–698. [CrossRef]
10. Marr, D.; Hildreth, E. Theory of edge detection. *Proc. R. Soc. Lond. B* **1980**, *207*, 187–217.
11. Kovesi, P. Image features from phase congruency. *Videre J. Comput. Vis. Res.* **1999**, *1*, 1–26.
12. Hu, G.; Gao, Q. A non-parametric statistics based method for generic curve partition and classification. In Proceedings of the 2010 IEEE International Conference on Image Processing, Hong Kong, China, 26–29 September 2010; IEEE: New York, NY, USA, 2010; pp. 3041–3044.
13. Arbelaez, P.; Maire, M.; Fowlkes, C.; Malik, J. Contour detection and hierarchical image segmentation. *IEEE Trans. Pattern Anal. Mach. Intell.* **2010**, *33*, 898–916. [CrossRef]
14. Zhou, C.; Huang, Y.; Xiang, M.; Ren, J.; Ling, H.; Zhang, J. Generative Edge Detection with Stable Diffusion. *arXiv* **2024**, arXiv:2410.03080. [CrossRef]
15. Xie, S.; Tu, Z. Holistically-nested edge detection. In Proceedings of the 2015 IEEE International Conference on Computer Vision (ICCV), Santiago, Chile, 7–13 December 2015; pp. 1395–1403.
16. He, J.; Zhang, S.; Yang, M.; Shan, Y.; Huang, T. Bi-directional cascade network for perceptual edge detection. *IEEE Trans. Pattern Anal. Mach. Intell.* **2019**, *44*, 3828–3837.
17. Hu, G.; Saeli, C. Enhancing deep edge detection through normalized Hadamard-product fusion. *J. Imaging* **2024**, *10*, 62. [CrossRef]
18. Liu, N.; Zhang, N.; Wan, K.; Shao, L.; Han, J. Visual saliency transformer. In Proceedings of the 2021 IEEE/CVF International Conference on Computer Vision (ICCV), Montreal, QC, Canada, 10–17 October 2021; pp. 4722–4732.
19. Pu, M.; Huang, Y.; Liu, Y.; Guan, Q.; Ling, H. Edter: Edge detection with transformer. In Proceedings of the 2022 IEEE/CVF Conference on Computer Vision and Pattern Recognition (CVPR), New Orleans, LA, USA, 18–24 June 2022; pp. 1402–1412.
20. Jie, J.; Guo, Y.; Wu, G.; Wu, J.; Hua, B. EdgeNAT: Transformer for Efficient Edge Detection. *arXiv* **2024**, arXiv:2408.10527. [CrossRef]
21. Cetinkaya, B.; Kalkan, S.; Akbas, E. Ranked: Addressing imbalance and uncertainty in edge detection using ranking-based losses. In Proceedings of the 2024 IEEE/CVF Conference on Computer Vision and Pattern Recognition (CVPR), Seattle, WA, USA, 16–22 June 2024; pp. 3239–3249.
22. Liufu, X.; Tan, C.; Lin, X.; Qi, Y.; Li, J.; Hu, J.F. SAUGE: Taming SAM for Uncertainty-Aligned Multi-Granularity Edge Detection. In Proceedings of the 2025 AAAI Conference on Artificial Intelligence, Philadelphia, PA, USA, 25 February–4 March 2025; Volume 39, pp. 5766–5774.
23. Liu, Y.; Cheng, M.M.; Hu, X.; Wang, K.; Bai, X. Richer convolutional features for edge detection. *IEEE Trans. Pattern Anal. Mach. Intell.* **2017**, *41*, 3000–3009.
24. Xavier, S.; Riba, E.; Sappa, A. Dense extreme inception network: Towards a robust cnn model for edge detection. In Proceedings of the 2020 IEEE Winter Conference on Applications of Computer Vision (WACV), Snowmass, CO, USA, 1–5 March 2020; pp. 1923–1932.
25. Xu, D.; Ouyang, W.; Alameda-Pineda, X.; Ricci, E.; Wang, X.; Sebe, N. Learning deep structured multi-scale features using attention-gated crfs for contour prediction. *Adv. Neural Inf. Process. Syst.* **2017**, *30*, 3964–3973.
26. Simonyan, K.; Zisserman, A. Very deep convolutional networks for large-scale image recognition. *arXiv* **2014**, arXiv:1409.1556.
27. He, K.; Zhang, X.; Ren, S.; Sun, J. Deep residual learning for image recognition. In Proceedings of the 2016 IEEE Conference on Computer Vision and Pattern Recognition (CVPR), Las Vegas, NV, USA, 27–30 June 2016; pp. 770–778.
28. Vaswani, A.; Shazeer, N.; Parmar, N.; Uszkoreit, J.; Jones, L.; Gomez, A.N.; Kaiser, Ł.; Polosukhin, I. Attention is all you need. *Adv. Neural Inf. Process. Syst.* **2017**, *30*.
29. Dosovitskiy, A.; Beyer, L.; Kolesnikov, A.; Weissenborn, D.; Zhai, X.; Unterthiner, T.; Dehghani, M.; Minderer, M.; Heigold, G.; Gelly, S.; et al. An image is worth 16 × 16 words: Transformers for image recognition at scale. *arXiv* **2020**, arXiv:2010.11929.
30. Wang, L.; Lu, H.; Wang, Y.; Feng, M.; Wang, D.; Yin, B.; Ruan, X. Learning to detect salient objects with image-level supervision. In Proceedings of the 2017 IEEE Conference on Computer Vision and Pattern Recognition (CVPR), Honolulu, HI, USA, 21–26 July 2017; pp. 136–145.
31. Liu, Z.; Lin, Y.; Cao, Y.; Hu, H.; Wei, Y.; Zhang, Z.; Lin, S.; Guo, B. Swin transformer: Hierarchical vision transformer using shifted windows. In Proceedings of the 2021 IEEE/CVF International Conference on Computer Vision (ICCV), Montreal, QC, Canada, 10–17 October 2021; pp. 10012–10022.
32. Kirillov, A.; Mintun, E.; Ravi, N.; Mao, H.; Rolland, C.; Gustafson, L.; Xiao, T.; Whitehead, S.; Berg, A.C.; Lo, W.Y.; et al. Segment Anything. *arXiv* **2023**, arXiv:2304.02643.
33. Ye, Y.; Xu, K.; Huang, Y.; Yi, R.; Cai, Z. Diffusionedge: Diffusion probabilistic model for crisp edge detection. In Proceedings of the 38th Annual AAAI Conference on Artificial Intelligence, Vancouver, BC, Canada, 20–27 February 2024; Volume 38, pp. 6675–6683.

34. Ho, J.; Jain, A.; Abbeel, P. Denoising diffusion probabilistic models. *Adv. Neural Inf. Process. Syst.* **2020**, *33*, 6840–6851.
35. Isola, P.; Zhu, J.Y.; Zhou, T.; Efros, A.A. Image-to-image translation with conditional adversarial networks. In Proceedings of the 2017 IEEE Conference on Computer Vision and Pattern Recognition (CVPR), Honolulu, HI, USA, 21–26 July 2017; pp. 1125–1134.
36. Goodfellow, I.; Pouget-Abadie, J.; Mirza, M.; Xu, B.; Warde-Farley, D.; Ozair, S.; Courville, A.; Bengio, Y. Generative adversarial networks. *Commun. ACM* **2020**, *63*, 139–144. [CrossRef]
37. Silberman, N.; Hoiem, D.; Kohli, P.; Fergus, R. Indoor segmentation and support inference from rgbd images. In Proceedings of the Computer Vision–ECCV 2012: 12th European Conference on Computer Vision, Florence, Italy, 7–13 October 2012; Proceedings, Part V 12; Springer: Berlin/Heidelberg, Germany, 2012; pp. 746–760.
38. Martin, D.R.; Fowlkes, C.C.; Malik, J. Learning to detect natural image boundaries using local brightness, color, and texture cues. *IEEE Trans. Pattern Anal. Mach. Intell.* **2004**, *26*, 530–549. [CrossRef]
39. Lou, C.; Jia, Z.; Zheng, Z.; Tu, K. Sparser is faster and less is more: Efficient sparse attention for long-range transformers. *arXiv* **2024**, arXiv:2406.16747. [CrossRef]
40. Mei, Y.; Fan, Y.; Zhou, Y. Image super-resolution with non-local sparse attention. In Proceedings of the 2021 IEEE/CVF Conference on Computer Vision and Pattern Recognition (CVPR), Nashville, TN, USA, 20–25 June 2021; pp. 3517–3526.
41. Khan, A.; Rauf, Z.; Sohail, A.; Khan, A.R.; Asif, H.; Asif, A.; Farooq, U. A survey of the vision transformers and their CNN-transformer based variants. *Artif. Intell. Rev.* **2023**, *56*, 2917–2970. [CrossRef]
42. Tang, H.; Chen, Y.; Wang, T.; Zhou, Y.; Zhao, L.; Gao, Q.; Du, M.; Tan, T.; Zhang, X.; Tong, T. HTC-Net: A hybrid CNN-transformer framework for medical image segmentation. *Biomed. Signal Process. Control.* **2024**, *88*, 105605. [CrossRef]
43. Tamkin, A.; Liu, V.; Lu, R.; Fein, D.; Schultz, C.; Goodman, N. Dabs: A domain-agnostic benchmark for self-supervised learning. *arXiv* **2021**, arXiv:2111.12062.
44. Graikos, A.; Malkin, N.; Jojic, N.; Samaras, D. Diffusion models as plug-and-play priors. *Adv. Neural Inf. Process. Syst.* **2022**, *35*, 14715–14728.
45. Wan, Z.; Zhang, P.; Wang, Y.; Yong, S.; Stepputtis, S.; Sycara, K.; Xie, Y. Sigma: Siamese mamba network for multi-modal semantic segmentation. In Proceedings of the 2025 IEEE/CVF Winter Conference on Applications of Computer Vision (WACV), Tucson, AZ, USA, 26 February–6 March 2025; IEEE: New York, NT, USA, 2025; pp. 1734–1744.
46. Chen, Q.; Huang, G.; Wang, Y. The weighted cross-modal attention mechanism with sentiment prediction auxiliary task for multimodal sentiment analysis. *IEEE/ACM Trans. Audio Speech Lang. Process.* **2022**, *30*, 2689–2695. [CrossRef]
47. Weerakoon, K.; Sathyamoorthy, A.J.; Liang, J.; Guan, T.; Patel, U.; Manocha, D. Graspe: Graph based multimodal fusion for robot navigation in outdoor environments. *IEEE Robot. Autom. Lett.* **2023**, *8*, 8090–8097. [CrossRef]

mathematics

MDPI

Article

ConvNext as a Basis for Interpretability in Coffee Leaf Rust Classification

Adrian Chavarro [1,†], Diego Renza [1,†] and Ernesto Moya-Albor [2,*,†]

1 Facultad de Ingeniería, Universidad Militar Nueva Granada, Carrera 11 101-80, Bogotá 110111, Colombia; est.adrian.chavarro@unimilitar.edu.co (A.C.); diego.renza@unimilitar.edu.co (D.R.)
2 Facultad de Ingeniería, Universidad Panamericana, Augusto Rodin 498, Ciudad de México 03920, Mexico
* Correspondence: emoya@up.edu.mx; Tel.: +52-55-5482-1600 (ext. 5210)
† These authors contributed equally to this work.

Abstract: The increasing complexity of deep learning models can make it difficult to interpret and fit models beyond a purely accuracy-focused evaluation. This is where interpretable and eXplainable Artificial Intelligence (XAI) come into play to facilitate an understanding of the inner workings of models. Consequently, alternatives have emerged, such as class activation mapping (CAM) techniques aimed at identifying regions of importance for an image classification model. However, the behavior of such models can be highly dependent on the type of architecture and the different variants of convolutional neural networks. Accordingly, this paper evaluates three Convolutional Neural Network (CNN) architectures (VGG16, ResNet50, ConvNext-T) against seven CAM models (GradCAM, XGradCAM, HiResCAM, LayerCAM, GradCAM++, GradCAMElementWise, and Eigen-CAM), indicating that the CAM maps obtained with ConvNext models show less variability among them, i.e., they are less dependent on the selected CAM approach. This study was performed on an image dataset for the classification of coffee leaf rust and evaluated using the RemOve And Debias (ROAD) metric.

Keywords: class activation mapping (CAM); deep learning; model evaluation; model interpretation; rust classification; RemOve And Debias (ROAD); eXplainable Artificial Intelligence (XAI); green; environment

MSC: 68T05; 68T45; 92B20

1. Introduction

Agriculture has assumed a significant role in contemporary society due to the increasing global population and climate change, both of which can impact the worldwide food supply. Consequently, promoting the use of technology to optimize production processes is essential for enhancing productivity, which can be vulnerable to diseases. Many crop supervision tasks require intensive labor and considerable time for expert analysis to determine the current state of crops.

Recent advancements in computer vision and artificial intelligence techniques offer promising solutions for automating the detection and classification of leaf diseases. Deep-learning methods have shown impressive performance across various crops. However, despite their effectiveness, these models often lack interpretability, making it challenging to understand the reasoning behind their decisions [1]. Some efforts have focused on enhancing models for detecting plant diseases, with a particular emphasis on the critical role of visualization [2]. These studies have demonstrated significant success in accurately localizing infected regions in leaf images and highlighting them with heat map overlays. Despite the promising outcomes of using eXplainable Artificial Intelligence (XAI) methods in plant disease detection, it is recognized that the field of XAI is still developing. Current explanation maps remain inadequate for fully supporting decision-making processes in this domain [1].

Interpretability in Convolutional Neural Networks (CNNs) refers to the ability to understand, explain and attribute the decisions and predictions that the network makes. It is a crucial aspect for analyzing why a model makes certain decisions, which can be of particular importance in critical applications.

The initial visualization and interpretation works on CNNs were developed for AlexNet CNN to explore and find architectures that outperform a specific problem [3] and to provide tools to visualize the neurons of a CNN in response to specific inputs, as well as to analyze activation patterns in response to different types of activation [4]. Local Interpretable Model-agnostic Explanations (LIME) is then introduced as a solution based on the approximation of the original model by identifying a model that is locally faithful to the classifier on an interpretable representation, such as a binary vector indicating the presence or absence of a contiguous zone of similar pixels [5]. On the other hand, using game theory, SHapley Additive exPlanations (SHAP) is presented as a unified measure of feature importance that is approximated by several methods and, according to its authors, best matches human intuition [6].

In order to identify the areas of an image that represent greater importance when classifying with a CNN-based model, recent initiatives have been oriented towards methods based on class activation mapping (CAM) [7]. One of the most representative methods is Gradient-Weighted CAM (GradCAM), which computes the local importance from the gradient of the output of a given convolutional layer (usually the last one) with respect to the class predicted by the model [8]. GradCAM++ is presented as an enhancement to GradCAM that uses second-order gradients to improve object localization or multiple occurrences of the same class in an image [9], while XGradCAM is presented as an improvement to GradCAM in terms of both sensitivity and conservation, where sensitivity is understood as the consistency between the removal of input features and the respective change in output, and conservation refers to the magnitude of the output equivalent to the sum of the explanation responses [10].

Other recent examples of class activation mapping methods include HiResCAM, LayerCAM and EigenCAM. HiResCAM starts from the computation of the gradient with respect to the feature maps, and then obtains an attention map from the sum along the feature dimension of the element-wise products of the gradient and the feature maps. In order to reflect the model computations, such operations are performed when the gradients indicate that some elements of the feature map should be scaled or have their sign reversed [11]. A variant of the latter method, GradCAMElementWise, performs an element-wise multiplication of activations with gradients and then applies a ReLU (Rectified Linear Unit) operation before summation [12]. In LayerCAM, to obtain the class activation map of a layer, the activation value of each location in the feature map is multiplied by a weight that depends on the backward class-specific gradients; for locations with positive gradients, their gradients are used as the weight, while locations with negative gradients are assigned zero [13]. An alternative that, according to its authors, does not rely on feature weighting such as gradient backpropagation or class relevance scoring, is the so-called EigenCAM method, which uses the principal components of the learned representations of the convolutional layers to create the visual explanations [14].

The purpose of the methods related so far is primarily designed to illustrate which parts of an image influence the decision of a convolutional neural network. These methods have been validated on different CNNs architectures, including Inception (LIME), ResNet (SHAP, GradCAM, XGradCAM) and VGG16 (SHAP, GradCAM, GradCAM++, XGrad-CAM, LayerCAM, EigenCAM), with VGG16 and ResNet being the most commonly used architecture types when validating feature visualization methods using class maps. In addition to the above, it is important to keep in mind that the great diversity of methods for the identification of key regions for a classifier makes the selection of a specific method not straightforward [15]. Also, the diversity of CNN architectures and their evolution can hinder the process of selecting and analyzing CAM-type methods.

As is well known, CNNs revolutionized computer vision research by involving feature extraction learning in training without the need for manual extraction processes [16,17]. In addition to this, Vision Transformers (ViT), derived from the original work on natural language processing tasks, have become a competitive option thanks to their scalability in relation to model and dataset size [18]. Under this scenario, standard CNNs were also modernized towards the structure of a hierarchical Vision Transformer, without introducing any attention-based modules [16]. In this way, CNNs can be as scalable as vision transformers, highlighted especially in scenarios requiring low complexity. This family has been oriented to maintain the simplicity and efficiency of CNNs and has been named ConvNext [16], and its evolution, ConvNext2, is based on self-supervised learning [18].

As previous research has pointed out, the quality of heat maps depends both on the methods by which they are calculated and on the performance of the classifier (model+data) [19]. Consequently, this paper evaluates the interpretability of CNN models in the classification of coffee leaf rust. In particular, three CNN architectures (VGG16, Resnet50 and ConvNext-T) are evaluated via the seven CAM techniques both qualitatively and quantitatively using a ROAD metric.

The rest of the paper is structured as follows: Section 2 provides the fundamentals of ConvNext networks, the selected CAM methods, and the ROAD metric. Section 3 describes the coffee leaf dataset, the pre-trained models and the evaluation method. Section 4 presents preliminary qualitative results for some example data and qualitative results evaluated on the test dataset. Finally, Section 5 concludes the research results.

2. Background

In this section, the fundamentals of the ConvNext architecture, its detailed specifications and its variants are explained. Moreover, the theory of CAM methods and its evolution are given. Finally, this section details the ROAD metric and its calculation.

2.1. ConvNext

The ViT model [20], launched on 2021, quickly overcame the performance of the state-of-the-art CNN models for recognition tasks. It uses patches of an image in a similar way as the natural language processing uses tokens from structured text. Thus, by using CNNs or ConvNet modules in ViT, it was shown that the accuracy increased whereas the computational cost was reduced [21], thereby enabling the transformer to be used in various visual tasks, such as image classification, object detection and semantic segmentation problems. On the other hand, the Swim Transformer [22], a new transformer model that uses a shifted windowing scheme, has also been also used to improve the performance of models. This is because the shifted windows limit the self-attention computation to non-overlapping local windows, but allows the cross-window connections.

In contrast to transformers, in Liu et al. [16] was presented ConvNext, a natural CNN network that uses a ResNet-50 network containing ConvNet modules and is based on the design of the Swim Transformer. ConvNext overcomes the performance of the Swim Transformer, e.g., in accuracy and scalability, whereas it holds the simplicity and efficiency of standard ConvNet modules.

A ConvNext network is composed by a stem unit, four stages of ConvNext blocks, a Global Average Pooling (GAP) layer followed a Layer Normalization and a last linear layer. The typical structure of a ConvNext network is shown in Figure 1 and its internal structure on Figure 2 [23].

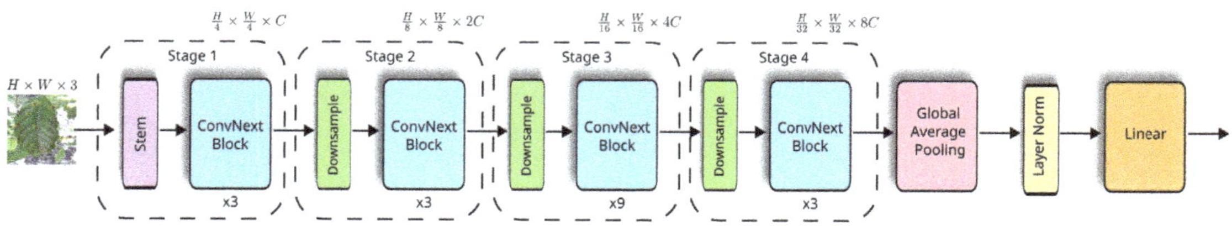

Figure 1. ConvNext architecture.

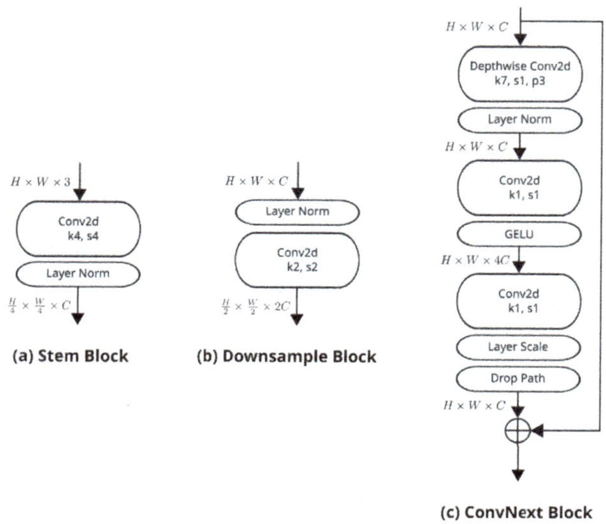

Figure 2. (**a**) The internal structure of Stem Block. (**b**) The internal structure of Downsample Block. (**c**) The internal structure of ConvNext Block.

The Stem Block in Figure 1 processes the input images via a patchify structure composed by a convolutional layer and a layer normalization, as can be seen in Figure 2a. Thus, for an input image that $H \times W \times 3$, with H and W as the height and width of the image, respectively, the convolution layer downsampled it through C kernels of 4×4 ($k = 4$) and stride of 4 ($s = 4$), generating features of a quarter of the input size with C channels for each. Finally, a layer normalization is applied to the generated features.

Moreover, the four stages of the ConvNext network in Figure 1 are each composed of a Downsample Block and the so-called ConvNext Block . In addition, each of the ConvNext blocks cycles for a different number of times, as indicated at the bottom of the block, e.g., $\times 3$ or $\times 9$ cycles.

On the one hand, the Downsample Block (Figure 2b) generates hierarchical features at multi-scale by performing convolutions with kernels of 2×2 ($k = 2$) and stride of 2 ($s = 2$), which reduces the resolution of the input feature maps and doubles the number of channels ($2C$). On the other hand, each ConvNext block (Figure 2c) contains three convolutional layers, a depth-wise convolution ($k = 7, s = 1$), two 1×1 convolution layers ($k = 1, s = 1$), and a Gaussian Error Linear Unit (GELU) as the activation function between them. At the output of the second convolutional layer a Layer Scale and a Drop Path are applied. Finally, a Residual Connection is used.

In applying the depth-wise convolutions in the ConvNext architecture, a kind of group-wise convolution that groups channels is performed. Thus, joining the depth-wise convolution and the 1×1 convolution layers generates a similar effect as the shared property in vision transformers [23].

Finally, there are several ConvNext versions. They follow the notation ConvNext-variant, where "variant" could be Tiny (T), Small (S), Base (B), Large (L) and X-Large (XL). These variants differ by the number of channels in each stage and the number of cycles of each ConvNext block (see Table 1) [23]. Thus, the ConvNext network shown in Figure 1 corresponds to the Tiny variant or ConvNext-T depending on the number or cycles of each ConvNext block (3, 3, 9, 3). The detailed specifications of the ConvNext-T architecture are shown in Table 2 [10].

Table 1. Variants of ConvNext architecture.

Variant ConvNext	Number of Channels in Each Stage	Number of Cycles of Each ConvNext Block
ConvNext-T	(96, 192, 384, 768)	(3, 3, 9, 3)
ConvNext-S	(96, 192, 384, 768)	(3, 3, 27, 3)
ConvNext-B	(128, 256, 512, 1024)	(3, 3, 27, 3)
ConvNext-L	(192, 384, 768, 1536)	(3, 3, 27, 3)
ConvNext-XL	(256, 512, 1024, 2048)	(3, 3, 27, 3)

Table 2. ConvNext-T architecture detailed specifications. The specifications in each Res stage indicate the type of convolution (d for depth-wise convolutions), the size of the convolution kernel, the number of channels and the number of blocks.

Stage	Output Size	Details
Stem	56×56	4×4, stride 4
Res2	56×56	$\begin{bmatrix} d7 \times 7 & 96 \\ 1 \times 1 & 384 \\ 1 \times 1 & 96 \end{bmatrix} \times 3$
Res3	28×28	$\begin{bmatrix} d7 \times 7 & 192 \\ 1 \times 1 & 768 \\ 1 \times 1 & 192 \end{bmatrix} \times 3$
Res4	14×14	$\begin{bmatrix} d7 \times 7 & 384 \\ 1 \times 1 & 1536 \\ 1 \times 1 & 384 \end{bmatrix} \times 9$
Res5	7×7	$\begin{bmatrix} d7 \times 7 & 768 \\ 1 \times 1 & 3072 \\ 1 \times 1 & 768 \end{bmatrix} \times 3$

2.2. CAM Methods

One of the weaknesses of deep learning-based classification models is that most of the time they are seen as "black boxes", since it is not intuitive to relate the weight of the parameters to the type of features identified by the model. Within the current state of the art, several methods have been proposed that precisely focus on the visualization and interpretation of the patterns learned by the CNN models. Thus, CAM methods provide a visual output of the most interesting areas found by the CNN models [8,24]. The main characteristics of the CAM methods used in this work are presented below.

GradCAM: The original CAM method generates localized visual explanations, but requires architecture adjustments for its application. In contrast, GradCAM generates visual explanations of the decisions made by CNNs, locating the regions of greatest importance to the classifier's decision-making [25]. This method calculates the gradient of the network output with respect to the feature maps of the last convolutional layer, indicating how a small change in each feature map would affect the network output for a specific class. The weighted average of the gradients of each neuron in the last convolutional layer is then calculated, where each value represents the relative importance of that region of the image for classification [8].

GradCAM++: Considering that GradCAM calculates the weighted average of the gradients in the last convolutional layer, which can lead to visualizations that are fuzzy or do not accurately capture the most important regions, GradCAM++ was proposed. To solve this problem, GradCAM++ is not limited to the last convolutional layer, but calculates gradients in multiple layers of the network and uses a more sophisticated weighting function to combine gradients from different layers [9].

XGradCAM: A method based on a linear combination of the feature maps of the last convolutional layer, whose weights are calculated in such a way that the visualizations are sensitive to changes in the input (sensitivity) and that the sum of the values in the activation map is constant and related to the confidence of the classification (conservation) [10].

HiResCAM: A technique proposed as an enhancement to GradCAM to generate more accurate and detailed visual explanations. HiResCAM is not limited to the last convolutional layer, but generates activation maps for multiple layers of the network, including the initial layers. These activation maps are combined using a weighting function that assigns higher weights to layers that contribute more to the classification. In addition, a refinement process is applied to the combined activation maps to improve their quality and reduce noise [11].

LayerCAM: LayerCAM is an attention method for generating reliable class activation maps in shallow layers with large spatial resolutions, not only in the final convolutional layer, where classic approaches such as GradCAM fail to obtain accurate subtle localization of objects [13]. This CAM method harvests reliable and complementary class activation maps for all layers to generate an individual weight for the spatial location in the feature map, benefiting weakly supervised tasks. In addition, LayerCAM is easily used in the common CNN models without modifying their architectures and backpropagation form.

EigenCAM: Unlike the previous methods, the EigenCAM method leverages the principal components, calculated from learned representations at the final convolutional layer, to generate visual and explicable representations of the CNN model [14]. It is a simple method to calculate CAM from output layers independent of the class relevance score. Furthermore, it has the advantage of locating objects in the presence of adversarial noise without modification, training, or backpropagating parameters across layers of the network architecture, yet overcomes gradient-based CAM methods, e.g., GradCAM.

2.3. ROAD

The RemOve And Retrain (ROAR) framework is an initiative to evaluate interpretability methods, testing how the performance of a re-trained model is affected as important input features are removed. In this method, the dataset images are modified (training and testing) such that the pixels identified as the most important are replaced by a fixed non-informative value. Using the modified datasets, new models are trained, repeating the process several times in order to minimize the variance in accuracy [15].

Based on the ROAR proposal, a new proposal called RemOve And Debias (ROAD) was proposed by Rong et al. [26]. It offers greater consistency and substantially reduces the computational cost since it does not require the retraining phase. In general, this method takes as input the result of an explanation method that indicates the most important pixels and by means of a selection operator discriminates the relevant features. These features go to an imputation operator that reduces mutual information and bias. This process, in turn, can balance both a strategy of eliminating the most relevant pixels first or the least relevant pixels first [26].

ROAD is a conceptual framework for evaluating attribution methods that is based on the idea of removing features from the input data and then compensating for possible information leakage. Generally speaking, ROAD proposes to remove a subset of features (e.g., pixels in an image) to quantify the degree of importance of the features identified by the attribution method. However, it does not rely on simple feature removal, as the removed area may unintentionally influence the remaining features, leading to misleading results. Consequently, ROAD proposes a technique called "Noisy Linear Imputation" which consists in replacing the removed features by values obtained from a linear interpolation [26].

To use ROAD to evaluate a feature attribution method, first, a prediction is performed with a classification model, keeping all the input features intact. The behavior of the classifier is analyzed through a feature attribution method (e.g., CAM), which allows assigning to each input dimension an importance value. These values in turn can be ordered from two points of view: descending (Most Relevant First, MoRF) and ascending (Least Relevant First, LeRF), and a given number of features are selected per instance that will form the subset of features that will be replaced in the input data (masked features).

For the calculation of the ROAD metric, both the most relevant and the least relevant features of the image are first removed independently. To eliminate the most relevant features, a threshold value is calculated, which is defined from a percentile given by the user. Subsequently a binary mask is calculated, given by Equation (1):

$$binary\_mask = mask < threshold, \tag{1}$$

where *mask* corresponds to the importance values given by the CAM method. In the case of no specifying percentile, the *threshold* value is calculated via Otsu's method. Next, an imputation of the input tensor is performed using the binary mask obtained according to Equation (2).

$$MoRF = imputer(input\_tensor, binary\_mask). \tag{2}$$

This imputer is responsible for filling in the missing values in the input tensor after applying the binary mask. For the case of an image, each pixel ($x_{i,j}$) is approximated by a weighted average of its neighbors according to Equation (3):

$$x_{i,j} = w_d\left(x_{i,j+1} + x_{i,j-1} + x_{i+1,j} + x_{i-1,j}\right) + w_i\left(x_{i+1,j+1} + x_{i-1,j+1} + x_{i+1,j-1} + x_{i-1,j-1}\right) + X, \tag{3}$$

where $w_d = 1/6$ and $w_i = 1/12$ are constant coefficients for the direct neighbors and the indirect neighbors (diagonals). X is a random variable whose purpose is to add random noise ($\sigma = 0.1$) to the solution.

To obtain the modified image after eliminating the less relevant features, a similar procedure is applied but previously inverting *mask* and applying similar equations:

$$mask_{lerf} = 1 - mask \tag{4}$$

$$binary\_mask_{lerf} = mask_{lerf} < threshold, \tag{5}$$

$$LeRF = imputer(input\_tensor, binary\_mask_{lerf}). \tag{6}$$

Since the outcome of evaluation strategies can be very sensitive to parameters such as ordering, (i.e., *MoRF*, *LeRF*) ROAD metric can consider the inclusion of these two approaches by semi-differencing them (Equation (7)).

$$ROAD_{combined} = (scores_{lerf} - scores_{morf})/2, \tag{7}$$

Thus, the higher the value of the ROAD combined metric ($ROAD_{combined}$), the more important the features identified by the attribution method are to the classifier.

For the calculation of the ROAD metric, a library available under MIT license and published on GitHub [12] has been used. This library has been implemented based on the article where this metric has been proposed [26], in which they point out that this method has significantly reduced the cost with respect to previous methods such as RemOve And Retrain (ROAR) [15]. It is important to note that as evidenced by Equations (1)–(7), the computational cost of the ROAD metric itself is not high. The computational cost of applying the metric depends primarily on the cost of the CAM method used.

3. Materials and Methods

3.1. Dataset

In this study, several datasets were leveraged to aggregate a comprehensive repository of images utilized across the training, validation and testing stages of CNN model design and evaluation for the detection of coffee leaf rust. These datasets include Rocole [27], Bracol [28], Digipathos [29], D&P [30] and Licole [31]. They encompass instances of both healthy leaves and various disease types such as rust (R: Hemileia vastatrix), leaf miner (LM: Leucoptera coffellum), phoma (P: Phoma spp.), cercospora (C: Cercospora coffeicola), spider mite (RSM: Oligonychus yothersi), bacterial blight (BB: Pseudomonas syringae), blister spot (BS: Colletotrichum gloeosporioides) and sooty mold (SM: Capnodium spp.).

The datasets under consideration demonstrate a notable breadth of diversity, encompassing various elements such as the presence of distractors, diverse illumination conditions (whether from artificial sources or natural sunlight), a spectrum of backgrounds and varying quantities of leaves per individual image. To prepare the dataset for analysis, meticulous preparatory procedures were executed. These tasks encompassed the precision cropping of images to isolate coffee leaves, the meticulous removal of distractors including undesired objects such as the hands of coffee growers, the exclusion of images featuring diseases other than rust, and the elimination of artificial backgrounds. The resultant dataset consists of 1686 images classified as healthy, 1723 images depicting rust, and 1928 images portraying other diseases (specifically, LM, P, C, RSM, BB, BS, and SM). The allocation for training, validation, and testing purposes involved distributing 70%, 20%, and 10% of the total images, respectively. Additionally, all the jpg images were uniformly resized to dimensions of 224 × 224 pixels to facilitate consistent computational processing and analysis using RGB color space.

3.2. Pre-Trained Models

The CAM maps were calculated from three CNN classifiers based on VGG, ResNet, and ConvNext. The first two correspond to the two most used architectures in the evaluation of CAM methods proposed in the literature. The third one corresponds to one of the most recent architectures designed to be competitive in the context of computer vision applications. In this case, the two most widely used CNN architectures in the literature for evaluating CAM methods (VGG and ResNet) were selected, and one of the most recent pre-trained CNN architectures (ConvNext) that are widely available (e.g., in frameworks such as Tensorflow or PytTorch) for transfer learning and/or fine tuning tasks was added.

The selected architectures were used from the pre-trained models and weights available in PyTorch. The version selected in the case of VGG was VGG16, in the case of ResNet it was ResNet50, and for ConvNext the lighter version of ConvNext was used (ConvNext-T). In all three cases, the architecture classification stage was adjusted to three classes.

A fine tuning process was applied to the three pre-trained models using the training and validation subsets of the dataset explained above. This pre-training was performed on all model layers for 50 epochs, with an Adam optimizer, and learning rate decay. A checkpoint was also used to save the model, only upon an improvement in validation accuracy. The performance achieved during model fine tuning is shown in Tables 3 and 4.

Table 3. Classifier performance in terms of accuracy for the three selected architectures when applying fine tuning on the rust dataset.

Model	Val. Loss	Val. Accuracy (%)	Test Accuracy (%)	Epoch	Size (MB)
VGG16	0.1998	96.46	93.33	16	512.3
ResNet50	0.2089	94.38	91.25	46	90
ConvNext-T	0.2825	94.45	92.08	23	106.2

Table 4. Evaluation metrics on test data derived from the confusion matrix for the models with fine tuning.

		VGG16		
Class	Accuracy (%)	Precision	Recall	F1-Score
1	96.25	0.91	0.98	0.94
2	95.42	0.97	0.91	0.94
3	95.00	0.92	0.92	0.92
		ResNet50		
Class	Accuracy (%)	Precision	Recall	F1-Score
1	94.38	0.86	0.98	0.91
2	95.42	0.97	0.91	0.94
3	93.00	0.91	0.85	0.88
		ConvNext-T		
Class	Accuracy (%)	Precision	Recall	F1-Score
1	94.58	0.89	0.95	0.92
2	96.04	0.96	0.94	0.95
3	93.54	0.92	0.87	0.89

3.3. Selected CAM Methods

To identify the regions of importance in the decision-making of the selected classification models, class activation mapping methods were used, starting from the pioneering GradCAM scheme towards those that have made proposals to improve different aspects of the interpretation of the model output. The selected CAM schemes are listed below:

- GradCAM;
- XGradCAM;
- HiResCAM;
- LayerCAM;
- GradCAM++;
- GradCAMElementWise;
- EigenCAM.

3.4. Evaluation Using ROAD

As discussed so far, feature attribution methods such as those based on CAM aim to identify the importance of input features for a model's decision, within what is known as XAI. As the supply of such methods has increased, so has the need to establish techniques to evaluate these methods [26].

This type of strategy for evaluating distribution methods is generally based on perturbing the input features that have been identified by the model as important/unimportant. Here, it is assumed that affecting the most important input features affects the performance of the model, while affecting less important features has little impact on the performance of the model [26]. An example of such features can be the pixels of the regions identified via the attribution method (e.g., CAM).

According to the proposed methodology, the output feature map of the selected CNN architectures was evaluated using a CAM-type method. Subsequently, this CAM map was evaluated using the ROAD metric and the results were compared in order to evaluate the consistency of the selected CNN architectures. The objective here was to address the veracity/fidelity of explanation with respect to model prediction, i.e., to assess correctness, in the context of a recent classification of quantitative evaluation of Explainable AI methods [32]. The type of evaluation method selected consisted of removing features from the input as a function of explanation (CAM) and measuring the change in predictive model output [32].

4. Results and Discussion

This section shows the results of the application of the proposed methodology on the dataset outlined above. These results are presented first for the visual analysis of the results, and then the consolidated quantitative results for the test data are shown.

4.1. Preliminary Results (Qualitative)

In order to illustrate the differences between the regions of importance identified by the CAM methods in any of the three architectures, three examples with different levels of rust severity are shown. Figures 3–5 first show the original image evaluated and then the seven CAM maps (GradCAM, XGradCAM, HiResCAM, LayerCAM, GradCAM++, GradCAMElementWise, and EigenCAM) obtained with each of the three architectures.

(**a**) Test image

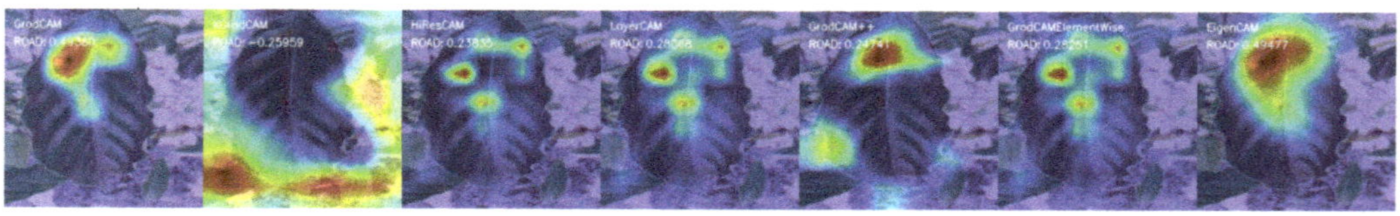

(**b**) VGG16

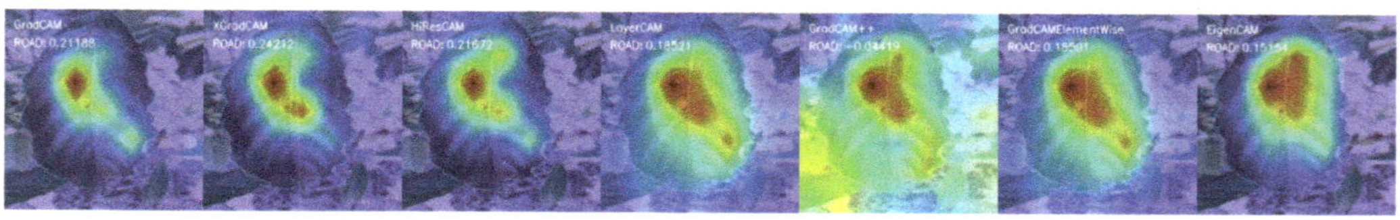

(**c**) ResNet50

(**d**) ConvNext-T

Figure 3. CAM maps obtained from a first test image, from left to right: GradCAM, XGradCAM, HiResCAM, LayerCAM, GradCAM++, GradCAMElementWise, and EigenCAM. (**a**) Test image. (**b**) CAD maps from VGG16 architecture. (**c**) CAM maps from ResNet50 architecture. (**d**) CAM maps from ConvNext-T architecture.

In the case of Figure 3, it is observed that the CAM maps obtained after applying the VGG16-based model show significant differences among them, mainly associated with the localization of the zones. However, some results show high similarity as in the case of HiResCAM, LayerCAM, and GradCAMElementWise. The results obtained using the ResNet50 architecture show a higher correlation between them, in terms of the localization and intensity of the most important zones (red tones). The differences found are mainly related to the size of the area. On the other hand, the CAM maps obtained with the ConvNext-T architecture present greater consistency among themselves in terms of localization and intensity, particularly for the first six methods shown. In this case, only the last method (EigenCAM) differs significantly from the other six.

(**a**) Test image

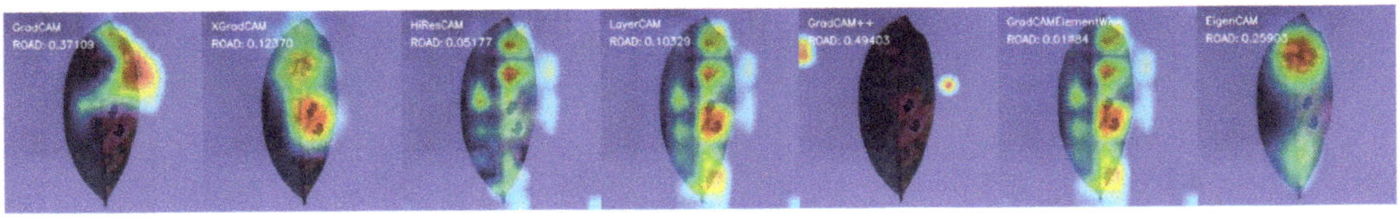

(**b**) VGG16

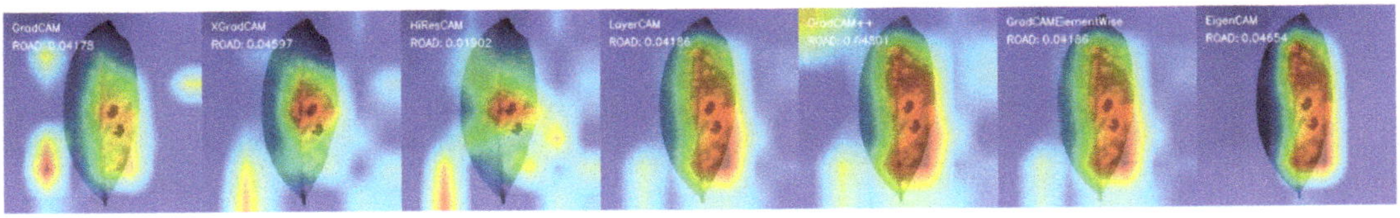

(**c**) ResNet50

(**d**) ConvNext-T

Figure 4. CAM maps obtained from a second test image, from left to right: GradCAM, XGradCAM, HiResCAM, LayerCAM, GradCAM++, GradCAMElementWise, and EigenCAM. (**a**) Test image. (**b**) CAD maps from VGG16 architecture. (**c**) CAD maps from ResNet50 architecture. (**d**) CAM maps from ConvNext-T architecture.

(a) Test image

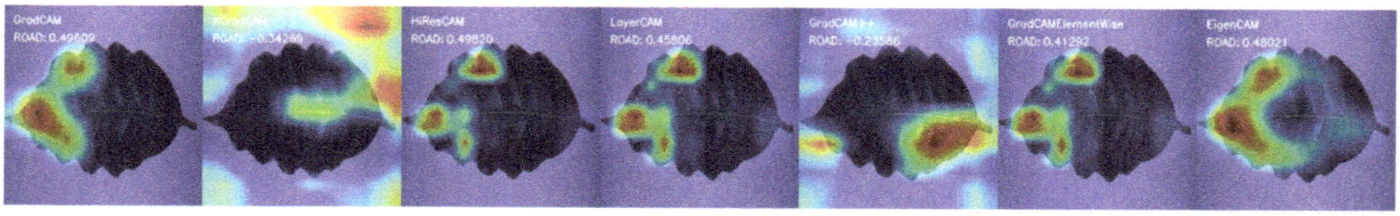

(b) VGG16

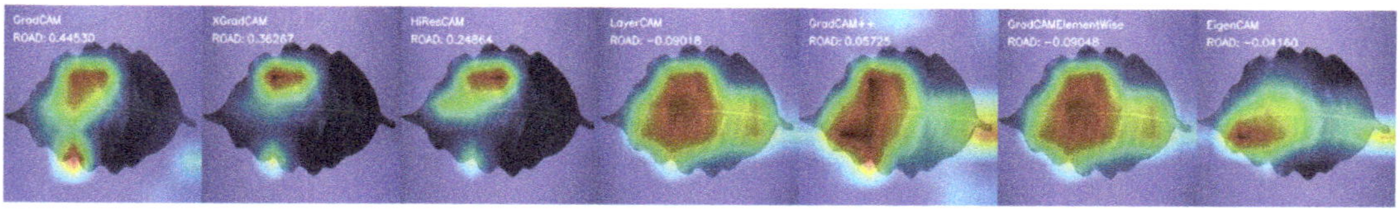

(c) ResNet50

(d) ConvNext-T

Figure 5. CAM maps obtained from a third test image, from left to right: GradCAM, XGradCAM, HiResCAM, LayerCAM, GradCAM++, GradCAMElementWise, and EigenCAM. (**a**) Test image. (**b**) CAD maps from VGG16 architecture. (**c**) CAD maps from ResNet50 architecture. (**d**) CAM maps from ConvNext-T architecture.

The second test image shows a higher number of areas with rust presence (Figure 4). First, the results for VGG16 show a great diversity of results among the different CAM methods evaluated. For example, there are methods that identify a single zone (GradCAM), methods that identify a large number of regions (HiResCAM, LayerCAM, GradCAMElementWise) or methods that identify regions outside the affected zone (GradCAM++). In the case of ResNet50-based results, from a similarity point of view, it is possible to group the results into two groups: The first identifies a small region of great importance inside the leaf, but also identifies regions outside the leaf (GradCAM, XGradCAM, HiResCAM). The second group highlights a wider area inside the leaf, covering a larger number of affected areas, but also identifies areas outside the leaf (LayerCAM, GradCAM++, GradCAMElementWise, EigenCAM). Something similar happens with the results obtained by

ConvNext-T. In this case, the first group shows a small region near the leaf tip with the presence of rust, which is cataloged as being of great importance (GradCAM, XGradCAM, HiResCAM, LayerCAM). The second group identifies much of the leaf, with the leaf tip being of greatest importance, although the heat map includes areas that fall outside the leaf.

The third example (Figure 5) shows a leaf with little presence of rust in specific areas. Again, some of the results obtained with the VGG16 architecture are similar to each other, but others vary substantially. In the case of ResNet50, the results show a similar trend to the two previous examples, where they can be framed in two groups that vary mainly by the size of the identified area. On the other hand, the results obtained with ConvNext-T show a great consistency, being very similar to each other.

4.2. Consolidated Results (Quantitative)

The proposed methodology was applied on the entire test dataset in order to evaluate the consistency of the results. In this regard, for each of the three selected CNN architectures (VGG16, ResNet50, and ConvNext-T), tests were performed with each of the seven CAM methods evaluated on the 480 images. In other words, in each architecture we have $480 \times 7 = 3360$ ROAD metric results for each type of CNN architecture, that is, a total of 10,080 results.

Since the visual results show considerable similarities between the maps obtained with different CAM methods, especially for the ConvNext-T architecture, scatter plots were made to show the relationship between the ROAD metrics of two given CAM methods. For comparison purposes, the ROAD metric obtained with the fundamental CAM method, i.e., GradCAM, was fixed on the X-axis. In this way, the Y-axis varies with each of the other six selected CAM methods.

Figure 6 shows the scatter plots after calculating the ROAD metric on the test dataset evaluated with the VGG16 architecture. The results in this plot show that there is no correspondence between the results obtained with different CAM methods For example, in methods such as XGradCAM, GradCAM++ or EigenCAM, it is observed that in a large number of cases, positive metric values in GradCAM correspond to negative metric values for the other CAM method and vice versa. Another aspect that stands out in this figure is that there is no proportionality between the values of the two axes, i.e., there is no relationship that illustrates that if one of the two metrics increases, the other also increases and vice versa.

Figure 7 shows the scatter plots after calculating the ROAD metric on the test dataset evaluated with the ResNet50 architecture. There is a fundamental aspect that stands out in the scatter plots of this graph: the proportionality of the results. In this case, a trend of proportionality is observed around a line of slope 1. That is, in these results, although there is a large variability around the hypothetical line of slope 1, there is a trend that as the X-axis values increase, the Y-axis values increase. This characteristic is an important difference with respect to the results obtained with VGG16.

Finally, Figure 8 shows the scatter plots after calculating the ROAD metric on the test dataset evaluated with the ConvNext-T architecture. The first thing that stands out in this plot is that the results of the XGradCAM and HiResCAM methods are highly similar to the results obtained with GradCAM. The values in these two plots, although they look like a perfect straight line, have slight differences as illustrated by the ROAD values shown in Figures 3–5d. However, what is corroborated here is that the importance of the regions identified through the GradCAM, XGradCAM, and HiResCAM methods have a nearly equal impact according to the ROAD metric values. Thirdly, the results obtained with the LayerCAM method show a trend proportional to the results obtained with GradCAM, i.e., it articulates around a straight line of slope 1, but with much less variation compared to the results of the ResNet50 architecture. In the case of the other three methods evaluated, the results do not present high similarities. However, the results using ConvNext-T are similar for four of the seven CAM methods evaluated, as shown in Figures 3–5d.

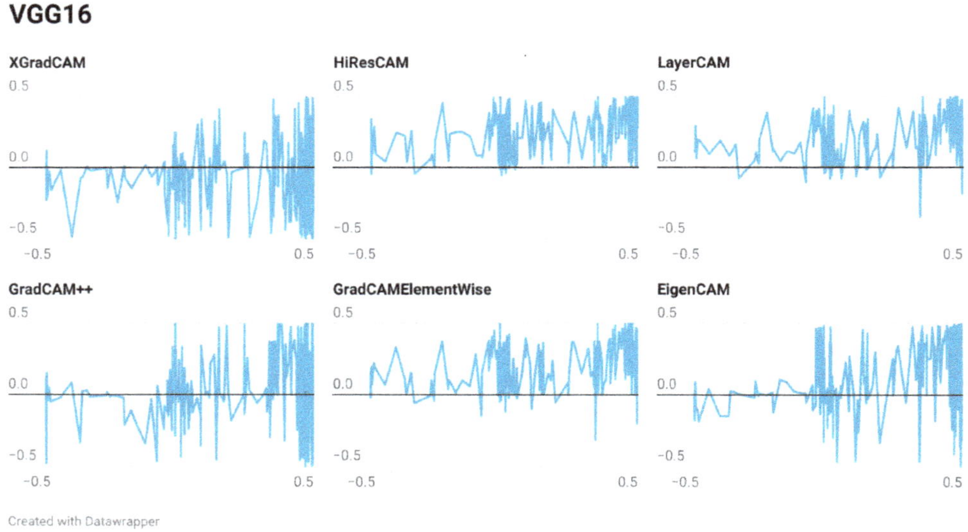

Figure 6. GradCAM ROAD metric versus ROAD metrics from six different CAM methods. CAM maps evaluated on the last convolutional layer of a VGG16 architecture adapted for three classes, on 480 test data. The X-axis corresponds to the ROAD metric evaluated on the GradCAM map, while the Y-axis corresponds to the ROAD metric evaluated on the respective CAM map.

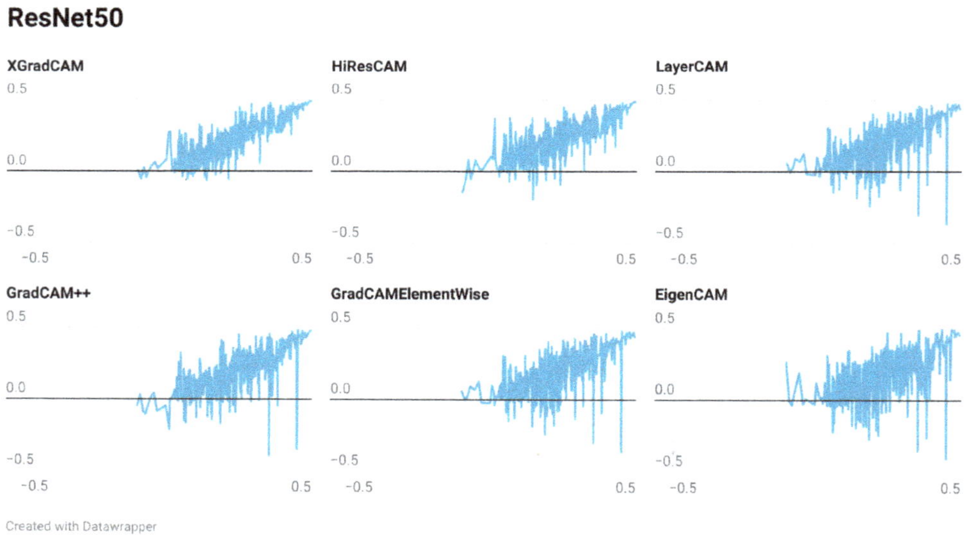

Figure 7. GradCAM ROAD metric versus ROAD metrics from six different CAM methods. CAM maps evaluated on the last block of a ResNet50 architecture adapted for three classes, on 480 test data. The X-axis corresponds to the ROAD metric evaluated on the GradCAM map, while the Y-axis corresponds to the ROAD metric evaluated on the respective CAM map.

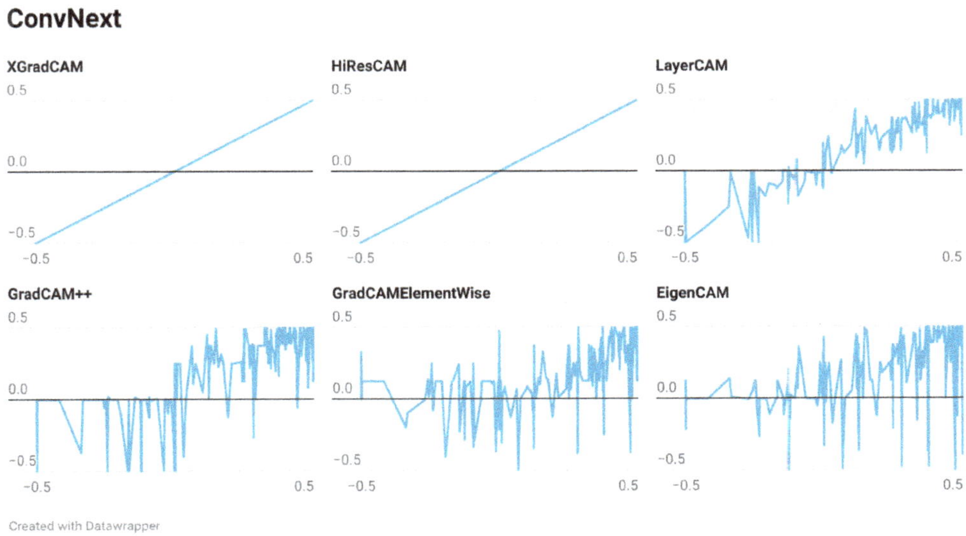

Figure 8. GradCAM ROAD metric versus ROAD metrics from six different CAM methods. CAM maps evaluated on the last block of a ConvNext tiny architecture adapted for three classes, on 480 test data. The X-axis corresponds to the ROAD metric evaluated on the GradCAM map, while the Y-axis corresponds to the ROAD metric evaluated on the respective CAM map.

According to the results shown, the identification of regions of importance for CNN-based classifiers presents more stable results when using the ConvNext-T architecture. As shown in Figures 3–5d, the locations of the regions identified by all CAM methods in the ConvNext architecture are very similar to each other, and their importance in the classifier decision process evaluated with the ROAD metric shows the same trend for four of the seven CAM methods evaluated.

These results show a large variability between the regions identified by either method in two of the most commonly used architectures when evaluating CAM methods. In contrast, the regions of importance identified by the methods in the ConvNext-T architecture are highly similar in most cases.

The consistency in the output CAM maps delivered by the ConvNext architecture represents a competitive advantage thanks to the great diversity of evaluation methods that support this type of explanation. Although the present study was focused on evaluating correctness, in order to describe the fidelity of the explanation with respect to the model, by masking input features and the respective evaluation of the change in the output of the predictive model, it is possible to use heat maps to evaluate other properties of the quality of the explanation. For example, it may be possible to evaluate completeness, continuity, contrastivity, covariate complexity, compactness or coherence, as described in a recent study on Explainable AI [32].

In terms of the type of architecture, an aspect that can make a difference has to do with the use of depth-wise convolutions. This ensures that in each operation performed by the network, information is only mixed across one dimension (either spatial or channel), but not both at the same time [16]. This validation can be further studied by considering architectures that use this type of convolution such as Xception-[33] or MobileNet [34]-based architectures.

Another aspect to consider here has to do with the type of data or specific problem on which the proposed methodology has been evaluated. Although the present study has focused on coffee leaf rust classification, a first alternative may be to consider extending these results to the detection of rust on other crops, or other leaf diseases. Beyond this, it is possible to apply similar studies on other classification problems or even object detection.

5. Conclusions

This paper presented an analysis of CNN architectures against the interpretation of model behavior by class activation maps. In particular, it was found that the two most frequent architectures when evaluating CAM methods, VGG16 and ResNet50, present a large variability in the CAM maps obtained. In other words, the interpretation of the results of these models depends to a large extent on the CAM method selected. This behavior is particularly evident in the VGG16 architecture, such that both the zones identified by the CAM method vary in location and intensity, as well as in the value of the ROAD metric. In the case of ResNet50, the results show an improvement with respect to VGG16 in terms of location and intensity, but may vary with respect to the area identified. Regarding the values of the ROAD metric with ResNet50, they show a more stable trend with respect to VGG16, but an important variability around the mean. Finally, the behavior of the CAM maps obtained with the ConvNext-T architecture presents results with a greater similarity between the different CAM methods, and even at the level of interpretability, the ROAD metric is highly similar for four of the seven CAM techniques evaluated. The importance of the above lies in the fact that the interpretation and adjustment of CNN models in terms of rust classification is simplified by the use of ConvNext-based models, in such a way that the selection of a type of CAM technique loses relevance, giving way to the adjustment of the model or dataset.

Author Contributions: Conceptualization, A.C. and D.R.; methodology, A.C.; software, D.R.; validation, E.M.-A.; formal analysis, D.R.; investigation, A.C. and E.M.-A.; resources, A.C. and D.R.; data curation, E.M.-A.; writing—original draft, D.R.; writing—review and editing, A.C. and E.M.-A.; visualization, A.C.; supervision, D.R. and E.M.-A.; project administration, D.R.; funding acquisition, E.M.-A. All authors have read and agreed to the published version of the manuscript.

Funding: This research has been funded by Universidad Panamericana under the Program "Fomento a la Investigación UP 2023" grant UP-CI-2023-MX-10-ING.

Data Availability Statement: The image datasets used in this work are publicly available from: Rocole dataset [27]: https://data.mendeley.com/datasets/c5yvn32dzg/2 (accessed on 10 August 2023); Bracol dataset [28]: https://data.mendeley.com/datasets/yy2k5y8mxg/1/ (accessed on 10 August 2023); Digipathos dataset [29]: https://www.redape.dados.embrapa.br/dataset.xhtml?persistentId=doi:10.48432/XA1OVL (accessed on 27 August 2024); D&P dataset [30]: https://data.mendeley.com/datasets/vfxf4trtcg/4/ (accessed on 10 August 2023); Licole dataset [31]: https://github.com/francismontalbo/swatdcnn/ (accessed on 14 November 2022).

Acknowledgments: Adrian Chavarro and Diego Renza thank the Universidad Militar Nueva Granada, since their contribution to this research is a product of the academic practice within this university. Ernesto Moya-Albor thanks Universidad Panamericana for all support in this work.

Conflicts of Interest: The authors declare no conflicts of interest.

Abbreviations

The following abbreviations are used in this manuscript:

CNN	Convolutional Neural Network
CAM	Class Activation Mapping
GAP	Global Average Pooling
GELU	Gaussian Error Linear Unit
GradCAM	Gradient-Weighted CAM

LeRF	Least Relevant First
LIME	Local Interpretable Model-agnostic Explanations
MoRF	Most Relevant First
ReLU	Rectified Linear Unit
ROAD	RemOve And Debias
ROAR	RemOve And Retrain
SHAP	SHapley Additive exPlanations
ViT	Vision Transformers
XAI	eXplainable Artificial Intelligence

References

1. Sagar, S.; Javed, M.; Doermann, D.S. Leaf-Based Plant Disease Detection and Explainable AI. *arXiv* **2023**, arXiv:2404.16833 .
2. Yebasse, M.; Shimelis, B.; Warku, H.; Ko, J.; Cheoi, K.J. Coffee disease visualization and classification. *Plants* **2021**, *10*, 1257. [CrossRef]
3. Zeiler, M.D.; Fergus, R. Visualizing and understanding convolutional networks. In Proceedings of the Computer Vision—ECCV 2014: 13th European Conference, Zurich, Switzerland, 6–12 September 2014; Proceedings, Part I 13; Springer, Cham: Berlin/Heidelberg, Germany, 2014; pp. 818–833.
4. Yosinski, J.; Clune, J.; Nguyen, A.; Fuchs, T.; Lipson, H. Understanding neural networks through deep visualization. *arXiv* **2015**, arXiv:1506.06579.
5. Ribeiro, M.T.; Singh, S.; Guestrin, C. "Why should i trust you?" Explaining the predictions of any classifier. In Proceedings of the 22nd ACM SIGKDD International Conference on Knowledge Discovery and Data Mining, San Francisco, CA, USA, 13 April 2016; pp. 1135–1144.
6. Lundberg, S.M.; Lee, S.I. A unified approach to interpreting model predictions. *Adv. Neural Inf. Process. Syst.* **2017**, *30*, 4768–4777.
7. Renza, D.; Ballesteros, D. Sp2PS: Pruning Score by Spectral and Spatial Evaluation of CAM Images. *Informatics* **2023**, *10*, 72. [CrossRef]
8. Selvaraju, R.R.; Cogswell, M.; Das, A.; Vedantam, R.; Parikh, D.; Batra, D. Grad-cam: Visual explanations from deep networks via gradient-based localization. In Proceedings of the IEEE International Conference on Computer Vision, Venice, Italy, 22–29 October 2017; pp. 618–626.
9. Chattopadhay, A.; Sarkar, A.; Howlader, P.; Balasubramanian, V.N. Grad-cam++: Generalized gradient-based visual explanations for deep convolutional networks. In Proceedings of the 2018 IEEE Winter Conference on Applications of Computer Vision (WACV), Lake Tahoe, NV, USA, 15 March 2018; IEEE: Piscataway, NJ, USA, 2018; pp. 839–847.
10. Fu, R.; Hu, Q.; Dong, X.; Guo, Y.; Gao, Y.; Li, B. Axiom-based grad-cam: Towards accurate visualization and explanation of cnns. *arXiv* **2020**, arXiv:2008.02312.
11. Draelos, R.L.; Carin, L. Use HiResCAM instead of Grad-CAM for faithful explanations of convolutional neural networks. *arXiv* **2020**, arXiv:2011.08891.
12. Gildenblat, J.; contributors. PyTorch Library for CAM Methods. 2021. Available online: https://github.com/jacobgil/pytorch-grad-cam (accessed on 10 August 2023).
13. Jiang, P.T.; Zhang, C.B.; Hou, Q.; Cheng, M.M.; Wei, Y. Layercam: Exploring hierarchical class activation maps for localization. *IEEE Trans. Image Process.* **2021**, *30*, 5875–5888. [CrossRef]
14. Muhammad, M.B.; Yeasin, M. Eigen-cam: Class activation map using principal components. In Proceedings of the 2020 International Joint Conference on Neural Networks (IJCNN), Glasgow, UK, 19 July 2020; IEEE: Piscataway, NJ, USA, 2020; pp. 1–7.
15. Hooker, S.; Erhan, D.; Kindermans, P.J.; Kim, B. A benchmark for interpretability methods in deep neural networks. *Adv. Neural Inf. Process. Syst.* **2019**, *32*, 9737–9748.
16. Liu, Z.; Mao, H.; Wu, C.Y.; Feichtenhofer, C.; Darrell, T.; Xie, S. A convnet for the 2020s. In Proceedings of the IEEE/CVF Conference on Computer Vision and Pattern Recognition, New Orleans, LA, USA, 24 June 2022; pp. 11976–11986.
17. Pachón, C.G.; Ballesteros, D.M.; Renza, D. An efficient deep learning model using network pruning for fake banknote recognition. *Expert Syst. Appl.* **2023**, *233*, 120961. [CrossRef]
18. Woo, S.; Debnath, S.; Hu, R.; Chen, X.; Liu, Z.; Kweon, I.S.; Xie, S. Convnext v2: Co-designing and scaling convnets with masked autoencoders. In Proceedings of the IEEE/CVF Conference on Computer Vision and Pattern Recognition, Vancouver, BC, Canada, 24 June 2023; pp. 16133–16142.
19. Samek, W.; Binder, A.; Montavon, G.; Lapuschkin, S.; Müller, K.R. Evaluating the visualization of what a deep neural network has learned. *IEEE Trans. Neural Netw. Learn. Syst.* **2016**, *28*, 2660–2673. [CrossRef] [PubMed]
20. Dosovitskiy, A.; Beyer, L.; Kolesnikov, A.; Weissenborn, D.; Zhai, X.; Unterthiner, T.; Dehghani, M.; Minderer, M.; Heigold, G.; Gelly, S.; et al. An Image is Worth 16x16 Words: Transformers for Image Recognition at Scale. *arXiv* **2021**, arXiv:2010.11929.
21. Xiao, T.; Singh, M.; Mintun, E.; Darrell, T.; Dollár, P.; Girshick, R.B. Early Convolutions Help Transformers See Better. *arXiv* **2021**, arXiv:2106.14881.
22. Liu, Z.; Lin, Y.; Cao, Y.; Hu, H.; Wei, Y.; Zhang, Z.; Lin, S.; Guo, B. Swin Transformer: Hierarchical Vision Transformer using Shifted Windows. In Proceedings of the 2021 IEEE/CVF International Conference on Computer Vision (ICCV), Montreal, BC, Canada, 17 October 2021; pp. 9992–10002. [CrossRef]

23. Zhang, Y.; Xu, A.; Lan, D.; Zhang, X.; Yin, J.; Goh, H.H. ConvNeXt-based anchor-free object detection model for infrared image of power equipment. *Energy Rep.* **2023**, *9*, 1121–1132. [CrossRef]
24. Alzubaidi, L.; Zhang, J.; Humaidi, A.J.; Al-Dujaili, A.; Duan, Y.; Al-Shamma, O.; Santamaría, J.; Fadhel, M.A.; Al-Amidie, M.; Farhan, L. Review of deep learning: concepts, CNN architectures, challenges, applications, future directions. *J. Big Data* **2021**, *8*, 1–74.
25. Zhou, B.; Khosla, A.; Lapedriza, A.; Oliva, A.; Torralba, A. Learning deep features for discriminative localization. In Proceedings of the IEEE Conference on Computer Vision and Pattern Recognition, Las Vegas, NV, USA, 30 June 2016; pp. 2921–2929.
26. Rong, Y.; Leemann, T.; Borisov, V.; Kasneci, G.; Kasneci, E. A consistent and efficient evaluation strategy for attribution methods. *arXiv* **2022**, arXiv:2202.00449.
27. Parraga-Alava, J.; Cusme, K.; Loor, A.; Santander, E. RoCoLe: A robusta coffee leaf images dataset for evaluation of machine learning based methods in plant diseases recognition. *Data Brief* **2019**, *25*, 104414. [CrossRef] [PubMed]
28. Krohling, R.A.; Esgario, J.; Ventura, J.A. BRACOL—A Brazilian Arabica Coffee Leaf Images Dataset to Identification and Quantification of Coffee Diseases and Pests. Mendeley Data. 2019. Available online: https://data.mendeley.com/datasets/yy2k5y8mxg/1 (accessed on 10 August 2023)
29. Barbedo, J.G.A.; Koenigkan, L.V.; Halfeld-Vieira, B.A.; Costa, R.V.; Nechet, K.L.; Godoy, C.V.; Junior, M.L.; Patricio, F.R.A.; Talamini, V.; Chitarra, L.G.; et al. Annotated plant pathology databases for image-based detection and recognition of diseases. *IEEE Lat. Am. Trans.* **2018**, *16*, 1749–1757. [CrossRef]
30. Brito Silva, L.; Cavalcante Carneiro, A.L.; Silveira Almeida Renaud Faulin, M. Rust (Hemileia vastatrix) and leaf miner (Leucoptera coffeella) in coffee crop (Coffea arabica). Mendeley Data. 2020. Available online: https://data.mendeley.com/datasets/vfxf4trtcg/4/ (accessed on 10 August 2023)
31. Montalbo, F.J.P.; Hernandez, A.A. Classifying Barako coffee leaf diseases using deep convolutional models. *Int. J. Adv. Intell. Inform.* **2020**, *6*, 197–209. [CrossRef]
32. Nauta, M.; Trienes, J.; Pathak, S.; Nguyen, E.; Peters, M.; Schmitt, Y.; Schlötterer, J.; Van Keulen, M.; Seifert, C. From anecdotal evidence to quantitative evaluation methods: A systematic review on evaluating explainable ai. *ACM Comput. Surv.* **2023**, *55*, 1–42. [CrossRef]
33. Chollet, F. Xception: Deep learning with depthwise separable convolutions. In Proceedings of the IEEE Conference on Computer Vision and Pattern Recognition, Honolulu, HI, USA, 26 July 2017; pp. 1251–1258.
34. Howard, A.G.; Zhu, M.; Chen, B.; Kalenichenko, D.; Wang, W.; Weyand, T.; Andreetto, M.; Adam, H. Mobilenets: Efficient convolutional neural networks for mobile vision applications. *arXiv* **2017**, arXiv:1704.04861.

mathematics

MDPI

Article

Deep Reinforcement-Learning-Based Air-Combat-Maneuver Generation Framework

Junru Mei, Ge Li and Hesong Huang *

College of Systems Engineering, National University of Defense Technology, Changsha 410073, China; meijunru@nudt.edu.cn (J.M.); lige@nudt.edu.cn (G.L.)
* Correspondence: huanghesonghhs@126.com

Abstract: With the development of unmanned aircraft and artificial intelligence technology, the future of air combat is moving towards unmanned and autonomous direction. In this paper, we introduce a new layered decision framework designed to address the six-degrees-of-freedom (6-DOF) aircraft within-visual-range (WVR) air-combat challenge. The decision-making process is divided into two layers, each of which is addressed separately using reinforcement learning (RL). The upper layer is the combat policy, which determines maneuvering instructions based on the current combat situation (such as altitude, speed, and attitude). The lower layer control policy then uses these commands to calculate the input signals from various parts of the aircraft (aileron, elevator, rudder, and throttle). Among them, the control policy is modeled as a Markov decision framework, and the combat policy is modeled as a partially observable Markov decision framework. We describe the two-layer training method in detail. For the control policy, we designed rewards based on expert knowledge to accurately and stably complete autonomous driving tasks. At the same time, for combat policy, we introduce a self-game-based course learning, allowing the agent to play against historical policies during training to improve performance. The experimental results show that the operational success rate of the proposed method against the game theory baseline reaches 85.7%. Efficiency was also outstanding, with an average 13.6% reduction in training time compared to the RL baseline.

Keywords: air combat; deep reinforcement learning; SAC; recurrent neural network

MSC: 93-10

1. Introduction

In recent years, along with the continuous advancement of UAV technology, the Unmanned Combat Aerial Vehicle (UCAV) is exerting an increasingly crucial role on the battlefield. However, in order to fully exploit the performance of UCAVs and achieve high-intensity air combats, UCAVs must be liberated from ground control and have a high degree of autonomous decision-making. Simultaneously, the combination of deep neural networks and machine learning has triggered a new research upsurge in artificial intelligence. In 2016, AlphaGo [1] defeated the human Go champion, drawing worldwide attention. On the one hand, AlphaGo's victory placed reinforcement learning on the historical stage, and on the other hand, it also verified the application potential of reinforcement learning in game issues. Reinforcement learning has had some successful applications in the military. In the DARPA (Defense Advanced Research Projects Agency) Subterranean Challenge, which began in 2018, competing unmanned ground vehicles used reinforcement-learning techniques to learn to navigate and perform tasks in complex environments, such as opening doors or crossing obstacles. The U.S. Army is developing a Synthetic Training Environment that incorporates RL to create adaptive virtual enemies for training purposes. These AI-controlled adversaries learn and adapt strategies based on the actions of human soldiers. The U.S. Navy has experimented with RL in coordinating drone swarms for maritime reconnaissance, where individual drones learn to work together efficiently in reconnaissance missions. In the DARPA AlphaDogfight trials in August 2020, the intelligent flight autonomous decision-making system designed by the American Heron Company successfully defeated the F-16's flight instructor with a score of 5:0, attracting the attention of military powers worldwide. Intelligent air-combat decision-making has become a research hotspot for all military powers presently.

So far, numerous studies have been conducted on the within-visual-range (WVR) air combat of UCAVs. The mainstream intelligent air-combat decision algorithms can be roughly divided into three categories. The first method is based on game theory [2–4], which includes two branches: differential game and matrix game. This method has a clear mathematical form, but the solution of the differential game is more complicated, and there are more strict conditions for solving the problem. Matrix games can only obtain the optimal decision in a short time, but cannot guarantee to generate the global optimal decision sequence. The second approach is based on expert knowledge [5,6]. This method has the advantages of rigorous logical thinking, clear and reasonable interpretation decisions, and low consumption of computing resources, but it still has several shortcomings. For example, the process of transforming expert knowledge into a knowledge base is complicated, and it is difficult to establish a complete rule base. Expert knowledge comes from the experience of human pilots and cannot guarantee the global optimality of decision-making strategies. The learning performance of the expert system is poor. After the design of the expert system is completed, it is difficult to extend the learning of new knowledge. The third method is based on reinforcement learning, which has the advantages of high control precision, fast response speed, potential optimal solutions that can be searched, long-term tactical planning ability, and strong relearning ability. However, at the same time, there are also shortcomings, such as the design of the reward function being highly subjective and the training of the algorithm requires a large amount of data and a long time. Considering the advantages and disadvantages of the above methods, we decide to adopt the reinforcement-learning method to solve the WVR air-combat problem of UCAVs.

Early studies were carried out in simplified environmental and dynamic models. The current aerial combat setting can be classified as either a two-dimensional or a three-dimensional space. Some studies, such as the work in [7], ignore the altitude aspect of UAVs, effectively reducing the combat space from three dimensions to a two-dimensional surface. These researchers have utilized platforms like DCS World to simulate one-on-one aerial combat scenarios and have endowed UAVs with the ability to make intelligent maneuver decisions. In an attempt to simulate the aerial combat scenario more accurately, various scholars have delved into the study of 3D combat situations. For instance, the work in [8,9] introduced a deep reinforcement-learning (DRL) algorithm for UAVs to make intelligent decisions, leveraging virtual reality technology. They trained UAVs in a 3D virtual environment using the Unity3D platform, enabling the visualization of the combat process in three dimensions.

Numerous studies on aircraft maneuvers for air combat assume comprehensive knowledge of both one's own aircraft and the target's information [10–14]. However, in real-world scenarios, specific details such as the position, orientation, and velocity of the target can be imprecise due to limitations in sensor capabilities, measurement errors, and fluctuating weather conditions. Moreover, pursuing an opponent is not always successful, leading to situations where target information becomes unobtainable. Therefore, for an air-combat model to be applicable to actual aircraft operations, it is crucial to incorporate the constraints and inaccuracies associated with the aircraft's integrated sensor systems.

In this paper, we propose a hierarchical DRL framework for constructing an air-combat model equipped with a high-fidelity 6-DOF dynamics model, which can conduct air combat in partially observable Markov decision process (POMDP) environments with imperfect information. The framework is segregated into two layers and solved independently using reinforcement learning. The upper layer formulates the combat policy based on the current combat situation and generates the autopilot instructions for the lower layer. The lower layer employs the control policy to directly manipulate the engine and rudder surfaces of the aircraft in accordance with the upper-layer instructions. Considering the issue of information imperfection in real-world air combat, we focus on two distinct sensor types: radar and visual sensors. Regarding the radar system, it is assumed that it operates with high accuracy and is capable of precisely detecting the opponent in the forward direction of one's own ship. Conversely, the vision sensor can detect opponents from all directions but has a shorter detection range and is mainly utilized in close combat scenarios. Nevertheless, it is pertinent to note that neither of these sensors is infallible and can, at times, be subject to constraints and errors. Critically, these may result in situations where there is a complete lack of information about the opponent. To robustly cope with such unpredictable circumstances, we incorporate the Gate Recurrent Unit (GRU), a type of recurrent neural network, into our system architecture. Additionally, we apply the renowned soft actor–critic (SAC) algorithm to our system. This design enables us to effectively address the challenge of handling sensor errors, ensuring that our system is resilient and reliable, even in the presence of inherent limitations in the sensors' functionality.

Furthermore, to solve the low efficiency of exploration in the early training stage, we utilize curriculum learning, a technique that restricts state space exploration. Similar to the human learning process, it is an effective tool to gradually learn from simple concepts to complex problems. It breaks down complex knowledge by providing a series of progressively more difficult learning tasks. It has been proven to reduce training time and induce reward signal feedback in environments with sparse reward, therefore balancing action and exploration.

Finally, sufficient experimentation shows that the proposed method can effectively deal with problems caused by missing information while also addressing the exploration difficulty in a high-fidelity simulation environment. Compared to the baselines, our agent has demonstrated exceptional combat performance. The contributions of this paper are as follows:

- Proposal of a novel two-layer hierarchical decision framework using the upper layer to generate combat policy based on the current combat situation and the lower layer to control aircraft maneuvers according to upper layer instructions.
- A decision method for partially observable environments is proposed, utilizing the learning algorithm SAC and GRU to extract temporal context information and effectively address problems arising from sensor limitations and errors.
- A series of reward functions based on expert knowledge are designed to effectively enhance the efficiency and stability of early training of agents. A virtual self-game mechanism is proposed to improve combat performance by fighting against its historical policy and iterating. Experimental results demonstrate the effectiveness of this mechanism.

The rest of this article is organized as follows. In Section 2, we review our work on air-combat solutions. Section 3 describes our approach in detail, including a partially observable Markov decision process (POMDP) model for WVR air combat and a DRL decision framework. In Section 4, we conducted a series of experiments and discussed the results. Finally, we conclude Section 5.

2. Related Work

In recent years, Deep Reinforcement Learning (DRL) has emerged as a key focus in the field of artificial intelligence. It demonstrates extraordinary decision-making capabilities in games, navigation, multi-agent cooperation, autonomous driving, and Robot control [15]. For instance, Alpha Go triumphed over the world Go champion with a score of 4:1 [1]. Alpha Star defeated the StarCraft professional with a score of 10:1 [16]. OpenAI Five defeated DOTA2's highest competition level TI champion OG team with a score of 2:0 [17]. These examples show that reinforcement learning has surpassed the average human level in decision-making speed and precision. Consequently, numerous researchers have applied DRL to air-combat scenarios and achieved remarkable accomplishments.

Yang et al. [18] utilized deep Q networks to train an agent that selects from a collection of discrete maneuvers in a customized 3D environment. Hu et al. [19] modeled one-to-one beyond-visual-range air combat by establishing the attack envelope and kill envelope of air-to-air missiles. The results validate that it can effectively represent the motion of the UAV in space and achieve intelligent decision-making. In the work of [20], a UAV dynamics model was established, where continuous throttle, angle of attack, and bank angle were taken as control variables. Different air-combat intentions, such as attack, escape, and pursuit, were designed, and an intelligent decision framework for MAV/UAV was established. The results show that the UAV can enhance the intelligent level under the guidance of MAV intention. The work in [21] presented a semi-realistic flight simulation environment known as Harfang3D Dog-Fight Sandbox for UAVs. The work in [7,18,19,22,23] established realistic air-combat environments with UAV and missile models. They used DRL algorithms such as TD3 to train the agent. Yuan et al. [24] proposed a novel heuristic deep deterministic policy gradient (DDPG) algorithm to enhance the exploration efficiency in the continuous action space. The aforementioned methods assumed 3-DOF aircraft dynamics for the air-combat task, disregarding the attack and sideslip angles by presuming that the velocity vector aligns with the nose direction. This oversimplification makes these methods unsuitable for more complex and realistic 6-DOF aircraft dynamics.

To enhance realism in simulating real-world scenarios, recent studies have utilized 6-DOF (six degrees of freedom) aircraft models for close-range air combat, referred to as WVR (within-visual-range) engagements. As the complexity of the dynamics escalates, so does the complexity of the simulation methods. The work in [25] presents a hierarchical deep reinforcement learning (DRL) policy that encompasses a high-level policy selector along with a set of specialized low-level policies, each trained to deal with different domains within the state space. Both the high-level and low-level components are trained through off-policy, maximum entropy techniques, and expert knowledge is incorporated by shaping the reward structure. This approach was conspicuously successful, achieving second place in the AlphaDogfight Trials and surpassing a human expert. While this method yields remarkable results, crafting reward functions that ensure optimal performance for subpolicy training continues to be a formidable challenge. Additionally, it demands considerable computational resources, involving tens of thousands of neurons at each layer of the neural network. The study referred to as [26] presents an alternative hierarchical framework with the aim of tackling the challenges of one-on-one WVR air combat in the context of 6-DOF dynamics. The method divides the decision-making process into inner and outer loops. However, In this work, the depiction of the missile is overly simplified. They employ a pre-defined weapon engagement zone (WEZ) and operate under the assumption that an adversary can be neutralized as soon as they are within this zone. This simplification can hardly explain the agent's evasion tactics compared to the behavior of real-world air combat.

The neural network used by the algorithm in the above research only has a fully connected layer and cannot extract historical information, so it has poor performance in the environment of lack of information. Yang et al. [27] compared the performance of off-policy reinforcement-learning algorithms such as DDPG, TD3, and SAC with cyclic neural networks such as LSTM and GRU. The experimental results show that SAC and recurrent neural networks (RNN) have the best combined ability. Among them, the combination of SAC and GRU has the best performance, which is comparable to the combination of SAC and LSTM, but has advantages in convergence speed. Therefore, we use the SAC+GRU algorithm, which improves the SAC algorithm based on GRU networks and utilizes historical temporal data. Detailed descriptions of the air-combat scenario, the related theories, and our proposed decision-making framework are presented as follows.

3. Preliminaries

3.1. Air-Combat Scenario Description

In this paper, one-to-one (red versus blue) WVR air combat is investigated. The aircraft adopts a 6-DOF motion model, and its motion is controlled by the forces and moments generated by each control surface. Both Red and Blue use the JSBSim [28] open-source environment to simulate the aircraft model, the flight dynamics model (FDM) based on wind tunnel test data, with high fidelity. This system model is also used by the Defense Advanced Research Projects Agency (DARPA) AlphaDogfight project. Detailed equations of motion can be found in [29].

The operating frequency of the simulation environment is 60 Hz. The action space consists of continuous signal inputs from the flight control system (ailerons, elevators, rudder, and throttle). WEZ is defined as a spherical cone with a 2° aperture extending from the nose of an aircraft at 200–1000 m (Figure 1) with a range of 200–1000 m. Either side takes sustained damage while in the opponent's WEZ. Damage per second in WEZ can be calculated using the following formula:

$$d_{wez} = \begin{cases} 0 & d > 1000 \text{ m} \\ \frac{1000-d}{800} & 200 \text{ m} < d < 1000 \text{ m} \\ 0 & d < 200 \text{ m} \end{cases} \quad (1)$$

where d is the distance between the two aircraft.

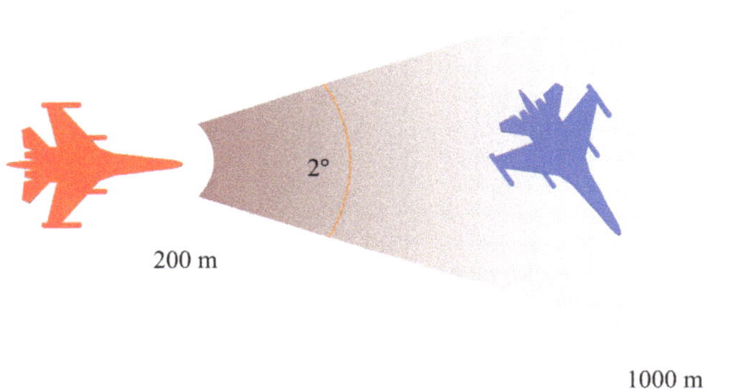

Figure 1. Weapon Engagement Zone (WEZ).

The engagement ends when the simulation time reaches 300 s or the health of either aircraft reaches zero. The agent can win by causing enough damage to the opponent or forcing him to break the minimum height limit of 100 m. If both sides have not cleared HP after 300 s, the engagement is considered a draw.

3.2. Partially Observable Markov Decision Processes (POMDP)

In real-world settings, an agent often can only have access to a reflection of the underlying state of the environment due to noise, occlusions, limited measurement resolutions, and so on. Under partial observability, a decision-making problem can be effectively modeled as a POMDP [30]. We can formally specify a finite-horizon POMDP by the tuple $\langle S, A, T, R, \Omega, O, \gamma \rangle$, where S is the state space, A is the action space, $T(s_{t+1} \mid s_t, a_t)$ is the transition probability, $R(s_t, a_t, s_{t+1})$ is the reward function, Ω is the observation space, $O(o_t \mid s_{t+1}, a_t)$ is the observation function, and $\gamma \in [0,1]$ is the discount factor. At each time period, the environment is in a certain state $s \in S$. The agent takes action $a \in A$, which results in an environment transition to state s' with probability $T(s' \mid s, a)$. At the same time, the agent receives observation $o \in Omega$, which depends on the new state of the environment with probability $O(o \mid s', a)$. Finally, the agent receives a reward r equal to $R(s, a)$. Then, repeat the process. The goal is to have the agent select the action that maximizes its expected future discount reward at each time step: $\mathbb{E}\left[\sum_{t=0}^{\infty} \gamma^t r_t\right]$. To achieve optimal performance in a partially observable Markov decision process (POMDP), an agent must base its policy on the complete history of observations and actions it has encountered thus far. However, the dimensionality of the history space grows exponentially with the length of this history. Consequently, recurrent neural networks are frequently employed to encapsulate this historical information, allowing the policy to be conditioned on the fixed-size hidden states produced by the RNN.

3.3. Maximum Entropy Reinforcement Learning

The idea of most deep reinforcement-learning methods is to find a deterministic notion of optimality, and although some random policies can increase agent exploration, most of them are heuristic, such as adding random noise or initializing a random policy with high entropy. But in some cases, we prefer to learn random policies, which can increase exploration in tasks with different goals and can also gain different combinations through pre-training. Previous studies have shown that when we consider optimal control versus probabilistic inference, random policies may be the best choice, i.e., the solution to the maximum entropy learning problem. Intuitively, we do not just want to capture a single deterministic behavior with a minimum cost (maximum cumulative expected return) but a set of behaviors with a low cost, i.e., we want to learn policies that can accomplish a task in multiple ways. If we can learn such a random policy, it can be a very good initialization policy when we want to fine-tune it to fit a more specific task.

Our goal is to learn a policy $\pi(a_t \mid s_t)$. The learning goal of traditional reinforcement learning is:

$$\pi_{std}^* = \arg\max_{\pi} \sum_t \mathbb{E}_{(s_t, a_t) \sim \rho_\pi}[r(s_t, a_t)] \tag{2}$$

The maximum entropy RL adds a policy entropy to the end of the reward term, which aims to not only maximize the reward at every moment but also combine with the policy entropy at every moment to increase the uncertainty of the policy.

$$\pi_{MaxEnt}^* = \arg\max_{\pi} \sum_t \mathbb{E}_{(s_t, a_t) \sim \rho_\pi}[r(s_t, a_t) + \alpha \mathcal{H}(\pi(\cdot \mid s_t))] \tag{3}$$

Where α is the temperature parameter that controls the randomness of the optimal policy, ρ_π denotes the state-action marginals of the trajectory distribution induced by the policy, $r(s_t, a_t)$ is the reward obtained at time t, and $\mathcal{H}(\pi)$ represents the entropy $\mathbb{E}[-\log \pi(a \mid s)]$ of the policy. This approach helps balance development and exploration. When the environment reward is high, the policy entropy is low, and the agent tends to choose the optimal action. When the environmental reward is high, the policy entropy is higher, which leads to greater exploration. If we wish to extend the maximum entropy RL to the infinite horizon problem, then it is convenient to introduce the discount factor γ to ensure that the expected reward and entropy are bounded. The entropy term makes the exploration range of the policy wider and eventually converges to the optimal action. Previous studies have confirmed the advantages of maximum entropy RL in the exploration of multimodal problems and learning speed [31,32].

3.4. The Soft Actor–Critic Algorithm

Soft actor–critic (SAC) [33] is an off-policy reinforcement-learning algorithm that combines maximum entropy learning with an actor–critic framework. Common reinforcement-learning algorithms tend to become more and more deterministic in the learning process, which greatly weakens the exploration ability of the algorithm in the late training and easily converges to the local optimal solution. In SAC, we want not only to maximize the reward of the environment but also to maximize the entropy of the policy, which makes the algorithm have better exploration ability and, therefore, show better performance on various tasks.

In SAC, we want to find a policy that maximizes the reward and, at the same time, maximizes the entropy of the policy's action distribution in each state. Therefore, the reward function is expanded:

$$r_{soft}(s_t, a_t) = r(s_t, a_t) + \alpha \mathop{\mathbb{E}}_{s_{t+1}\sim p}[\mathcal{H}(\pi(\cdot \mid s_{t+1}))] \tag{4}$$

Where α is the entropy temperature coefficient, which determines the degree of emphasis on entropy maximization; p is the state transition distribution, and $r(s_t, a_t)$ is the expectation that performing s_t in a_t will result in a reward. By bringing (4) into the Bellman equation, we obtain the soft action value function:

$$Q_{soft}(s_t, a_t) = r(s_t, a_t) + \mathop{\mathbb{E}}_{s_{t+1}, a_{t+1}}[Q_{soft}(s_{t+1}, a_{t+1}) - \alpha \log(\pi(a_{t+1} \mid s_{t-1}))] \tag{5}$$

Using (5), we can calculate the $Q_{soft}^{\pi_{old}}$ value of the current policy π_{old}. Then, solve the following optimization problems to find a new policy π_{new} that is better than the current policy:

$$\pi_{new} = \arg\min_{\pi'} D_{KL}\left(\pi'(\cdot \mid s_t) \,\|\, \frac{\exp(\alpha^{-1} Q_{soft}^{\pi_{old}}(s_t, \cdot))}{Z^{\pi_{old}}(s_t)}\right) \tag{6}$$

In (6), we first exponential $Q_{soft}^{\pi_{old}}$ to $\exp(\alpha^{-1}Q_{soft}^{\pi_{old}})$ and then normalize it to the distribution $\frac{\exp(\alpha^{-1}Q_{soft}^{\pi_{old}}(s_t,\cdot))}{Z^{\pi_{old}}(s_t)}$, where $Z^{\pi_{old}}(s_t) = \int \exp(\alpha^{-1}Q_{soft}^{\pi_{old}}(s_t, a_t))da_t$ is the normalized function. We hope to find a policy π' in the state s_t with minimum KL-divergence between its action distribution and $\frac{\exp(\alpha^{-1}Q_{soft}^{\pi_{old}}(s_t,\cdot))}{Z^{\pi_{old}}(s_t)}$ distribution, i.e., the distribution $\pi'(\cdot \mid s_t)$ and $\frac{\exp(\alpha^{-1}Q_{soft}^{\pi_{old}}(s_t,\cdot))}{Z^{\pi_{old}}(s_t)}$ are expected to be as similar as possible.

We use θ and ϕ to represent the parameters of action value function $Q_\theta(s, a)$ and policy $\pi_\theta(a \mid s)$, respectively. Similar to TD3 [34] algorithm, we use Clip Double Q learning to mitigate overestimating so that we will study θ_1 and θ_2 networks, respectively. In addition, in order to ensure the stability of learning, we will also maintain two target networks, $\hat{\theta}_1$ and $\hat{\theta}_2$.

According to (5), the training of parameter a of q can be achieved by minimizing the soft Bellman residual:

$$\begin{aligned} L_Q(\theta_i) = \mathop{\mathbb{E}}_{(s_t,a_t,r_t,s_{t+1})\sim D}[(Q_{\theta_i}(s_t, a_t) - (r_t + \mathop{\mathbb{E}}_{a_{t+1}\sim\pi_\phi}[\min_{\hat{\theta}_i} Q_{\hat{\theta}_i}(s_{t+1}, a_{t+1}) \\ - \alpha \log(\pi_\phi(a_{t+1} \mid s_{t+1}))]))^2] \end{aligned} \tag{7}$$

where D denotes the replay buffer.

Further simplifying (6), The training of parameter ϕ of $\pi_\phi(a_{t+1} \mid s_{t+1})$ can be achieved by minimizing the following loss function:

$$L_\pi(\phi) = \mathop{\mathbb{E}}_{s_t\sim D}[\mathop{\mathbb{E}}_{a_t\sim\pi_\phi(\cdot|t)}[\alpha \log \pi_\phi(a_t \mid s_t) - \min_{\theta_i} Q_{\theta_i}(s_{t+1}, a_{t+1})]] \tag{8}$$

The SAC algorithm sets a target entropy for each task. When the entropy of the policy is greater than the target entropy, reduce α to reduce the degree of maximum entropy of the algorithm, thus reducing the entropy of the current policy. When the entropy of the policy is less than the target entropy, increasing α increases the degree of the algorithm to maximize the entropy, thus increasing the entropy of the current policy. The loss function of design entropy is as follows:

$$\begin{aligned} L(\alpha) &= \underset{s_t \sim D, a_t \sim \pi}{\mathbb{E}}[-\alpha \log \pi_\phi(a_t \mid s_t) - \alpha \hat{\mathcal{H}}] \\ &= \alpha \underset{s_t \sim D}{\mathbb{E}}[H(\pi(\cdot \mid t)) - \hat{\mathcal{H}}] \end{aligned} \tag{9}$$

4. Method

Due to the continuous high-dimensional state and action space, as well as the complex calculation of aircraft dynamics by JSBSim, it is difficult to directly train the combat policy of the aircraft. The idea of hierarchical reinforcement learning is to decompose a complex problem into several smaller subtasks for easy solutions. In this paper, we propose a hierarchical decision framework as shown in Figure 2. The upper level outputs maneuvering instructions based on the current battle situation (the position, attitude, and speed of the warring parties) and acts as input to the lower level. The lower level determines the output signals of each control part of the aircraft (ailerons, elevators, rudder, and throttle) according to the instructions of the upper level and the current state of the aircraft to complete the maneuver. At the same time, based on how well the control policy completes the mission, the combat policy plans its behavior more rationally. The upper and lower layers have different decision frequencies. At the micro level, the decision frequency is high because the control policy needs to track the target with high-precision and keep the aircraft stable. On the macro level, in order to avoid the trajectory shock caused by frequent policy changes and reduce the consumption of computing resources, the combat policy operates at a lower decision frequency.

The training of the overall framework is divided into two phases. In the first phase, we trained the control policy using the SAC algorithm to perform tasks such as navigation and tracking with high performance. In the second stage, we use the trained control policy to supplement the training of combat policy while introducing the self-game mechanism. This training method improves the training efficiency and improves combat performance. The specific description is as follows.

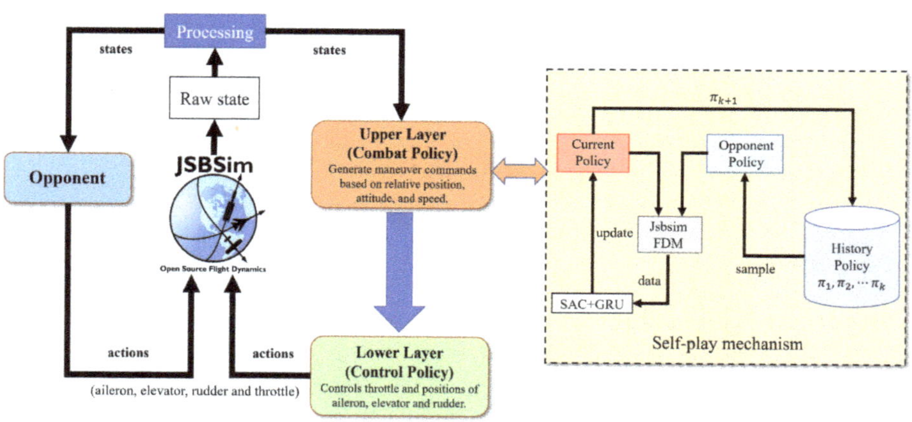

Figure 2. The hierarchical decision frame is constructed with two layers: the upper (combat policy) and the lower (control policy).

4.1. The 6-DOF Dynamic Control Policy

Reinforcement-learning algorithms interact with the environment based on the current state, move to the next state, and receive a reward, a process often modeled as MDP. In this work, JSBSim serves as a simulation environment for aircraft dynamics modeling and air-combat maneuvers. It can simulate the movement of the aircraft with high precision and return a variety of desired variables. This will generate a state space with position and velocity, etc., which we define as s_{basic}. The definition of each state is shown in Table 1.

Table 1. Basic state definitions of 6-DOF aircraft.

State	Definition	State	Definition
x	x-axis position	α	angle of attack
y	y-axis position	β	sideslip angle
h	altitude	δ_a	aileron position
u, v, w	linear velocity	δ_e	elevator position
ϕ, θ, ψ	roll, pitch and yaw angles	δ_r	rudder position
p, q, r	roll, pitch and yaw rate	δ_t	throttle

The control policy realizes motion control by controlling the pitch, roll, and yaw angles of the aircraft. Because the three Euler angles are coupled, the yaw angle can be controlled indirectly by the pitch angle and the roll angle. The control policy takes the pitch angle, roll angle, and velocity difference of the target as the tracking target and outputs the deflection angle instructions of ailerons, elevators, rudders, and throttle valves to directly control the aircraft. The control process can be established as A Markov decision process (MDP): $M = \langle S, A, R, P, \gamma \rangle$, where S is the state space, A is the action space, R is the reward function, P is the state transition, and γ is the discount factor. The details are as follows.

4.1.1. State Space

State s_t describes the control of the aircraft at time t. It can be divided into two parts. The first part is the basic state of the aircraft. It is used to describe the current state of the vehicle for attitude control. The second part describes the situation of angle and velocity tracking. It contains the tracking errors of pitch angle, roll angle, and velocity: $[\Delta\theta_t, \Delta\phi_t, \Delta v_t]$, where $\Delta\theta_t, \Delta\phi_t, \Delta v_t$ are, respectively, the difference between and target pitch angle, target roll angle and speed. In summary, the state s_t is $[s_{basic}, \Delta\theta_t, \Delta\phi_t, \Delta v_t]$. All bounded variables are normalized.

4.1.2. Action Space

The t time action includes the aileron c_a, the elevator c_e, the rudder c_r deflection angle command, and the throttle c_t, which, respectively, controls the aircraft roll, pitch, yaw, and engine thrust. The bound of each variable is shown in Table 2.

Table 2. Action and bound.

Action	Bound	Action	Bound
c_a	$[-21.5°, 21.5°]$	c_e	$[-25°, 25°]$
c_r	$[-30°, 30°]$	c_t	$[0, 1]$

4.1.3. Reward

The positive/negative feedback signal is obtained by the agent after performing an action, and the reward set is all the feedback information available to the agent. However, due to the sparsity of the reward signal, the model training may have high variance, resulting in unstable performance. Therefore, the design of reward functions is important for successful learning. For the training of control policy, we design an effective reward function:

$$r = \beta_1 r_{track} + \beta_2 r_{action} + r_{stable} + r_{crash} \tag{10}$$

The first part is the target tracking reward, which is determined by the errors between the current pitch angle, roll angle, and target angles:

$$r_{track} = -\sqrt{(\Delta\theta)^2 + (\Delta\phi)^2 - \exp(-|\Delta v|) + 1} \tag{11}$$

The second part is the action reward, which is summed up by 4 parts:

$$r_i = \exp(-\frac{0.5|c_i|}{k_i}), i = a, e, r, t \tag{12}$$

where c_i represents the action value. As $|c_i|$ increases, r_i monotonically decreases from 1 to 0. This helps provide the agent with smooth feedback on changes and reduces system fluctuations. The scale factor $|k_i|$ can be set to an appropriate value according to the unit of measure $|c_i|$. β_1 and β_2 in (10) are constants and can be adjusted to change the effect of the corresponding reward.

The third part is related to aircraft stability. If the aircraft exceeds the limit angle of attack or sideslip angle or is below the critical speed will lose control.

$$r_{position} = \begin{cases} 1 & stable \\ 0 & instable \end{cases} \tag{13}$$

If it crashes, the penalty $r_{crash} = -100$ is given, and the simulation ends.

We use the SAC algorithm to train the control policy. Initialize the network parameters ϑ_1, ϑ_2, and φ and initialize a replay buffer $\mathcal{D}$ at the beginning of each epoch. Then, start the simulation until the replay buffer is full. We select the position signal of the random target as the tracking target and re-initialize it regularly to enhance the exploration ability of the agent. The trace data are stored in the buffer $\mathcal{D}$. When enough data are collected, we update the parameters of the actor network and critic network according to the loss function described in Section 3.3. Through training, the control policy finally converges and obtains good tracking performance, which will be verified in the experiment. The training process can be described by Algorithm 1:

Algorithm 1: SAC-Based Control Policy Training

Randomly initialize the parameter φ of the actor network and the parameters ϑ_1, ϑ_2 of the critic network, initialize the experience replay buffer $\mathcal{D}$
Initialize the target networks $\bar{\vartheta}_1 \leftarrow \vartheta_1$, $\bar{\vartheta}_2 \leftarrow \vartheta_2$, set the capacity N of the experience replay buffer
for *each iteration* **do**
    for *each environment step* **do**
        Obtain the observation s_t and select an action a_t with policy $\pi_l(\cdot|s_t)$ to follow random target signals
        Perform action a_t, obtain next observation s_{t+1}, receive reward r_t
        $\mathcal{D} \leftarrow \mathcal{D} \cup \{(s_t, a_t, r(s_t, a_t), s_{t+1}\}$
    end
    for *each gradient step* **do**
        Update Q network using gradient descent method $\vartheta_i \leftarrow \vartheta_i - \lambda_Q \hat{\nabla}_{\vartheta_i} J_Q(\vartheta_i)$ for $i \in \{1,2\}$
        Update policy network using gradient descent method $\varphi \leftarrow \varphi - \lambda_\pi \hat{\nabla}_\varphi J_\pi(\varphi)$
        $\alpha \leftarrow \alpha - \lambda \hat{\nabla}_\alpha J(\alpha)$
        Update target network $\bar{\vartheta}_i \leftarrow \tau \bar{\vartheta}_i + (1-\tau)\bar{\vartheta}_i$ for $i \in \{1,2\}$
    end
end
Store the converged policy in $\mathcal{B}$
Output: $\vartheta_1, \vartheta_2, \varphi$

After the training of the control policy is complete, we freeze its parameters and encapsulate them into the decision framework. This facilitates the training of the combat policy and allows the control policy to be modularized for reuse in other missions.

4.2. The SAC+GRU-Based Combat Policy

In this section, we train the upper-layer combat policy based on the control policy introduced in the previous section. Including state space design, reward design, and action space design. As shown in Figure 2, the combat policy outputs macro commands based on the state of the aircraft and the relative position and speed of the opponent. Considering the sensor characteristics of the configuration, the simulator uses state to generate reward r_t and observations o_t. Trajectories (o_t, a_t, r_t) are stored in the replay buffer, and a fixed-length sequence of trajectories is sampled for the critic network.

4.2.1. Observation Space

The combat policy observation space contains information about air combat, including information such as the position and posture of the self relative to the opponent. Figure 3 shows the basic geometry between the two aircraft and represents the components of the relative target information. Relative target information at time step t includes: Aspect Angle (AA_t), Antenna Train Angle (ATA_t), Heading Crossing Angle (HCA_t), relative distance (d_t), relative speed ($\Delta V_t = [\Delta u_t, \Delta v_t, \Delta w_t]$), ownship HP ($oHP_t$), and target HP ($tHP_t$):

$$\Omega_t = \{AA_t, ATA_t, HCA_t, d_t, \Delta V_t, oHP_t, tHP_t\} \tag{14}$$

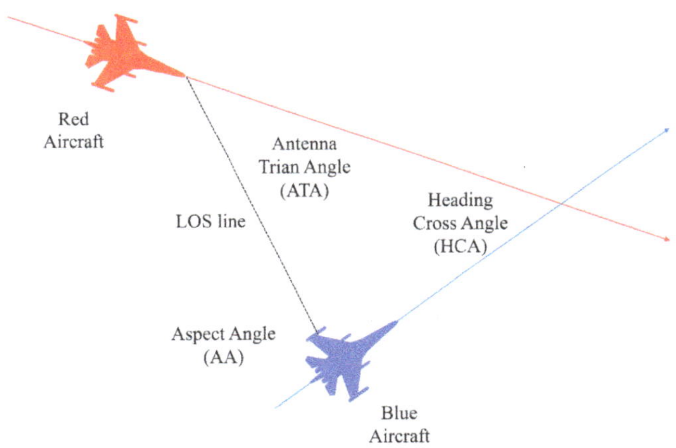

Figure 3. Geometry of air combat.

4.2.2. Action Space

The upper layer generates the target tracking track in real time according to the combat situation and determines the optimal sailing target point according to the position, attitude, and speed vector of both sides. The output of the upper layer is the expected roll angle $\dot{\phi}_t$, pitch angle $\dot{\theta}_t$ and speed $\dot{v}_t$. Its action space is:

$$a_t = (\dot{\phi}_t, \dot{\theta}_t, \dot{v}_t) \tag{15}$$

4.2.3. The Design of POMDP Environment

In the real environment, it is often difficult to obtain complete target information due to the resolution limitation and noise of the sensor. The accumulation of errors in the training process will eventually lead to an overestimation of the Q value, which may result in training converging to sub-optimal policies or even training failure. In practice, RNNS can be used to model from the time series information of the target to predict the trajectory of the target beyond the detection range of the sensor and reduce the impact of noise. In this paper, we designed a POMDP environment to reflect the real environment characteristics. We used two types of sensors for POMDP environments, as shown in Figure 4:

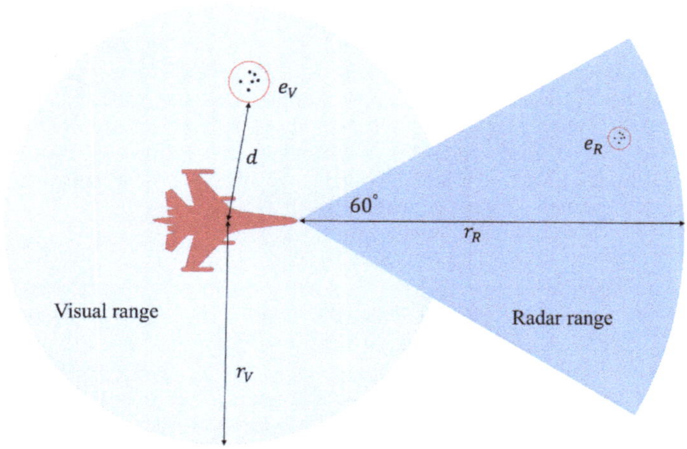

Figure 4. POMDP environment design.

The first is radar: fixed search 60° ahead. The search distance is far. The second is the vision sensor: the search distance is close, and the error is large, but can cover a 360-degree range. r_V and e_V are the detection distance and error rate of the vision sensor, while r_R and e_R are the detection distance and error rate of the radar sensor, respectively. When the target is within the detection range of the sensor, the error e_t in the observation space is proportional to the relative distance d. When the target is in the overlapping area of two sensors, $e_t = \min\{e_V, e_R\}$. If the target is out of the detection range, the target-related information is set to 0.

4.2.4. Reward

To improve the exploration difficulties caused by sparse reward problems, signal delay, local optimal, and high variance, we design a series of reward functions based on expert knowledge to form dense reward signals. The reward function of combat policy can be divided into five types: shoot-down, WEZ, control zone, crash, and control stability. Details for each item are as follows:

$$r_{ATA} = -\overline{ATA}^{\frac{1}{2}} \tag{16}$$

$$r_{AA} = -\overline{AA} \tag{17}$$

r_{ATA} and r_{AA} are related to the tracking angle of the target and penalize the agent when there is a non-zero angle.

$$r_{position} = \begin{cases} e^{-\overline{ATA}} \tanh(1000 - d) & 200 \text{ m} < d < 1000 \text{ m} \\ -2e^{-\overline{ATA}} \tanh(1000 - d) & d < 200 \text{ m} \end{cases} \tag{18}$$

$r_{position}$ encourages the agent to move to the opponent's six o'clock direction and approach the opponent at the same time to take advantage of the position. At the same time, curb the tendency to move too close so as not to overtake the opponent.

$$r_{wez} = \begin{cases} \exp(\frac{1000-d}{800})^2 & 200 \text{ m} < d < 1000 \text{ m} \; and \; ATA < 2^\circ \\ 0 & otherwise \end{cases} \tag{19}$$

$$r_{damaged} = \begin{cases} -\exp(\frac{1000-d}{800})^2 & 200 \text{ m} < d < 1000 \text{ m} \; and \; ATA > 178^\circ \\ 0 & otherwise \end{cases} \tag{20}$$

r_{shoot} rewards the agent for attacking an opponent in WEZ, and $r_{damaged}$ punishes the agent for entering the opponent's WEZ.

If either side is shot down within the time limit, the simulation ends, and the agent receives a reward $r_{win} = 100$ or $r_{lose} = -100$ based on the outcome of the combat.

4.2.5. Network Architecture

The diagram of the actor network and critic network in this article is shown in Figure 5. The network architecture of the SAC+GRU algorithm consists of two parts: actor network and critic network. The actor network takes the observed feature vector o as input, passes through the fully connected layer FC1, and uses the ReLU activation function to enhance feature extraction. The subsequent layer is a GRU layer that enables the algorithm to extract timing information and improve decisions based on the current state. After the GRU layer is the fully connected layer FC2, which uses ReLU activation functions to process the output of the GRU layer. Finally, a full connection layer produces the average μ and σ, used to resampling from Gaussian distribution $N = (\mu, \sigma)$. The resulting action a is obtained by the tanh activation function.

The critic network takes the observation vector o and the action vector a as inputs. The FC1 and GRU layers are the same as the actor network. The FC2 layer is a fully connected layer, and features are extracted from A using the ReLU activation function. The FC3 layer takes features from the GRU layer and the FC2 layer as input. Finally, the output layer consists of a neuron that outputs Q values to update the actor network.

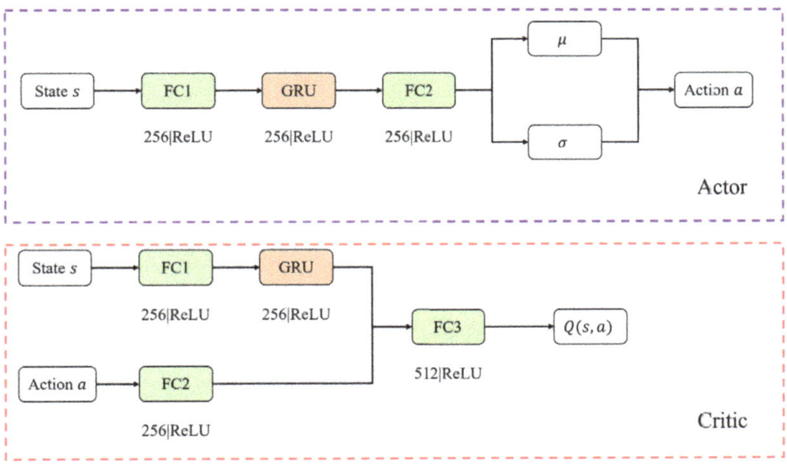

Figure 5. Network architecture of SAC+GRU.

4.2.6. The Self-Play Training Mechanism

In the early stages of training, the agent's opponents are created using a rule-based approach. We build the agent's historical policy pool $\mathcal{B}$, record the historical policy and update it after each episode. We adopt the self-game course-learning method. When the winning rate of the agent in training exceeds 50%, we will optimize the current policy π_h by making the agent fight against its own historical policy according to the game theory method. The network parameters stored in the pool are frozen and uniformly sampled as the opponent policy in the self-game mechanism. The training process is shown in Algorithm 2. It has been shown that self-game-based methods can converge to Nash equilibrium.

The rule for opponent updates is to sample opponents from the historical policy pool according to a certain probability distribution. This distribution is inversely proportional to the winning percentage of the last 100 games against each opponent and is dynamically updated after each episode. The sampling process is designed to allow the agent to play more against more intelligent opponents and reduce redundant games to avoid performance degradation. The policy can be iterated and converged to Nash equilibrium policy by self-evolving training.

Algorithm 2: SAC+GRU Combat Policy Training

Initialize a historical policy pool $\mathcal{B}$
Get a control policy π_l by Algorithm 1
Randomly initialize the parameter φ of the actor network and the parameters ϑ_1, ϑ_2 of the critic network, initialize the experience replay buffer $\mathcal{D}$
Initialize the target networks $\bar{\vartheta}_1 \leftarrow \vartheta_1$, $\bar{\vartheta}_2 \leftarrow \vartheta_2$, set the capacity N of the experience replay buffer
for *each generation* $n = 1, 2, 3, \cdots$ **do**
    for *each iteration* **do**
        for *each environment step* **do**
            Obtain the observation s_t and select an action a_t using the current policy network
            Perform action a_t, obtain next observation s_{t+1}, receive reward r_t
            $\mathcal{D} \leftarrow \mathcal{D} \cup \{(s_t, a_t, r(s_t, a_t), s_{t+1}\}$
        end
        for *each gradient step* **do**
            Update Q network using gradient descent method $\vartheta_i \leftarrow \vartheta_i - \lambda_Q \hat{\nabla}_{\vartheta_i} J_Q(\vartheta_i)$ for $i \in \{1, 2\}$
            Update policy network using gradient descent method $\varphi \leftarrow \varphi - \lambda_\pi \hat{\nabla}_\varphi J_\pi(\varphi)$
            $\alpha \leftarrow \alpha - \lambda \hat{\nabla}_\alpha J(\alpha)$
            Update target network $\bar{\vartheta}_i \leftarrow \tau \bar{\vartheta}_i + (1 - \tau) \bar{\vartheta}_i$ for $i \in \{1, 2\}$
        end
    end
    Store the converged policy in $\mathcal{B}$
    $\mathcal{B}_{n+1} = \mathcal{B} \cup \pi_h$
end
Output: $\vartheta_1, \vartheta_2, \varphi$

5. Experiment

This section discusses the training progress and results of the control policy and the combat policy. This experiment was performed on an Intel Core I9-13900K CPU (24 cores, 3.00 GHz), an Nvidia RTX 4090 GPU (24 GB) and 64 GB RAM. The Pytorch version is v1.10.1. The interaction frequency of the JSBSim environment is set to 60 Hz, the decision frequency of the control policy is set to 30 Hz, and the decision frequency of the combat policy is set to 10 Hz. Due to the large number of parameters to calculate in the JSBSim environment, training in control policy took 16 days, and combat policy took 19 days. We also set up a control group where the GRU layer of the combat policy was replaced with a fully connected layer, and the control policy remained the same. It takes about 22 days to train. The main parameters in the experiment are shown in Table 3.

Table 3. Parameter settings.

Parameter	Setting
Discount factor γ	0.99
Worker number	64
Optimizer	Adam
Learning rate	0.0001
Replay buffer size	1×10^6
Iteration per epoch	10
Batch size	12,800
Activation function	ReLU
Sequence length	64

5.1. Experiments on Control Policy

In the control policy experiment, the coefficient of r_{track} term and r_{action} term in the reward will affect the performance of the control policy. To study the influence of the coefficient, we first limited the coefficient β_1 of r_{track} between 0.1 and 2 and the coefficient β_2 of r_{action} between 0.1 and 0.5 according to experience.

We adjusted two coefficient pairs with a step size of 0.01 to train the control policy and conducted 200 random target signal tests for each control policy. If the training is convergent, the final average reward is normalized. Then, ultimately, reward the highest combination of coefficients as our control policy. The results in Figure 6 show that different combinations of coefficients lead to performance differences. When β_1/β_2 is too large, the stability of the aircraft will be affected, and the angle of attack and sidescale angle of the aircraft will be greatly shaken, which will lead to the aircraft stall or tailspin. When β_1/β_2 is too small, the plane's rudder angle is too small, which reduces aircraft maneuverability.

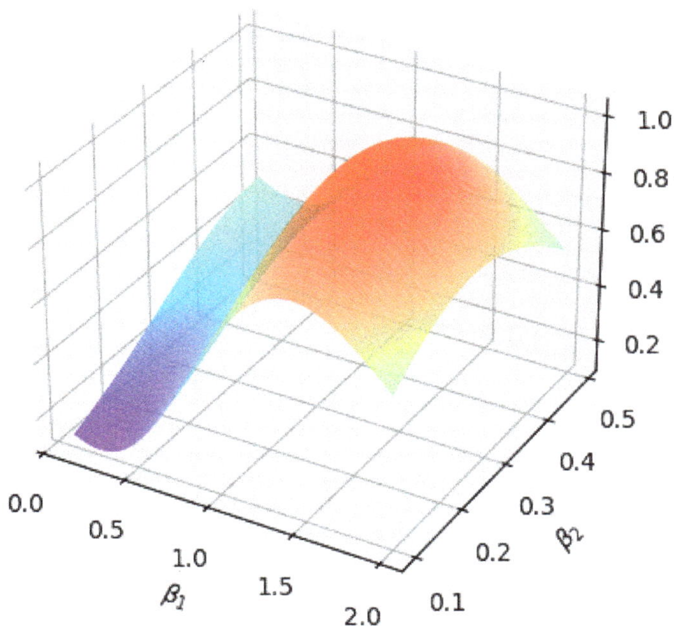

Figure 6. Performance of control policy under different combinations of coefficients. The maximum value is obtained when $\beta_1 = 1.22$ and $\beta_2 = 0.27$.

According to the experimental results, we finally determined the coefficients $\beta_1 = 1.22$ and $\beta_2 = 0.27$ as our control policy. The learning curves of the control policy are shown in Figure 7, which is obtained by averaging the results of 5 repeated experiments, where the blue curve represents the average reward, and the yellow curve represents the tracking error. The tracking error is determined by the sum of roll angle, pitch angle, and speed difference. At the initial stage of training, the reward increases rapidly, and the average tracking error decreases. After about 200 episodes of training, the two curves gradually converge. The average reward per episode converges to about 5000, and the average tracking error converges to about 0.01. This shows that the control stability and tracking performance are improved with the iteration of training. The convergence of the tracking error curve to 0.01 indicates that the tracking error angle of the convergent agent is less than 1 degree and the speed difference is close to 0.

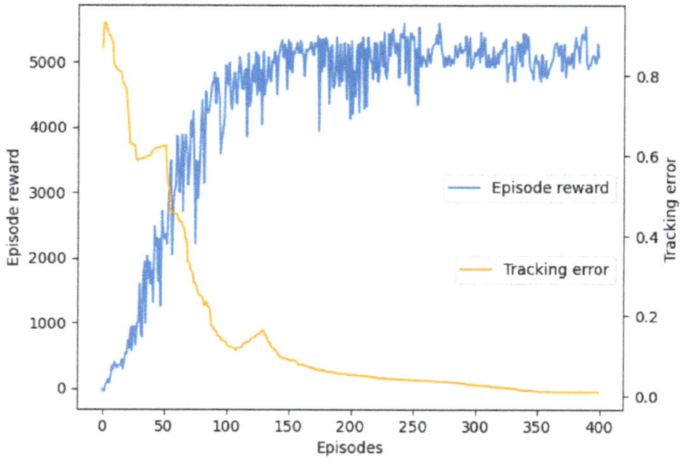

Figure 7. Average episode reward and tracking error of the control policy.

In the experiments to test the tracking performance of a control policy, we compare it with a traditional proportional integral derivative (PID) controller. It is the most widely used automatic controller. It has the advantages of simple principle, easy realization, wide application, independent control parameters, and simple selection of parameters. Details can be found in [35].

We tested the tracking performance of SAC and PID in two different scenarios: (a) sinusoidal signals and (b) Random signals. Figure 8 shows the tracking results of two control policies under two target signals. Both can stably track target course, altitude and speed. However, compared with the PID controller, the SAC control policy has advantages in overshoot, response speed, stability time and steady-state error. The control policy based on reinforcement learning can track the target signal quickly and smoothly, while the PID controller has obvious oscillation. In addition, model-free reinforcement-learning algorithms do not require complex analysis of aircraft dynamics to find superior policies that can be applied to a wider range of scenarios.

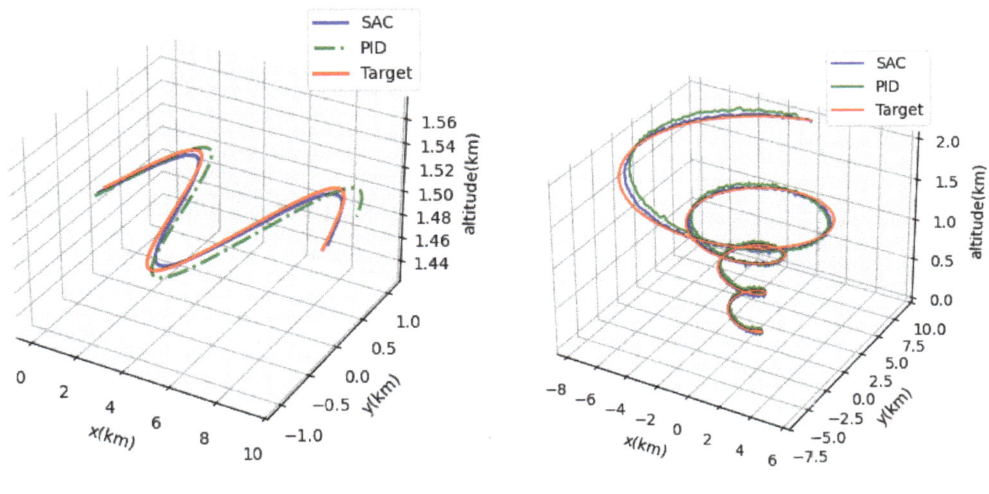

(**a**) (**b**)

Figure 8. Performance of SAC-based and PID-based control policies tracking two different target signals. (**a**) Sinusoidal signal. (**b**) Random signal.

5.2. Experiments on Combat Policy

5.2.1. Evaluation of SAC+GRU Algorithm

The adversary policy in this experiment is modeled according to the method in [2], which is based on game theory and modified by us to apply to the aircraft dynamic model. We assume that the adversary has access to all our agent's information to prevent performance degradation, limit the agent's inefficient search of the state space, and speed up early training.

To examine the role of the GRU layer, we set up a control group trained only with SAC, with both layers having previously trained control policies. Set the battlefield range to 10 km $\times$ 10 km, and randomly initialize the positions of both sides after each combat. We examined the effect on the training of both groups of agents by varying the size of the r_V (10 km, 3 km, and 1 km). In the experiment, the noise e_R of the radar is set to Gaussian noise $N(0, 0.01^2)$, and the noise e_V of the vision sensor increases with the increase of distance, which is expressed as $e_V = \frac{\sqrt{d}}{2r_V}$. Figure 9 shows the learning curves of SAC+GRU and SAC agents, where the solid line represents the average reward. When r_V = 10 km, SAC+GRU and SAC performed similarly, with SAC+GRU averaging 4.86% higher rewards than SAC but converging faster. The reason for this is that most of the time vision sensors can cover the entire battlefield, and the decision-making process in this case is similar to MDP. When r_V = 5 km, both SAC+GRU and SAC end up with a lower average reward. SAC+GRU was down 4.19%, while SAC was down 11.70%. Through visual analysis, we found that the SAC agent has a certain probability of losing the target after intersecting with the enemy aircraft, especially at higher speeds, because the turning radius is larger at this time. When r_V = 1 km, in the worst case, the opponent cannot be seen until he is in the six o'clock direction and within his WEZ. Compared to an MDP-like environment, the average reward for SAC+GRU dropped by 12.85%, while SAC dropped even more, to 37.11%. It is also observed that SAC+GRU curves converge faster, and oscillation amplitude is smaller in all scenarios. This may be because SAC is more sensitive to noise. The accumulation of errors in the training process eventually leads to an overestimation of the Q value, which makes the training effect worse.

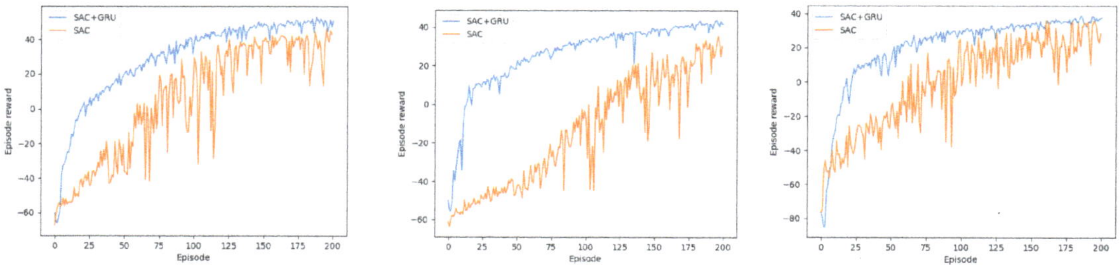

(a) (b) (c)

Figure 9. Average episode reward of SAC+GRU and SAC in restricted observation space. (**a**) r_V = 10 km; (**b**) r_V = 3 km; (**c**) r_V = 1 km.

Figure 10 shows the results of 100 battles with the trained SAC+GRU and SAC versus the game theory-based approach, respectively. SAC+GRU-based agents maintained a winning rate of 82%, 76%, and 69% in all three cases, respectively. SAC-based agents won 79%, 70%, and 55% of the time, respectively. The experimental results show that the SAC+GRU algorithm is superior to the traditional SAC algorithm in the POMDP environment because of its ability to extract historical information.

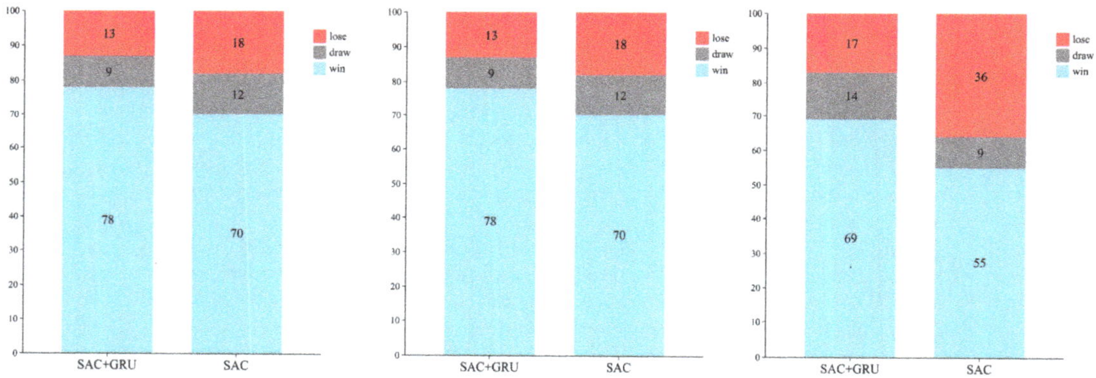

(a) (b) (c)

Figure 10. Combat results of SAC+GRU and SAC in restricted observation space. (**a**) r_V = 10 km; (**b**) r_V = 3 km; (**c**) r_V = 1 km.

5.2.2. Evaluation of Self-Play Mechanism

In the training process of the agent, we design a course-learning method based on self-game, and the agent constantly confronts the historical policy to improve the current policy. The training generation ends when the average reward increase within 50 consecutive iterations is less than 5%. This generation policy is then stored in the history policy pool, and the parameters are frozen. When training the next generation policy, sample from the historical policy pool as described in Section 4.2.6. The first-generation opponent policy is the same as in the previous section, and the agents used in the previous section are all first-generation. Here, we use the r_V = 5 km agent because this is similar to the pilot's perspective.

We ran a total of 5 iterations, as shown in Figure 11, with each curve representing the average reward of the current generation versus the previous generations. All learning curves grow with the number of training steps until they converge. This suggests that each generation's policy has an advantage over the previous one. In addition, with the increase of algebra, the growth rate of the curve begins to slow down, the convergence value decreases, and the standard deviation increases. This is because the improvement of the policy allows the new policy to accumulate absolute advantages over a long period of time and magnifies the impact of wrong actions. After the training was completed, we conducted 100 tests between every two generations, the results of which are shown in Table 4. Our older generation combat policies have a higher win rate than our younger generation combat policies. This indicates that the course-learning method based on self-game can achieve a more stable policy learning process by interacting with its own past version. Statistics show that each generation has a higher victory rate in battles with the previous generation. But at the same time, as the number of iterations increases, the advantages begin to decrease until they converge to the optimal policy.

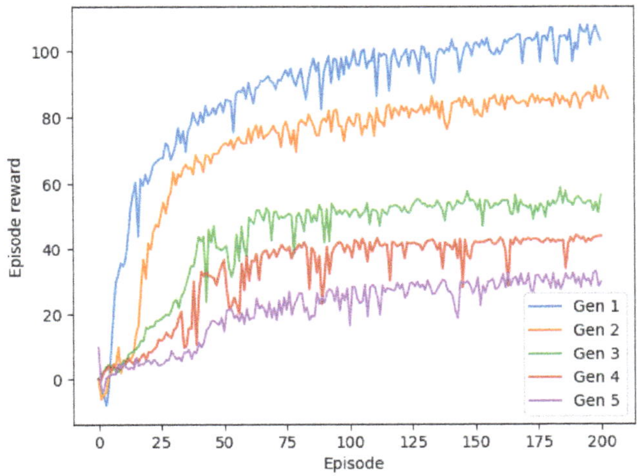

Figure 11. Learning curves of 5 generations. Each curve represents the change in the average episode reward when the current policy battles the previous policy.

Table 4. Combat results between different generations.

Red\Blue	Gen 1	Gen 2	Gen 3	Gen 4
Gen 2	78/12/10	-	-	-
Gen 3	80/13/7	73/16/11	-	-
Gen 4	76/9/15	77/10/13	70/14/16	-
Gen 5	81/12/14	75/11/14	72/12/14	69/13/18

5.2.3. Behavior Analysis

After the fifth iteration, we test the performance of the agent (red) against the baseline in the three typical scenarios shown in Figure 12. They are advantage scenarios, disadvantage scenarios, and neutral scenarios. A total of 100 episodes were tested in each scenario, and the specific results are shown in Table 5.

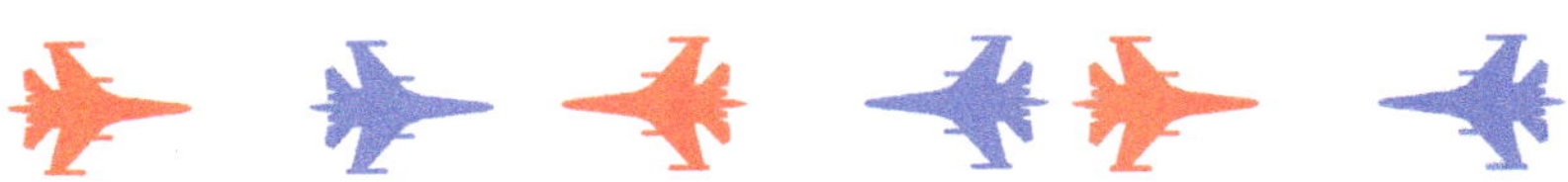

(**a**) (**b**) (**c**)

Figure 12. Testing scenarios. (**a**) Advantage scenario; (**b**) Disadvantage scenario; (**c**) Neutral scenario.

Table 5. Test results in three typical scenarios.

Scenarios	Win	Lose	Draw
Advantage	95%	3%	2%
Disadvantage	76%	19%	5%
Neutral	86%	8%	6%
Total	85.7%	10.0%	4.3%

Under favorable initial conditions, the agent can usually react quickly to the opponent's behavior and track the opponent stably. Therefore, we mainly analyze the agent's behavior in neutral and unfavorable scenarios. We used visual tools to analyze the two-party maneuvering trajectories stored during the test.

Figures 13 and 14 show two policies for the agent under neutral initial conditions. In the first, as shown in Figure 13, the agents choose to control the turning speed after the intersection in order to maintain the energy at a high level while maneuvering in the vertical direction, further depleting the energy of the opponent. Eventually, the opponent's mobility decreases due to low energy, and the agent has a higher speed, thus taking advantage and moving into the offensive position. In the second scenario, as shown in Figure 14, the agent chooses to convert speed into height after the horizontal intersection, thus obtaining a certain increase in mobility in the short term. At the same time, the reduction of speed reduces the turning radius so that the agent occupies a favorable position in the one-circle maneuver.

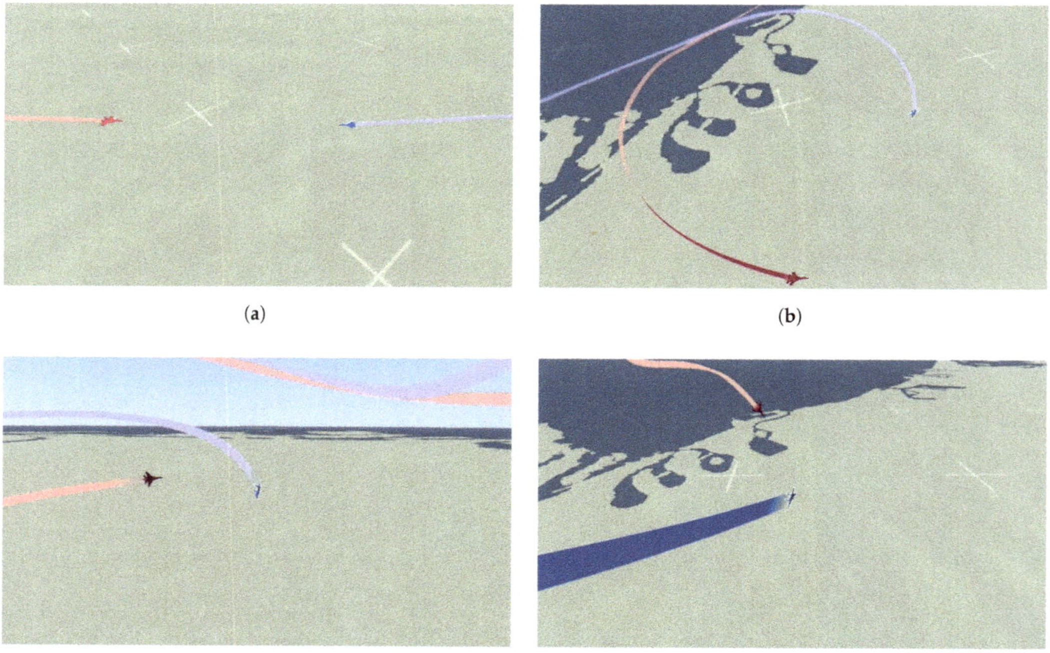

(a) (b) (c) (d)

Figure 13. *Cont.*

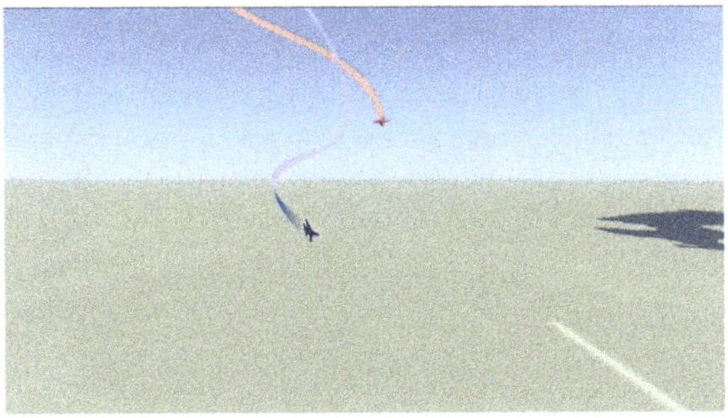

(e)

Figure 13. Neutral scenario. Win by high energy. (**a**) Beginning; (**b**) Maintain speed; (**c**) Boom and Zoom; (**d**) Take advantage; (**e**) Track.

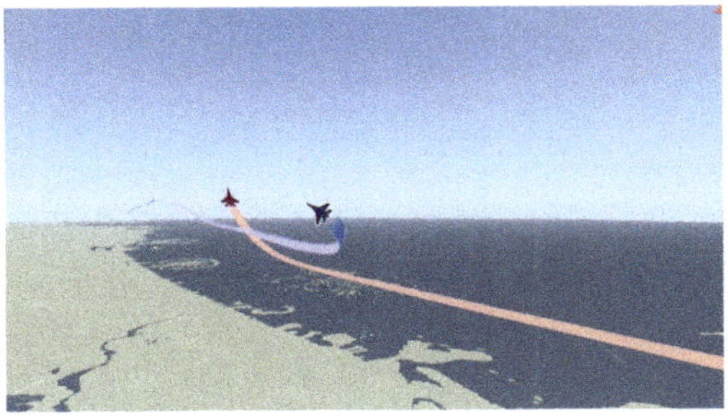

(a)

(b)

(c)

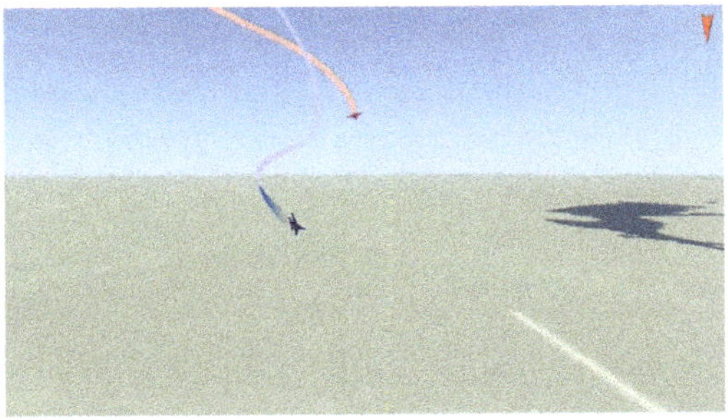

(d)

(e)

Figure 14. Neutral scenario. Win by a faster turn. (**a**) Beginning; (**b**) Pull up; (**c**) One-circle maneuver; (**d**) Inside the opponent's turn path; (**e**) Track.

Figure 15 shows the agent's policy under adverse initial conditions. The initial height, speed, and position of the agent are all at a disadvantage. The agent first tries to induce the opponent to overshoot by rolling scissors and barrel-rolling maneuvers while diving to recover energy. After that, the opponent chooses to pull up to avoid overshooting. At this point, the agent and the opponent enter the scissors maneuver in the vertical direction and take advantage of the smaller turning radius.

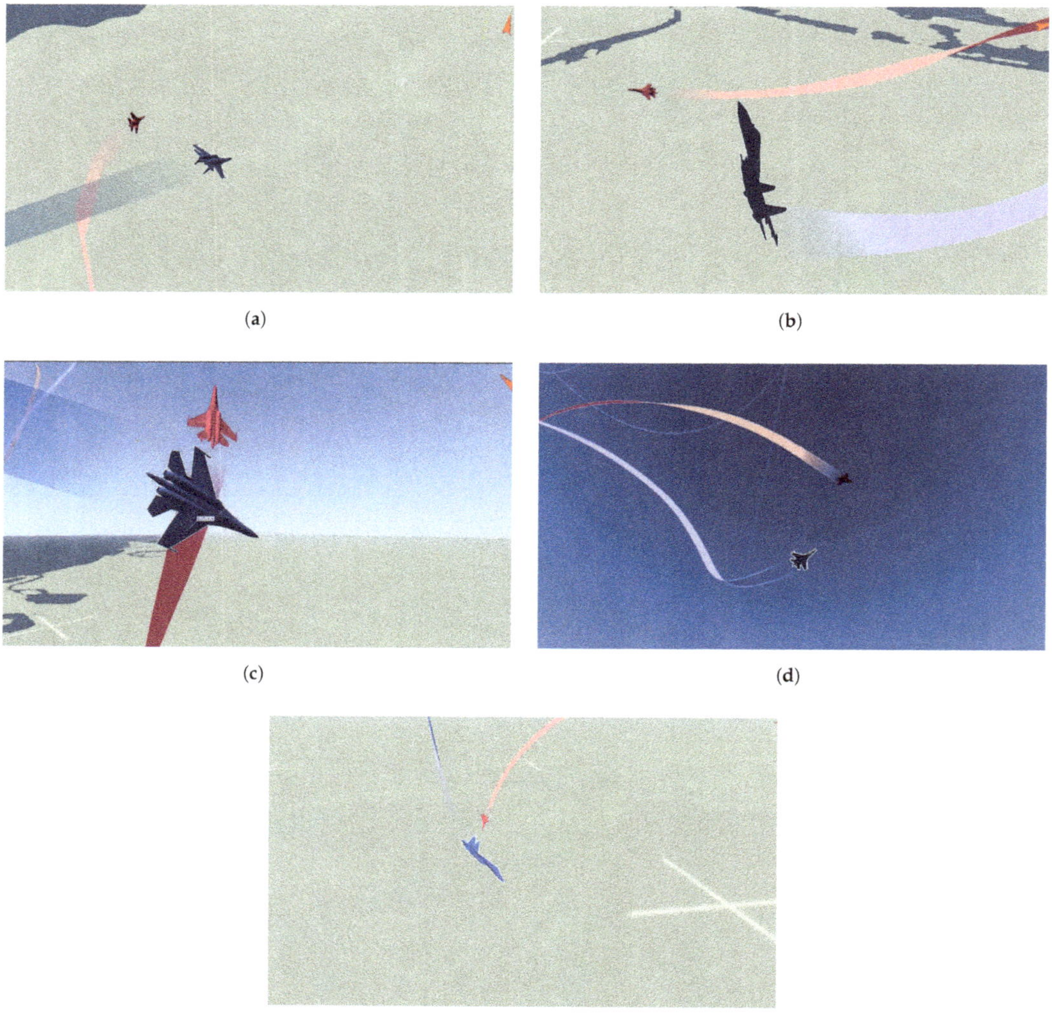

(a) (b) (c) (d) (e)

Figure 15. Disadvantage scenario. Win by defense and counterattack. (**a**) Beginning; (**b**) Induce the opponent to overshoot; (**c**) Vertical Rolling Scissors; (**d**) Take advantage of the smaller turning radius; (**e**) Track.

6. Conclusions

In this paper, we present a layered framework for air-combat-maneuver decision-making in a high-fidelity environment, taking into account the sensor limitations and errors in the real world. We designed a two-stage training mechanism, with the lower layer directly exporting control signals to various parts of the aircraft and the upper layer introducing a GRU network to generate maneuvering commands based on combat situation and historical information. In the training of combat policy, we also introduce the course-learning method based on self-game. The experimental results show that the control policy can track the target quickly, stably and accurately, and the effect is better than the traditional control method. The combat policy shows better performance and robustness to noise than the baseline method. We also analyze the maneuvering trajectories during the training process and demonstrate the strong learning ability of the agent. In future studies, we plan to further examine the applicability of the proposed method to real machines. In addition, future research will no longer be limited to single-aircraft confrontation, introducing high-precision radar missile models and increasing over-the-horizon air-combat scenarios.

Author Contributions: Conceptualization, J.M. and H.H.; methodology, H.H.; writing—original draft preparation, J.M. and G.L.; writing—review and editing, J.M. and H.H.; visualization, H.H. and G.L.; supervision, G.L.; project administration, J.M. and H.H.; funding acquisition, H.H. All authors have read and agreed to the published version of the manuscript.

Funding: This work was supported in part by the Hunan Natural Science Foundation under Grant 2024JJ6478.

Data Availability Statement: The data that support the findings of this study are available from the corresponding author upon reasonable request.

Conflicts of Interest: The authors declare no conflict of interest.

References

1. Silver, D.; Huang, A.; Maddison, C.J.; Guez, A.; Sifre, L.; Van Den Driessche, G.; Schrittwieser, J.; Antonoglou, I.; Panneershelvam, V.; Lanctot, M.; et al. Mastering the game of Go with deep neural networks and tree search. *Nature* **2016**, *529*, 484–489. [CrossRef] [PubMed]
2. Austin, F.; Carbone, G.; Falco, M.; Hinz, H.; Lewis, M. Game theory for automated maneuvering during air-to-air combat. *J. Guid. Control. Dyn.* **1990**, *13*, 1143–1149. [CrossRef]
3. Cruz, J.B.; Simaan, M.A.; Gacic, A.; Jiang, H.; Letelliier, B.; Li, M.; Liu, Y. Game-theoretic modeling and control of a military air operation. *IEEE Trans. Aerosp. Electron. Syst.* **2001**, *37*, 1393–1405. [CrossRef]
4. Poropudas, J.; Virtanen, K. Game-Theoretic Validation and Analysis of Air Combat Simulation Models. *IEEE Trans. Syst. Man, Cybern.-Part A Syst. Humans* **2010**, *40*, 1057–1070. [CrossRef]
5. Chai, R.; Tsourdos, A.; Savvaris, A.; Xia, Y.; Chai, S. Real-Time Reentry Trajectory Planning of Hypersonic Vehicles: A Two-Step Strategy Incorporating Fuzzy Multiobjective Transcription and Deep Neural Network. *IEEE Trans. Ind. Electron.* **2020**, *67*, 6904–6915. [CrossRef]
6. Huang, C.; Dong, K.; Huang, H.; Tang, S. Autonomous air combat maneuver decision using Bayesian inference and moving horizon optimization. *J. Syst. Eng. Electron.* **2018**, *29*, 86–97. [CrossRef]
7. Qiu, X.; Yao, Z.; Tan, F.; Zhu, Z.; Lu, J.G. One-to-one Air-combat Maneuver Strategy Based on Improved TD3 Algorithm. In Proceedings of the 2020 Chinese Automation Congress (CAC), Shanghai, China, 6–8 November 2020. [CrossRef]
8. Wang, L.; Wei, H. Research on Autonomous Decision-Making of UCAV Based on Deep Reinforcement Learning. In Proceedings of the 2022 3rd Information Communication Technologies Conference (ICTC), Nanjing, China, 6–8 May 2022. [CrossRef]
9. Xianyong, J.; Hou, M.; Wu, G.; Ma, Z.; Tao, Z. Research on Maneuvering Decision Algorithm Based on Improved Deep Deterministic Policy Gradient. *IEEE Access* **2022**, *10*, 92426–92445. [CrossRef]
10. Wang, L.; Wang, J.; Liu, H.; Yue, T. Decision-Making Strategies for Close-Range Air Combat Based on Reinforcement Learning with Variable-Scale Actions. *Aerospace* **2023**, *10*, 401. [CrossRef]
11. Wei, Y.; Zhang, H.; Wang, Y.; Huang, C. Maneuver Decision-Making through Automatic Curriculum Reinforcement Learning without Handcrafted Reward Functions. *Appl. Sci.* **2023**, *13*, 9421. [CrossRef]
12. Chen, R.; Li, H.; Yan, G.; Peng, H.; Zhang, Q. Hierarchical Reinforcement Learning Framework in Geographic Coordination for Air Combat Tactical Pursuit. *Entropy* **2023**, *25*, 1409. [CrossRef]
13. Wang, D.; Zhang, J.; Yang, Q.; Liu, J.; Shi, G.; Zhang, Y. An Autonomous Attack Decision-Making Method Based on Hierarchical Virtual Bayesian Reinforcement Learning. *IEEE Trans. Aerosp. Electron. Syst.* **2024** . [CrossRef]
14. Sun, L.; Qiu, H.; Wang, Y.; Yan, C. Autonomous UAV maneuvering decisions by refining opponent strategies. *IEEE Trans. Aerosp. Electron. Syst.* **2024**, *60*, 3454–3467. [CrossRef]
15. Arulkumaran, K.; Deisenroth, M.P.; Brundage, M.; Bharath, A.A. Deep Reinforcement Learning: A Brief Survey. *IEEE Signal Process. Mag.* **2017**, *34*, 26–38. [CrossRef]
16. Vinyals, O.; Babuschkin, I.; Czarnecki, W.M.; Mathieu, M.; Dudzik, A.; Chung, J.; Choi, D.H.; Powell, R.; Ewalds, T.; Georgiev, P.; et al. Grandmaster level in StarCraft II using multi-agent reinforcement learning. *Nature* **2019**, *575*, 350–354. [CrossRef]

17. Berner, C.; Brockman, G.; Chan, B.; Cheung, V.; Dębiak, P.; Dennison, C.; Farhi, D.; Fischer, Q.; Hashme, S.; Hesse, C.; et al. Dota 2 with large scale deep reinforcement learning. *arXiv* **2019**, arXiv:1912.06680.
18. Yang, Q.; Zhang, J.; Shi, G.; Hu, J.; Wu, Y. Maneuver Decision of UAV in Short-Range Air Combat Based on Deep Reinforcement Learning. *IEEE Access* **2020**, *8*, 363–378. [CrossRef]
19. Hu, D.; Yang, R.; Zuo, J.; Zhang, Z.; Wu, J.; Wang, Y. Application of Deep Reinforcement Learning in Maneuver Planning of Beyond-Visual-Range Air Combat. *IEEE Access* **2021**, *9*, 32282–32297. [CrossRef]
20. Li, B.; Gan, Z.; Chen, D.; Sergey Aleksandrovich, D. UAV maneuvering target tracking in uncertain environments based on deep reinforcement learning and meta-learning. *Remote Sens.* **2020**, *12*, 3789. [CrossRef]
21. Din, A.F.; Mir, I.; Gul, F.; Mir, S. Non-linear intelligent control design for unconventional unmanned aerial vehicle. In Proceedings of the AIAA SCITECH 2023 Forum, National Harbor, MD, USA, 23–27 January 2023; p. 1071.
22. Zhang, H.; Zhou, H.; Wei, Y.; Huang, C. Autonomous maneuver decision-making method based on reinforcement learning and Monte Carlo tree search. *Front. Neurorobotics* **2022**, *16*, 996412. [CrossRef]
23. Jiang, Y.; Yu, J.; Li, Q. A novel decision-making algorithm for beyond visual range air combat based on deep reinforcement learning. In Proceedings of the 2022 37th Youth Academic Annual Conference of Chinese Association of Automation (YAC), Beijing, China, 19–20 November 2022. [CrossRef]
24. Yuan, W.; Xiwen, Z.; Rong, Z.; Shangqin, T.; Huan, Z.; Wei, D. Research on UCAV Maneuvering Decision Method Based on Heuristic Reinforcement Learning. *Comput. Intell. Neurosci.* **2022**, *2022*, 1477078 [CrossRef]
25. Pope, A.P.; Ide, J.S.; Mićović, D.; Diaz, H.; Twedt, J.C.; Alcedo, K.; Walker, T.T.; Rosenbluth, D.; Ritholtz, L.; Javorsek, D. Hierarchical reinforcement learning for air combat at DARPA's AlphaDogfight trials. *IEEE Trans. Artif. Intell.* **2022**, *4*, 1371–1385. [CrossRef]
26. Chai, J.; Chen, W.; Zhu, Y.; Yao, Z.X.; Zhao, D. A hierarchical deep reinforcement learning framework for 6-DOF UCAV air-to-air combat. *IEEE Trans. Syst. Man Cybern. Syst.* **2023**, *53*, 5417–5429. [CrossRef]
27. Yang, Z.; Nguyen, H. Recurrent Off-policy Baselines for Memory-based Continuous Control. *arXiv* **2021**, arXiv:cs.LG/2110.12628.
28. Berndt, J. JSBSim: An Open Source Flight Dynamics Model in C++. In Proceedings of the AIAA Modeling and Simulation Technologies Conference and Exhibit, Providence, RI, USA, 16–19 August 2004. [CrossRef]
29. Nguyen, L.T. *Simulator Study of Stall/Post-Stall Characteristics of a Fighter Airplane with Relaxed Longitudinal Static Stability*; National Aeronautics and Space Administration: Washington, DC, USA, 1979; Volume 12854.
30. Kaelbling, L.P.; Littman, M.L.; Cassandra, A.R. Planning and acting in partially observable stochastic domains. *Artif. Intell.* **1998**, *101*, 99–134. [CrossRef]
31. Haarnoja, T.; Tang, H.; Abbeel, P.; Levine, S. Reinforcement Learning with Deep Energy-Based Policies. In Proceedings of the 34th International Conference on Machine Learning, Sydney, NSW, Australia, 6–11 August 2017.
32. Schulman, J.; Chen, X.; Abbeel, P. Equivalence Between Policy Gradients and Soft Q-Learning. *arXiv* **2017**, arXiv:1704.06440.
33. Haarnoja, T.; Zhou, A.; Abbeel, P.; Levine, S. Soft Actor-Critic: Off-Policy Maximum Entropy Deep Reinforcement Learning with a Stochastic Actor. In Proceedings of the 35th International Conference on Machine Learning, Stockholm, Sweden, 10–15 July 2018.
34. Fujimoto, S.; Hoof, H.; Meger, D. Addressing Function Approximation Error in Actor-Critic Methods. In Proceedings of the 35th International Conference on Machine Learning, Stockholm, Sweden, 10–15 July 2018.
35. Stevens, B.L.; Lewis, F.L.; Johnson, E.N. *Aircraft Control and Simulation: Dynamics, Controls Design, and Autonomous Systems*; John Wiley & Sons: Hoboken, NJ, USA, 2015.

mathematics

MDPI

Article

Predictive Maintenance Algorithms, Artificial Intelligence Digital Twin Technologies, and Internet of Robotic Things in Big Data-Driven Industry 4.0 Manufacturing Systems

Marek Nagy [1], Marcel Figura [1], Katarina Valaskova [1] and George Lăzăroiu [2,3,4,5,*]

[1] Faculty of Operation and Economics of Transport and Communications, University of Zilina, Univerzitna 1, 010 26 Zilina, Slovakia; marek.nagy@stud.uniza.sk (M.N.); marcel.figura@stud.uniza.sk (M.F.); katarina.valaskova@uniza.sk (K.V.)
[2] Faculty of Science and Engineering, Curtin University, Bentley, WA 6102, Australia
[3] Intelligent Communication and Computing Laboratory, Toronto Metropolitan University, Toronto, ON M5B 2K3, Canada
[4] The Creative Computing Research Centre, Cardiff Metropolitan University, Cardiff CF5 2YB, UK
[5] Department of Economic Sciences, Spiru Haret University, 030045 Bucharest, Romania
* Correspondence: george.lazaroiu@spiruharet.ro

Abstract

In Industry 4.0, predictive maintenance (PdM) is key to optimising production processes. While its popularity among companies grows, most studies highlight theoretical benefits, with few providing empirical evidence on its economic impact. This study aims to fill this gap by quantifying the economic performance of manufacturing companies in the Visegrad Group countries through PdM algorithms. The purpose of our research is to assess whether these companies generate higher operational profits and lower sales costs. Using descriptive statistics, non-parametric tests, the Hodges–Lehmann median difference estimate, and linear regression, the authors analysed data of 1094 enterprises. Results show that PdM significantly improves economic performance, with variations based on geographic scope. Regression analysis confirmed PdM as an essential predictor of performance, even after considering factors like company size, legal structure, and geographic scope. Enterprises with more effective cost management and lower net sales were more likely to adopt PdM, as revealed by decision tree analysis. Our findings provide empirical evidence of the economic benefits of PdM algorithms and highlight their potential to enhance competitiveness, offering a valuable foundation for business managers to make informed investment decisions and encouraging further research in other industries.

Keywords: predictive maintenance; internet of robotic things; industry 4.0 manufacturing; artificial intelligence digital twin technologies

MSC: 62H10; 62J05

1. Introduction

The concept of Industry 4.0 has significantly transformed the manufacturing sector in recent decades. The complexity of manufacturing equipment has increased due to the increasing number of components, which has also exacerbated the challenges associated with maintenance. In order to mitigate unanticipated failures or downtimes, it is advantageous for organisations to implement continuous monitoring of such intricate systems. The rapid technological advancements brought about by big data-driven Industry 4.0 manufacturing systems have not only introduced new possibilities in maintenance and technical condition prediction but have also provided companies with new ways to monitor the condition and diagnose the health of equipment. PdM is a significant trend that is currently emerging in this context [1]. Traditional maintenance methods, including reactive and preventive maintenance, were employed by companies until recently. Nevertheless, these conventional methods are linked to inefficiencies in the management of production processes. The risks of unexpected downtimes and financial losses are significantly increased by reactive maintenance, which concentrates on repairs only after a failure has occurred. On the other hand, preventive maintenance frequently results in high costs and unnecessary component replacements, despite its ability to prevent breakdowns. In this context, PdM is a solution that integrates predictive capabilities with efficiency [2,3]. The primary advantages of PdM algorithms are the reduction in overall operational costs and the optimisation of equipment utilisation, which results in extended equipment lifespans [4]. PdM, the most sophisticated maintenance approach, is built upon sophisticated machine learning algorithms that serve as the foundation for precise fault diagnostics and predictions [5]. The Internet of Robotic Things (IoRT) is also essential, as it enables the continuous monitoring of manufacturing processes through sensor networks and big data analytics [6]. An active area of research is the quantification of the economic performance of enterprises as a result of PdM algorithms and artificial intelligence digital twin technologies. Research has shown that PdM has a direct impact on business competitiveness by significantly reducing maintenance costs and improving operational efficiency [7,8]. Furthermore, PdM contributes to the reduction in defective products, which in turn results in an increase in customer satisfaction [9,10]. However, PdM implementation poses numerous substantial obstacles, despite its numerous benefits. High costs are the initial significant obstacle. Companies that implement PdM algorithms and artificial intelligence digital twin technologies must incur substantial upfront expenses, as it is imperative to not only acquire PdM solutions but also integrate them with existing systems. This frequently necessitates substantial investments in the development of a comprehensive sensor infrastructure. Additionally, organisations may encounter an absence of expertise among their personnel during the implementation phase. Finally, organisations must resolve concerns regarding predictive models' dependability [11–13]. In this investigation, the economic advantages of PdM algorithms in manufacturing enterprises that operate within the Visegrad Group (V4) countries are examined. Based on 1094 enterprise data, this investigation inspects the influence of PdM on the cost of goods sold and the operating profit. This study employs linear regression, descriptive analysis, non-parametric tests, and the Hodges–Lehmann median difference estimation to quantify this effect. This investigation also inspects the factors that may have influenced companies to implement PdM algorithms and contrasts those that have and those that have not. The results emphasise the significance of PdM in the manufacturing sector, offer managerial assistance in determining its implementation, and establish the foundation for additional academic research. The research gap this study addresses resides in the lack of empirical evidence on the economic benefits of PdM algorithms in manufacturing enterprises within the V4 countries. While PdM is widely acknowledged for its potential to enhance operational efficiency, the existing literature primarily focuses on case studies, theoretical models, or broader industry-level analyses without region-specific insights.

This paper is divided into several sections. The Section 2 summarises the most recent and relevant studies related to the research issue. The Section 3 describes the data acquisition and selection process as well as the description of the methodological steps of the analysis. The outputs of various methods and tests are presented in the Section 4, and the findings are further compared and analysed in the Section 5. A summary of the results and analysis is included in the Section 6, while the last section covers theoretical and practical implications, together with limitations and further directions of research.

2. Literature Review

In today's dynamic business environment, maintenance has become a widely discussed topic. It is not only addressed by the academic community but also by representatives from a wide range of industrial sectors. In terms of cost-saving potential, maintenance is one of the most critical factors for manufacturing [14]. Nevertheless, the concept of maintenance is delineated in a variety of ways. Maintenance contributes to increased customer satisfaction, enhances performance levels, affects production quality and volume, and helps preserve asset value [15]. This suggests that maintenance is no longer solely concerned with resolving machine malfunctions or failures. Rather, it is perceived as a strategic instrument for the enhancement of efficiency and the optimisation of processes [16]. The emergence of big data-driven Industry 4.0 manufacturing systems and the continuous development of innovative technologies have been the driving forces behind this transformation [17,18].

Enterprises were primarily characterised by reactive maintenance during the initial stages of industrialisation [19]. This method has been found to be inefficient as industrial complexity has increased, because it involves responding to machine issues only after they have occurred [20]. Reactive maintenance is primarily employed by manufacturing companies that are unable to retain a skilled maintenance workforce and lack sufficient staff [21]. Preventive maintenance emerged as a more proactive alternative to the costly and inefficient reactive maintenance approach. Its objective is to conduct routine inspections and servicing of equipment to avert potential failures before they occur [22]. This form of maintenance has been widely implemented by organisations due to its ability to facilitate the early identification and resolution of issues, thereby reducing the likelihood of equipment failures and extending its lifespan [23].

Maintenance has undergone a significant transformation as a result of the development of breakthrough technologies and the introduction of big data-driven Industry 4.0 manufacturing systems. PdM is currently regarded as the most sophisticated form of maintenance, having emerged as a consequence of the adoption of new technologies [19]. Technologies that facilitate PdM include artificial intelligence (AI), the IoRT, big data analytics, and digital twins. PdM is increasingly becoming a critical component of industrial strategies as a result of the utilisation of these intelligent tools [24]. The foundational layer is the IoRT, which is composed of networks of numerous sensors that collect real-time operational data from machines for digital entrepreneurial self-efficacy [25,26]. The data that are collected are subsequently transmitted through communication protocols for further analysis [27].

The massive influx of data from the IoRT is processed by big data analytics, which identifies patterns and anomalies (Figure 1). Potential failures or equipment issues that pose significant risks to businesses can be identified based on these patterns [28]. By employing machine learning algorithms, artificial intelligence significantly improves analytical processes. These algorithms make a substantial contribution to the advancement of predictive models by consistently enhancing their accuracy over time [29].

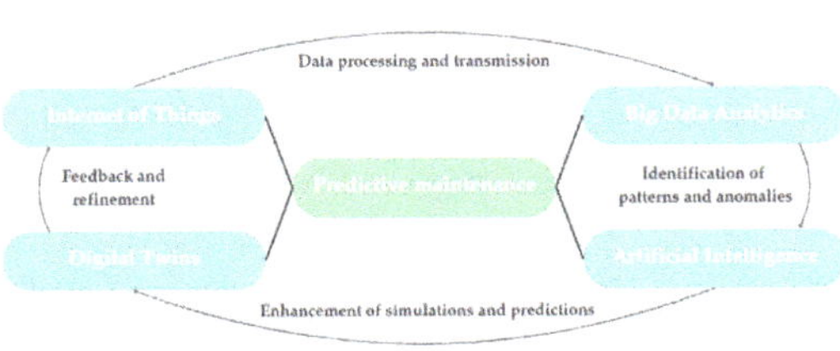

Figure 1. PdM schema in the context of Industry 4.0. Source: the authors' compilation.

Cognitive manufacturing systems are significantly influenced by digital twins [30] that facilitate real-time decision-making processes and enable simulations [31]. The integration of digital twins into cyber–physical systems furthers industries to utilise vast amounts of data for resource optimisation and decision-making, thereby enhancing predictive capabilities [32].

Businesses regard the principle of maintenance as an inevitability that must be executed solely when production ceases [33]. In an industrial setting, forced maintenance results in a substantial increase in operational costs, which can account for approximately 15% to 60% of total manufacturing expenses [34]. In this context, Meddaoui et al. [5] emphasise the benefits of PdM associated with reducing downtime and costs based on a case study and enhancements in product quality. Franco and de Figueiredo [35] and Scaife [36] focus on the continuous monitoring of equipment and the analysis of historical data, as these methods allow for the real-time prediction of potential failures prior to their occurrence. The risks associated with unplanned production shutdowns are reduced through the early detection of faults. Furthermore, the reliability of manufacturing equipment is improved by timely servicing [37]. A survey on the adoption of PdM among companies in Slovenia indicates that PdM increases customer satisfaction by extending the lifespan of equipment, reducing maintenance costs, and improving overall manufacturing efficiency. Ultimately, this results in higher-quality products [9]. Companies can reduce total maintenance costs by 5% to 10%, increase equipment uptime by 10% to 20%, and shorten maintenance planning time by 20% to 50% by implementing PdM [38,39]. The significance of PdM algorithms in the pursuit of manufacturing excellence is substantiated by the aforementioned advantages. A wide range of industries, such as aerospace, energy, healthcare, telecommunications, manufacturing, and the automotive sector, are interested in the significant potential of PdM [40,41]. This emphasises the transformative power and complexity of PdM algorithms in contemporary industrial strategies [42,43]. Nevertheless, it is imperative to recognise the obstacles and concerns that are linked to PdM algorithms, artificial intelligence digital twin technologies, and IoRT in big data-driven Industry 4.0 manufacturing systems, as they impede the widespread adoption of cutting-edge tools across various sectors. The processing of vast amounts of sensor-generated data are highly demanding [44]. This is due to the fact that, in many cases, these data contain noise or errors, which significantly affect the accuracy of equipment condition forecasts and negatively impact predictive models. Another obstacle is the phenomenon of "catastrophic forgetting" [45]. The data that represent the current state of equipment are used to train machine learning models that are used for PdM algorithms. Nevertheless, the usage patterns of equipment may evolve as it ages. This can result in inaccurate or misleading predictions [12]. In the final analysis, "catastrophic forgetting" entails that the model forgets the information it acquired from prior data and commences to commit errors [46]. PdM is also linked to economic concerns, in addition to technological obstacles. The high financial costs and organisational barriers that restrict its widespread adoption are underscored [47]. The technological infrastructure, which includes IoRT sensors, cloud systems, and artificial intelligence digital twin technologies, and the necessary capital to invest in PdM algorithms, may be lacking in many companies. Giada and Rossella [48] propose that companies should establish a model for assessing economic return prior to implementation, irrespective of the complexity of determining all costs and benefits, in response to this economic issue. Figure 2 illustrates the inconsistent market development across various regions and presents regional differences in the adoption of PdM algorithms. With a 34.81% market share in 2022, North America was the dominant region in the PdM industry. The widespread implementation of innovative and advanced technologies in North America is the cause of this strong dominance. In contrast, the Asia-Pacific region's PdM market is anticipated to reach the highest compound annual growth rate (CAGR) by 2030, as a result of the diverse expansion of small and medium-sized industrial sectors [49].

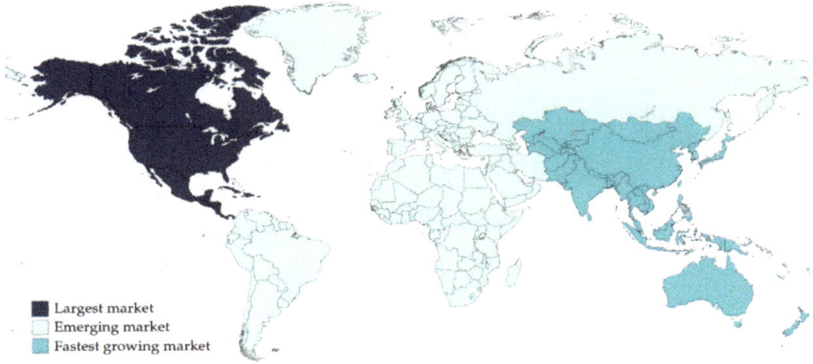

Figure 2. The inconsistent market development across a variety of regions. Source: the authors' compilation according to [49].

The authors have also discovered that, despite the substantial interest in PdM, there is a limited number of empirical studies that have been conducted to directly confirm the theoretically described benefits and challenges of PdM algorithms, artificial intelligence digital twin technologies, and IoRT in big data-driven Industry 4.0 manufacturing systems. This is based on the literature review that was conducted. The most significant empirical research conducted, as well as the key findings, are summarised in Table 1.

Table 1. The most significant empirical research on PdM.

Authors	Title of the Study	Number of Citations	Summarisation
Qi and Tao [44]	Digital Twin and Big Data Towards Smart Manufacturing and Industry 4.0: 360 Degree Comparison	819	The research highlights the significant roles of digital twin, PdM algorithms, and big data in advancing smart manufacturing.
Yan et al. [50]	Industrial Big Data in an Industry 4.0 Environment: Challenges, Schemes, and Applications for Predictive Maintenance	261	The research demonstrates that multisource heterogeneous data can provide new solutions for PdM, scheduling, and machining process optimisation aimed at energy saving.
Aivaliotis et al. [51]	The use of Digital Twin for predictive maintenance in manufacturing	196	The paper demonstrates that the digital twin approach can accurately estimate the remaining useful life of machinery (RUL), which is crucial for PdM in manufacturing settings.

Table 1. *Cont.*

Authors	Title of the Study	Number of Citations	Summarisation
Ayvaz and Alpay [52]	Predictive maintenance system for production lines in manufacturing: A machine learning approach using IoT data in real-time	189	The research presents a data-driven PdM system that utilises IoT sensor data to detect potential machinery failures in real-time, allowing for timely preventive actions to avoid production stops.
Jimenez et al. [53]	Towards multi-model approaches to predictive maintenance: A systematic literature survey on diagnostics and prognostics	141	The research highlights the increasing importance of PdM algorithms in research, indicating a shift towards multi-model approaches to address the complexity of technological systems.

Source: authors' compilation.

These studies underscore the importance of PdM algorithms in big data-driven Industry 4.0 manufacturing systems, which utilise technologies such as IoRT, big data, and digital twins to facilitate more intelligent manufacturing. Digital twin technology is emphasised as a critical instrument for estimating the remaining useful life (RUL) of machinery, thereby enhancing the accuracy of maintenance. Real-time fault detection is facilitated by multisource heterogeneous data and machine learning, which also optimises energy efficiency and scheduling in production. The research also emphasises a transition to multi-model approaches, which improve the diagnostics and prognostics of intricate industrial systems. In general, PdM algorithms, artificial intelligence digital twin technologies, and thee IoRT in big data-driven Industry 4.0 manufacturing systems are advancing through the implementation of sophisticated data-driven methodologies, which are designed to reduce downtime and enhance manufacturing efficiency.

3. Material and Methods

Data Acquisition: the Orbis Database, which contains comprehensive information on manufacturing companies in numerous countries, was the source of the data used in this study. In total, 5504 large and very large companies from V4 countries were included in the dataset: Slovakia (SK) with 598 companies, Czech Republic (CZ) with 1195 companies, Poland (PL) with 2922 companies, and Hungary (HU) with 789 companies. The size distribution of companies was also assessed prior to data processing in order to ensure the representativeness of the sample and the methodological consistency of the analysis. The companies were categorised based on the criteria of the Orbis database. The proportional representation of these size categories varied across the analysed countries and has been considered in the phase of sample selection.

Selection of Samples: a Python 3.12.7 script was employed to randomly select a sample of 1094 companies for analysis. This script was executed within the Jupyter Notebook 7.2.2 environment. A stratified random sampling method was employed to ensure the representativeness of the sample. The stratification was determined by two primary factors: the distribution of companies by country and by company size within each country. The country-level stratification was applied using the following proportions: 30% of Slovak companies, 20% of Czech companies, 15% of Polish companies, and 30% of Hungarian companies. Simultaneously, the proportional representation of large and very large companies within each country was preserved in accordance with their distribution in the overall population. As the Python script randomly selected company identification numbers for inclusion in the sample, it ensured that both country and size distributions accurately reflected the general structure of the analysed companies. The proportional distribution of companies in the sample was intended to ensure that it remained consistent with the overall distribution of manufacturing companies across the V4 countries. The direct proportional sampling approach would have generated an excessively large dataset, which would have increased the computational complexity and time required for the analysis, as the total number of companies in Poland is substantially greater than that of other V4 countries. In order to guarantee methodological feasibility and maintain the relative representation of corporations across countries, the sample was formed using predetermined proportions that correspond to the general structure of the population. This method was able to maintain the fundamental characteristics of the original distribution while allowing for a manageable dataset.

Utilisation of PdM Identification: a Python script was employed to search the annual reports for the years 2018, 2019, and 2020 in order to ascertain whether a company employed PdM. The search was conducted using keywords such as "Predictive Maintenance" and "PdM." This analysis does not incorporate annual reports from 2021 and 2022. The primary reason for the exclusion of these reports was to guarantee that the analysis accurately represented the impact of PdM algorithms over a minimum of three years. The results may have been distorted if a company implemented PdM in 2021 or 2022, as its effects may not have been fully observable by 2023. Consequently, the decision was made to concentrate on the 2018–2020 reports in order to guarantee a more comprehensive understanding of the long-term consequences of PdM implementation. The sample contains companies classified by country, company size, and the use of PdM.

Dataset, Variables, and Methods: the variables utilised in the analysis, along with their units of measurement and notation, are summarised in Table 2. The categorical variable Country (X_{14}) takes 4 variations, namely Czech Republic, Hungary, Poland, and Slovakia. There are three variations of the categorical variable Legal form (X_{15}): public limited companies, private limited companies, and partnerships. Company size (X_{16}) is regarded as a binary variable since the dataset contains only large and very large companies. Date of incorporation (X_{17}) is a nominal variable that contains only the year. PdM (X_{18}) specifies whether PdM algorithms have been implemented by the company. This variable is binary in nature, accepting only values of 0 and 1. Companies that have not implemented PdM are represented by a value of 0, while companies that have implemented PdM between 2018 and 2020 are represented by a value of 1. The data were collected from 2018 to 2023, and all analyses are based on this timeframe, with the exception of the decision tree model, which is based on the period from 2018 to 2020. This study implements a variety of statistical methodologies, utilising IBM SPSS 27 and Python 3.12.7 in the Jupyter Notebook 7.2.2 environment. For all statistical tests and regression analyses, the significance level was set at 5%, and all p-values were evaluated against this threshold.

Descriptive Analysis: the first step was descriptive analysis, which provided a basic comparison between companies using PdM and those that have not implemented PdM. The authors focused on two key indicators that best reflect company performance:

- Operating profit (reflects the company's efficiency in managing revenue and costs);
- Cost of goods sold (COGS) (indicator of the efficiency of production and logistics processes).

Each indicator was processed as follows: the average and median values, along with the coefficients of variation, were calculated for each country and for the groups of companies that use PdM and those that do not use PdM. The goal of the descriptive analysis was to provide an initial overview of the potential benefits of PdM algorithms in companies.

Table 2. Overview of variables.

Notation	Variable	Unit of Measurement
X_1	Operating profit	th. EUR
X_2	Costs of goods sold	th. EUR
X_3	Added value	th. EUR
X_4	Depreciation and amortisation	th. EUR
X_5	Net sales	th. EUR
X_6	Capital	th. EUR
X_7	Non-current assets	th. EUR
X_8	Current assets	th. EUR
X_9	Tangible fixed assets	th. EUR
X_{10}	Intangible assets	th. EUR
X_{11}	Other non-current assets	th. EUR
X_{12}	Other current assets	th. EUR
X_{13}	Total assets	th. EUR
Notation	**Variable**	**Type**
X_{14}	Country	Categorical
X_{15}	Legal form	Categorical
X_{16}	Company size	Binary
X_{17}	Date of incorporation	Nominal
X_{18}	PdM	Binary

Source: authors' compilation.

Normality test: next, it was necessary to test the statistical significance of these differences, after checking the normality of the data, i.e., whether the values of operating profit and COGS come from a normal distribution. To verify normality, the Kolmogorov–Smirnov test was used, as it is suitable for large datasets. The test confirmed that the data do not follow a normal distribution.

Mann–Whitney test: as the data do not follow a normal distribution, the Mann–Whitney test was used. To test the statistical significance of the differences between companies with PdM and without PdM, the Mann–Whitney test, which is suitable for comparing two independent groups with an unknown data distribution, was applied. This test was used for the following:

- Operating profit, where the results showed significantly higher values for companies with PdM;
- Cost of goods sold, where companies with PdM showed significantly lower values for this indicator.

The goal of this analysis was to statistically confirm whether there are significant differences in the operating profit and COGS between companies that implemented PdM and those that did not.

Kruskal–Wallis test: since the Mann–Whitney test only assessed global, overall differences, for a more detailed comparison between V4 countries, the Kruskal–Wallis test was used. This test allowed us to accomplish the following:

- Identify whether the effects of PdM differ across countries;
- Find out in which countries the differences between companies with PdM and without PdM are statistically significant.

Hodges–Lehmann median difference estimate: to obtain a precise expression of the magnitude of differences between companies with PdM and without PdM, the Hodges–Lehmann median difference estimate was applied. This method was selected for its robustness and suitability for non-normally distributed data, giving it an appropriate alternative to traditional parametric methods. The applied estimate provided the following:

- A quantitative measure of the impact of PdM on operating profit and COGS;
- A more accurate empirical determination of the differences between the countries analysed.

Regression analysis: to identify the factors that most significantly influence the operating profit of companies, the linear regression model was created. The dependent variable in the model was operating profit, and the set of independent variables consisted of six predictors, including an interaction effect between long-term tangible assets and added value. To ensure methodological rigor, a multicollinearity diagnostic check was incorporated. Variance inflation factors were computed to verify that the explanatory variables were not excessively correlated. The regression analysis was designed to assess the extent to which PdM algorithms affect company performance and to determine whether the impact remains substantial even after accounting for other variables.

Decision tree: as the final step, the decision tree method was conducted to analyse the key factors influencing companies' decisions to implement PdM. This technique is part of supervised machine learning methods that identify patterns in data. The objective of the analysis was to determine which combinations of economic factors were most strongly associated with the implementation of PdM in companies between 2018 and 2020. The target variable was a binary indicator of PdM implementation (0 = not implemented; 1 = implemented). A wide range of economic indicators were included that could influence a company's decision to implement PdM as explanatory variables. To train the decision tree, the CART (Classification and Regression Trees) algorithm was used, which splits the data based on optimal threshold values of explanatory variables.

The following section introduces step-by-step overview of the analysis to provide a structured and comprehensive explanation of the research process.

1. ***Data acquisition and sample selection***—data were extracted from the Orbis database, which covered 5504 large and very large companies from V4 countries. A final dataset of 1094 companies was obtained by employing a stratified random sampling method. The sample was proportionally stratified based on country and company size to maintain representativeness;
2. ***Utilisation of PdM identification***—the annual reports were examined using a Python script to identify any mentions of "Predictive Maintenance" or "PdM." Only annual reports from 2018 to 2020 were included to ensure that at least three years of data were available for assessing the impact of PdM. The classification of companies was determined by their adoption of PdM (0 indicates that it has not been implemented, while 1 indicates that it has been implemented);
3. ***Descriptive and statistical analysis***—key performance indicators (operating profit and COGS) were compared between PdM adopters and non-adopters. The Kolmogorov–Smirnov test verified that the data were not normally distributed. The economic performance of PdM and non-PdM companies was analysed through the Mann–Whitney test. The impact of PdM was evaluated across various V4 countries using the Kruskal–Wallis's test. The impact's magnitude was quantified using the Hodges–Lehmann median difference estimate;
4. ***Regression analysis***—in order to ascertain the primary economic variables that affect operating profit, a linear regression model was implemented. An interaction term between long-term tangible assets and added value was included, in addition to independent variables. To confirm that there was no multicollinearity, the variance inflation factor was checked. In order to determine whether PdM had a significant effect on company performance, the model was evaluated at a 5% significance level;

5. ***Decision tree model for PdM adoption analysis***—the CART algorithm was employed to develop a decision tree model. The target variable was the implementation of PdM (0 = no; 1 = yes). The most influential factors for the implementation of PdM were identified by incorporating a set of economic variables (X_1–X_{17}). The data were divided by the decision tree according to the most optimal threshold values, which enabled the identification of patterns in the decisions of companies.

4. Results

Table 3 presents a summary of the total number of manufacturing companies (NACE Rev. 2 Section C) in the Orbis database for V4 countries, as well as details regarding the sample of companies that were analysed for this study. The sample is stratified by the countries of operation (Slovakia, Czech Republic, Poland, and Hungary), size (large and very large), and whether PdM is implemented. Of the 1094 companies in the sample, 307 have implemented PdM, which accounts for approximately 28.06%. The proportion of companies that implemented PdM was highest among very large companies in Poland (34.45%), while the lowest proportion was observed among large companies in Hungary (22.67%). Initially, it is possible to discern minor variations in the implementation of PdM among countries and companies of varying sizes. These distinctions serve as the foundation for a more comprehensive examination of the variables that influence the implementation of PdM within organisations.

Table 3. Distribution of companies by size classification and application of PdM.

Country	No. of Companies in Orbis	No. of Companies Analysed	Distribution of Companies by Size Classification and Application of PdM			
Czechia (CZ)	1195	239	Very large		Large	
			69		170	
			PdM		PdM	
			Yes	No	Yes	No
			22	47	46	124
Hungary (HU)	789	237	Very large		Large	
			65		172	
			PdM		PdM	
			Yes	No	Yes	No
			17	48	39	133
Poland (PL)	2922	438	Very large		Large	
			119		319	
			PdM		PdM	
			Yes	No	Yes	No
			41	78	96	223
Slovakia (SK)	598	180	Very large		Large	
			44		136	
			PdM		PdM	
			Yes	No	Yes	No
			13	31	33	103
Total	5504	1094	x	x	x	x

Source: authors' compilation.

The descriptive analysis of two critical factors for companies, operating profit and cost of goods sold, was the primary focus of the initial phase. These factors were selected due to their relevance in evaluating the overall efficiency of operational processes within organisations. The descriptive statistics for the operating profit of companies are summarised in Table 4. It demonstrates that the average operating profits of companies that implement PdM algorithms are higher in all of the countries that have been analysed. In the Czech Republic, PdM has the most substantial impact, resulting in an operating profit increase of EUR 5602.31 thousand for companies that implement it in comparison to those that do not. Additionally, the median values, which provide a robust measure less affected by outliers, reveal a similar trend, highlighting higher operating profits for PdM adopters across all analysed countries. In other countries that were analysed, a comparable trend was observed, albeit with a slightly smaller impact of PdM usage. The utilisation of PdM had the least significant impact in Poland, where companies that employ PdM experience an average increase in operating profit of EUR 3318.96 thousand. This discovery implies that the utilisation of PdM has the capacity to substantially enhance company performance by increasing operating profits.

Table 4. Descriptive statistics—operating profit.

Country	PdM	Mean	Median	Std. Deviation	*n*
Czechia (CZ)	Yes	13,000.86	5067.24	25,481.02	68
	No	7398.54	3012.15	25,348.98	171
	Total	8992.50	3560.78	25,459.31	239
Hungary (HU)	Yes	7444.71	3397.43	12,955.96	56
	No	3574.20	1460.77	8119.58	181
	Total	4488.75	1717.01	9597.80	237
Poland (PL)	Yes	5588.88	1372.67	24,834.26	137
	No	2269.92	833.75	11,013.85	301
	Total	3308.04	955.99	16,660.92	438
Slovakia (SK)	Yes	4346.32	2008.06	6572.50	46
	No	3043.37	1752.73	4687.76	134
	Total	3376.35	1951.01	5245.23	180
Total	Yes	7382.96	2328.24	21,508.94	307
	No	3815.93	1386.89	14,419.33	787
	Total	4816.92	1600.11	16,781.22	1094

Source: authors' compilation.

The cost of goods sold was analysed using the same methodology. Table 5 displays the descriptive statistics that were generated for this factor. The table indicates that the average COGS of companies that employ PdM algorithms is lower in all of the countries that were analysed. The Czech Republic is the location where the most substantial distinction between companies with and without PdM is observed. In this country, the average cost of goods sold is EUR 32,513.93 thousand lower for companies that employ PdM than for those that do not. The second largest decrease in costs is observed in Poland, where companies with PdM report an average COGS of EUR 20,667.87 thousand lower. In Slovakia, the disparity is EUR 16,670.11 thousand. The median values confirm this trend, further demonstrating that companies with PdM consistently achieve lower costs of goods sold across all analysed countries. Nevertheless, Hungary's performance in international comparisons is the most subpar, as Hungarian companies that implement PdM have an average COGS that is only EUR 5226.98 thousand lower. This is the most minor discrepancy among the countries that were analysed. The variability of COGS is higher among non-PdM companies, as evidenced by the standard deviations.

Table 5. Descriptive statistics—cost of goods sold.

Country	PdM	Mean	Median	Std. Deviation	*n*
Czechia (CZ)	Yes	94,113.86	51,586.15	134,622.34	68
	No	126,627.48	59,476.11	204,884.33	171
	Total	117,376.74	57,119.63	187,888.25	239
Hungary (HU)	Yes	12,225.99	1575.87	32,768.19	56
	No	17,452.97	1832.32	51,333.16	181
	Total	16,217.90	1668.46	47,592.10	237
Poland (PL)	Yes	41,316.01	11,665.37	224,373.99	137
	No	61,983.88	13,029.54	246,542.93	301
	Total	55,519.28	12,669.21	239,765.25	438
Slovakia (SK)	Yes	50,284.06	13,856.64	64,368.45	46
	No	61,572.93	45,074.87	63,737.66	134
	Total	58,687.99	34,223.67	63,910.01	180
Total	Yes	49,048.07	13,478.50	166,948.70	307
	No	65,718.18	15,633.78	186,873.26	787
	Total	61,040.19	15,031.79	181,582.33	1094

Source: authors' compilation.

The descriptive analysis demonstrated that the average operating profit and cost of goods sold of companies with and without PdM were distinct. Consequently, the second phase of the analysis investigated the statistical significance of these company group differences. In order to ascertain the appropriate parametric or non-parametric test, it was imperative to evaluate the normality of the data distribution for both factors prior to conducting the test. The Kolmogorov–Smirnov test evaluated normality in relation to the size of the sample (Table 6). The data are not normally distributed, as the p-value for operating profit and cost of goods sold is less than 0.001 at 5%.

Table 6. Normality test of selected variables.

	Kolmogorov–Smirnov		
Operating profit	Statistic	df	Sig.
	0.315	1094	<0.001
	Kolmogorov–Smirnov		
Cost of goods sold	Statistic	df	Sig.
	0.315	1094	<0.001

Source: authors' compilation.

The non-parametric Mann–Whitney test was employed to evaluate the significance of the differences between the groups of companies with PdM and those without PdM, as our data do not follow a normal distribution. This test is an exceptional instrument for comparing the medians of two distinct groups. In this instance, the two distinct groups correspond to organisations that implement PdM and those that do not. The Mann–Whitney test is a reliable method for evaluating the differences between these independent groups and is resistant to violations of the normality assumption. Table 7 displays the operating profit test results. Companies that implement PdM (mean rank 632.97) exhibit a higher median operating profit than those that do not (mean rank 514.16). The statistical significance of this difference at the 5% significance level is confirmed by the test statistic $Z = -5.588$ and a p-value < 0.001. The results corroborate the hypothesis that companies can increase their operating profits by implementing PdM.

Table 7. Mann–Whitney test—operating profit.

	PdM	n	Mean Rank	Sum of Ranks
Operating profit	No	787	514.16	404,644.00
	Yes	307	632.97	194,321.00
	Total	1094		
	Operating profit			
Mann–Whitney U			94,566.00	
Wilcoxon W			404,644.00	
Z			−5.588	
Asymp. Sig. (2-tailed)			<0.001	

Source: Authors' compilation.

The Mann–Whitney test results for cost of goods sold are summarised in Table 8. It shows that companies that implement PdM achieve lower medians of COGS (mean rank 511.28) than those that do not (mean rank 561.23). In the same vein, the statistical significance of the difference at the 5% significance level is confirmed by the test statistic $Z = -2.368$ and a p-value of 0.018. The theory that companies can reduce their cost of goods sold by implementing PdM is further supported by the lower mean rank of the group of companies using PdM algorithms.

Table 8. Mann–Whitney test—cost of goods sold.

	PdM	n	Mean Rank	Sum of Ranks
Cost of goods sold	No	787	561.63	442,003.50
	Yes	307	511.28	156,961.50
	Total	1094		
	Cost of goods sold			
Mann–Whitney U			109,683.50	
Wilcoxon W			156,961.50	
Z			−2.368	
Asymp. Sig. (2-tailed)			0.018	

Source: Authors' compilation.

The results of the Mann–Whitney test are closely correlated with the results of descriptive statistics, which indicated that the average operating profit and average cost of goods sold were higher for companies that utilised PdM algorithms in all analysed countries. Nevertheless, this test did not ascertain whether the disparities in operating profit and cost of goods sold between companies with and without PdM are statistically significant within individual V4 countries. Consequently, the Kruskal–Wallis's test was chosen enabling the determination of statistical significance between the analysed V4 countries. Table 9 displays the operating profit results of the Kruskal–Wallis's test. The p-value was less than the established 5% significance level in three V4 countries: Czechia, Hungary, and Poland, indicating statistically significant differences. In contrast, Slovakia did not exhibit a statistically significant difference between companies that had PdM and those that did not (p-value = 0.282). This implies that there are no substantial distinctions between the various ethnic groups in this nation. The analysis may be somewhat restricted by the lower number of companies in Slovakia that utilise PdM, which may account for this absence. Conversely, it is possible that Slovak companies are not yet fully capitalised on the potential of PdM algorithms, or they may need a longer time frame to demonstrate more substantial economic disparities. The results on a global scale suggested that the relationship between PdM implementation and performance, as measured by operating profit, may be influenced by geographical factors and the local company environment.

Table 9. Kruskal–Wallis—operating profit.

Country	Null Hypothesis	Test	Sig.	Decision
CZ	The distribution of operating profit is the same across categories of PdM	Independent-Samples Kruskal–Wallis Test	<0.001	Reject the null hypothesis
HU	The distribution of operating profit is the same across categories of PdM	Independent-Samples Kruskal–Wallis Test	<0.001	Reject the null hypothesis
PL	The distribution of operating profit is the same across categories of PdM	Independent-Samples Kruskal–Wallis Test	<0.001	Reject the null hypothesis
SK	The distribution of operating profit is the same across categories of PdM	Independent-Samples Kruskal–Wallis Test	0.282	Retain the null hypothesis

Source: authors' compilation.

The Kruskal–Wallis test was also employed to investigate the costs of goods sold in the individual V4 countries between companies that utilised and did not utilise PdM. The statistical significance of the differences in the cost of goods sold in the Czech Republic (p-value = 0.032), Poland (p-value = 0.007), and Slovakia (p-value = 0.003) at the established significance level of 5% was confirmed by the results in Table 10. It is evident that PdM has a substantial effect on the reduction in sales costs in these three V4 countries. The only country in which statistically significant differences (p-value = 0.820) between companies with and without PdM were not confirmed was Hungary. This conclusion can also be drawn from the descriptive analysis, which revealed that Hungarian companies that implemented PdM did not exhibit any substantial disparities in their average sales costs when contrasted with those that did not. In general, the low significance of this economic indicator in the Hungarian business environment suggests that PdM has a weaker impact. The results indicate that the cost of goods sold in specific geographical areas can be significantly and positively impacted by the implementation of PdM from a global perspective. Nevertheless, this effect is also contingent upon the diverse cost structures of companies in various countries and the local conditions.

It was possible to determine the statistical significance of the differences in operating profit and cost of goods sold between companies that were using PdM and those that were not in the individual V4 countries using the Kruskal–Wallis's test. Nevertheless, it did not offer a precise quantitative representation of the distinctions between these two categories of companies. As a result, the Hodges–Lehmann estimate of the median difference was employed. This test is capable of numerical quantification of the effect of PdM implementation on the two selected key economic performance indicators. The Hodges–Lehmann method enables a more precise representation of the discrepancies in the centre of the distribution of both examined indicators between the analysed groups of companies. This method is renowned for its reliable estimates, as it is not affected by outliers or extreme values in the data.

Table 10. Kruskal–Wallis—cost of goods sold.

Country	Null Hypothesis	Test	Sig.	Decision
CZ	The distribution of cost of goods sold is the same across categories of PdM	Independent-Samples Kruskal–Wallis Test	0.032	Reject the null hypothesis
HU	The distribution of cost of goods sold is the same across categories of PdM	Independent-Samples Kruskal–Wallis Test	0.820	Retain the null hypothesis
PL	The distribution of cost of goods sold is the same across categories of PdM	Independent-Samples Kruskal–Wallis Test	0.007	Reject the null hypothesis
SK	The distribution of cost of goods sold is the same across categories of PdM	Independent-Samples Kruskal–Wallis Test	0.003	Retain the null hypothesis

Source: authors' compilation.

Table 11 presents the separate results of the Hodges–Lehmann median difference estimates for operating profit and cost of goods sold for each analysed country. The Czech Republic recorded the highest median difference (EUR 2198.17 thousand) in the operating profit estimates, underscoring the substantial and advantageous impact of PdM on this metric in that nation. The median difference in Slovakia is the lowest at EUR 403.10thousand, which is at the opposite end of the spectrum. These quantitative findings provide a novel perspective on the impact of PdM usage and substantiate the findings of the Kruskal–Wallis's test that revealed statistically significant differences in operating profit only in the Czech Republic, Hungary, and Poland, while no significant differences were observed in Slovakia (p-value = 0.282).

Table 11. Hodges–Lehmann—intervals.

Country	Hodges–Lehmann Estimate of the Median Difference for Operating Profit
CZ	2198.17
HU	1711.64
PL	572.49
SK	403.10
Country	**Hodges–Lehmann Estimate of the Median Difference for the Cost of Goods Sold**
CZ	−7581.25
HU	−51.39
PL	−1338.50
SK	−8355.49

Source: authors' compilation.

In the same vein, median interval estimates were implemented for the cost of goods sold. It was discovered that Slovakia and Czechia experienced the most significant median cost reductions for companies that implemented PdM (EUR −8355.49 thousand and EUR −7581.25 thousand, respectively). In Hungary, the most minimal and nearly negligible effect was observed, with a median reduction in cost of goods sold of EUR −51.39 thousand for companies that implemented PdM in comparison to those that did not. And again, the Hodges–Lehmann median difference estimates are consistent with the Kruskal–Wallis test results in this instance, as only the Czech Republic, Poland, and Slovakia exhibited statistically significant differences. These discrepancies were classified as statistically insignificant in Hungary (p-value = 0.820).

The Hodges–Lehmann estimates consistently complemented the Kruskal–Wallis's test results by providing direct quantitative evidence of the economic benefits of PdM algorithms, artificial intelligence digital twin technologies, and the IoRT in big data-driven Industry 4.0 manufacturing systems. The estimates not only illustrate the existence of disparities between countries and groups of companies with and without PdM, but they also supply a reliable representation of the effect of PdM usage on the operating profit and cost of goods sold. The findings of this study are a substantial addition to the current analysis and can also serve as an incentive for additional research on the effects of PdM implementation in organisations.

The quantification of the differences between companies that use and do not use PdM provided valuable insights into the dissimilarities in operating profit and cost of goods sold. Nevertheless, the objective of our analysis was to investigate the influence of individual factors on the performance of the company. Consequently, a linear regression model was developed. It could identify the factors that have the most substantial influence on operating profit as the key performance indicator. This implies that the dependent or target variable in our model was operating profit. Table 12 provides a concise summary of the model's quality.

Table 12. Regression—model summary.

R	R Square	Adjusted R Square	Std. Error of the Estimate
0.897	0.872	0.872	26,377.96

Source: authors' compilation.

The coefficient of determination was relatively high for the model (adjusted $R^2 = 0.872$). This suggests that the independent variables account for up to 87.2% of the variability in operating profit among the analysed companies. The statistical significance of the model (p-value = 0.000) at the 5% significance level was confirmed by the ANOVA test, as shown in Table 13. Consequently, the model is suitable for interpretation.

Table 13. Regression—ANOVA.

	Sum of Squares	df	Mean Square	F	Sig.
Regression	1.027×10^{14}	6	1.712×10^{13}	24,599.86	0.000
Residual	7.563×10^{11}	1087	6.958×10^{8}		
Total	1.035×10^{14}	1093			

Source: authors' compilation.

Table 14 illustrates the regression model that was generated. One of the six independent variables in the model is an interaction between long-term tangible assets and value added. The model does not exhibit multicollinearity between the independent variables, as the variance inflation factor (VIF) values are all less than 5. Additionally, all independent variables in the model are statistically significant, as they achieve p-values that are below the established 5% significance level. The reference categories for the dummy variables were determined as follows:

- ***Country***—the Czech Republic;
- ***Company legal structure***—public limited companies;
- ***Company size***—very large companies;
- ***Utilisation of PdM***—companies that implemented PdM.

Table 14. Regression—coefficients.

	Unstandardised Coefficients		Sig.	Collinearity Statistics	
	B	Std. Error		Tolerance	VIF
Constant	28,138.99	2561.00	<0.001		
Country = SK	−4806.18	2311.35	0.038	0.866	1.155
Country = PL	−5212.78	1795.29	0.004	0.822	1.217
Legal form = Private limited companies	−6917.01	1911.47	<0.001	0.916	1.091
Company size = Large company	−14,206.46	1829.42	<0.001	0.961	1.041
PdM = No	−5071.42	1780.26	0.004	0.994	1.006
Added Value*Tangible Fixed assets	0.227	0.001	0.000	0.921	1.127

Source: authors' compilation.

The structure of the regression equation is as follows:

$$\begin{aligned} Operating\ profit\ (Y) = \beta_0 - \beta_1 \cdot Country_{SK} - \beta_2 \cdot Country_{PL} - \beta_3 \cdot \\ Legal\ form_{Private\ limited\ companies} - \beta_4 \cdot Company\ size_{Large\ company} - \\ \beta_5 \cdot PdM_{No} + \beta_6 \cdot Added\ value * Tangible\ fixed\ assets \end{aligned} \quad (1)$$

Each of the coefficients in the regression equation represents the change in the dependent variable, which is operating profit, when the given independent variable changes, assuming that all other variables remain unchanged. The interpretation of the coefficients is as follows. The constant (28,138.99) indicates that if all independent variables take their reference values (i.e., it is a very large company using PdM, located in the Czech Republic with the legal form of a public limited company), the expected operating profit for this company would be EUR 28,138.99 thousand. For companies in Slovakia, the operating profit is, on average, EUR 4806.18 thousand lower compared to companies in the Czech Republic. Similarly, the operating profit of companies in Poland is, on average, EUR 5212.78 thousand, lower compared to companies in the Czech Republic. Companies with the legal form of private limited companies show, on average, EUR 6917.01 thousand lower operating profit compared to public limited companies. Large companies show, on average, EUR 14,206.46 thousand lower operating profit compared to very large companies. Companies that have not implemented and do not use PdM have, on average, EUR 5071.42 thousand lower operating profit compared to companies that have implemented PdM. Finally, an increase in the interaction effect between added value and fixed assets by one unit (EUR thousand) is associated with an increase in operating profit by EUR 0.227 thousand.

The results of the regression model confirmed the huge significance of implementing PdM for companies. Companies that do not use PdM have lower operating profits compared to companies that actively use PdM algorithms. The effect of PdM on the economic performance of companies remained consistent even after considering other factors. The model also revealed differences between the analysed countries and types of companies, reflecting that national and organisational characteristics play a significant role in determining the final impact of PdM on companies' economic performance.

The current analysis confirmed that PdM algorithms, artificial intelligence digital twin technologies, and the IoRT in big data-driven Industry 4.0 manufacturing systems have a positive effect on the economic performance of companies. However, answering the question of what motivates companies to use PdM algorithms is very challenging. It may be subjective motivation from managers or business owners, it may be a targeted management strategy, or it may simply be the desire to achieve better results than competitors. A questionnaire survey would provide the best answers to this question. It is believed that more can be revealed by data. For this purpose, the decision tree method, which is based on data from the period between 2018 and 2020 (the years in which PdM was implemented by companies), has been applied.

The decision tree shown in Figure A1 (Appendix A) provides a retrospective view of the economic characteristics of companies that implemented PdM in the period from 2018 to 2020. The main goal of this analysis was to identify combinations of factors that were more frequently associated with companies implementing PdM during this period. The root node of the tree identifies that net sales were the most important distinguishing factor between companies. Companies with net sales below EUR 11,947.07 thousand had a higher probability of implementing PdM during this period than companies with higher net sales. This finding may suggest that companies are motivated to use PdM in order to increase their net sales. The layers of the tree also reveal that companies with lower net sales were further characterised by costs of goods sold. Lower costs of goods sold (below EUR 8004.15 thousand) were more frequently observed in companies that implemented PdM. This may indicate that companies were already demonstrating efficient cost management before implementation and viewed PdM algorithms as tools to strengthen this efficiency. Alternatively, it could reflect companies' efforts to stabilise and maintain their advantageous cost structure amidst growing competition. For companies with higher net sales, factors such as company size or capital played a crucial role. The leaf nodes at the end of the decision tree present the final classifications and summarise the number of companies in each group based on the combination of the analysed factors. The resulting decision tree model provides valuable insight into the decision-making processes that may have led companies to implement PdM.

5. Discussion

The empirical study examining the economic consequences of PdM algorithms in manufacturing enterprises within the V4 countries provides a significant contribution to understanding the financial benefits and challenges of PdM implementation. The findings demonstrate that PdM algorithms, artificial intelligence digital twin technologies, and the IoRT in big data-driven Industry 4.0 manufacturing systems lead to increased operational profits and reduced costs of goods sold, highlighting their role as a strategic investment for companies aiming to enhance productivity and efficiency. However, this study also identifies regional disparities in PdM adoption and effectiveness, which raises important questions about the factors influencing its successful implementation [44]. PdM algorithms have gained widespread recognition for their ability to improve the financial and operational performance of manufacturing enterprises. By leveraging advanced technologies such as artificial intelligence, machine learning, and the IoRT, PdM enables companies to proactively monitor and assess equipment conditions, thereby minimising unplanned downtime and reducing maintenance costs [50]. This study's findings align with broader industry trends, confirming that companies implementing PdM experience higher operating profits and improved cost efficiency.

One key factor driving these benefits is PdM's ability to optimise resource allocation. Traditional maintenance approaches, such as reactive and preventive maintenance, often lead to either excessive downtime or unnecessary maintenance expenditures. In contrast, PdM algorithms, artificial intelligence digital twin technologies, and the IoRT in big data-driven Industry 4.0 manufacturing systems allows companies to perform maintenance only when necessary, reducing labour and material costs while extending the lifespan of machinery. The economic impact of these efficiencies is particularly evident in highly automated industries, where downtime can result in substantial financial losses [51]. This study's results indicate that the Czech Republic exhibited the most significant increase in operating profits among the V4 countries. This suggests that companies in this region have successfully integrated PdM algorithms into their operational strategies, likely benefiting from advanced digital infrastructure and government incentives promoting industrial modernisation [52]. However, the findings also highlight that Slovakia showed statistically insignificant improvements, raising concerns about potential barriers to PdM adoption in this region. The variations in PdM effectiveness across different V4 countries underscore the importance of considering geographical and industry-specific factors when analysing maintenance strategies [53]. Differences in digital maturity, workforce skills, regulatory environments, and economic conditions may all contribute to the disparities observed in this study. Countries with well-developed industrial infrastructure and high levels of digital adoption are more likely to achieve significant financial benefits from PdM [54]. In contrast, regions where businesses face challenges such as limited access to digital technologies, high initial investment costs, and a lack of technical expertise may struggle to realise PdM's full potential [55]. This study suggests that Hungarian companies, for instance, did not experience statistically significant reductions in COGS, which may indicate obstacles in implementing PdM at scale. Another key aspect influencing PdM's effectiveness is industry type. Manufacturing sectors that rely heavily on automated production lines, such as the automotive and electronics industries, are more likely to benefit from PdM algorithms, artificial intelligence digital twin technologies, and the IoRT in big data-driven Industry 4.0 manufacturing systems due to their reliance on complex machinery that requires continuous monitoring. Conversely, industries with lower automation levels may find the cost–benefit ratio of PdM less compelling, leading to slower adoption rates [56]. Despite the documented economic benefits, implementing PdM is not without challenges. There are also several barriers that may limit its adoption, including high upfront costs, integration complexities, and the need for skilled personnel to manage and interpret predictive analytics [57]. One major concern is the substantial investment required to deploy PdM solutions. Companies must invest in IoT sensors, data analytics platforms, and cloud computing infrastructure to enable real-time monitoring and predictive modelling. For small and medium-sized enterprises, these costs may be prohibitive, preventing them from fully adopting PdM. Another challenge is data management. PdM relies on vast amounts of sensor-generated data to predict equipment failures accurately [2]. However, collecting, storing, and analysing these data require robust data management systems and skilled personnel who can interpret complex analytical outputs. Companies that lack the necessary expertise may struggle to leverage PdM effectively, limiting its financial impact [58]. Additionally, PdM adoption requires organisational and cultural changes. Traditional maintenance teams accustomed to reactive strategies may resist transitioning to a predictive approach, necessitating training programs and change management initiatives to ensure a smooth integration process. Without adequate support and investment in workforce development, companies may face implementation difficulties [59]. The current study also highlights key factors that influence a company's likelihood of adopting PdM algorithms, artificial intelligence digital twin technologies, and the IoRT in big data-driven Industry 4.0 manufacturing systems, with net sales and cost structures playing a critical role. Companies with lower net sales and more efficient cost management strategies are more likely to implement PdM, suggesting that enterprises view PdM as a means of enhancing competitiveness and achieving financial stability. Interestingly, this study finds that PdM adoption is more common among smaller enterprises in the sample. This challenges conventional assumptions that larger corporations, with greater financial and technical resources, would be the primary adopters of PdM. One possible explanation is that smaller enterprises, facing higher financial pressures, may prioritise cost-saving initiatives such as PdM to remain competitive [60]. Another important measure influencing the PdM adoption is the company specification as also verified in the current study of V4 enterprises. Company-specific features and financial indicators play a crucial role in the successful implementation of PdM, as they influence a company's ability to invest in and benefit from advanced maintenance strategies [61]. These factors determine how well a company can integrate PdM algorithms, manage operational risks, and enhance overall efficiency. Company size and scale of operations play a significant role, as larger enterprises typically have more financial resources to invest in PdM technologies such as IoRT sensors, AI-driven analytics, and cloud computing. In contrast, small and medium-sized enterprises may face budget constraints but can still benefit from cost-effective PdM solutions tailored to their needs [62]. Additionally, scale impacts data availability, as larger companies generate vast operational datasets, improving prediction accuracy by production management algorithms. The industry sector and asset intensity also matter, as industries with high capital intensity, such as manufacturing, energy, and transportation, benefit most from PdM and digital twin-based cyber-physical manufacturing systems, given that unplanned downtime leads to significant financial losses. Enterprises operating in less asset-heavy sectors may find PdM less critical but can still use data-driven maintenance strategies to enhance efficiency in extended reality metaverse enterprise-based generative artificial intelligence economics [63]. Technological infrastructure and digital maturity, however, also influence the success of PdM adoption. Companies with strong digital capabilities, including automation, IoRT integration, and AI applications, can implement PdM more effectively, whereas enterprises with outdated systems may require higher initial investments to integrate PdM into existing workflows. Additionally, workforce skills and expertise are crucial, as companies need trained employees who understand data analysis, AI, and machine learning applications to interpret sensor data, optimise machine learning models, and make data-driven decisions [64]. Financial performance metrics directly impact the ability for a corporation to implement and sustain PdM systems. Operating profit is a key indicator, as companies with higher profits have more funds to invest in PdM solutions, which in turn further increases profits by reducing unplanned downtime, optimising asset use, and improving operational efficiency. The cost of goods sold is another critical factor, as enterprises with high COGS may benefit significantly from PdM algorithms, artificial intelligence digital twin technologies, and the IoRT in big data-driven Industry 4.0 manufacturing systems, since reducing machine failures lowers production costs [36], which is in line with the results of this study within the V4 countries. Moreover, other studies confirm that PdM optimises maintenance schedules, leading to lower repair costs and improved machine lifespan [65,66]. Capital expenditures and investment in technology also play a role, as those enterprises with higher expenditures can more easily invest in advanced sensors, cloud computing, and AI-driven PdM solutions, while companies with lower expenditures may opt for a gradual PdM adoption approach to balance cost-effectiveness. Net sales and revenue growth influence PdM adoption, as companies with strong revenue growth can scale PdM adoption across multiple production lines or facilities, while enterprises with declining sales may struggle to justify upfront PdM investments despite the potential for long-term savings [63]. Liquidity and cash flow are also crucial, as enterprises with strong cash flow can finance PdM implementation without disrupting other critical operations. Enterprises with liquidity constraints might need to explore leasing options, partnerships, or phased PdM implementation [47]. Another important factor is return on assets (ROA) and asset utilisation, as PdM directly improves asset utilisation, leading to higher ROA. Companies with capital-intensive assets benefit more from PdM by maximising equipment uptime and ensuring continuous production [67].

Based on the analysis of enterprises within the V4 countries and the results of other relevant studies, it can be concluded that company-specific features and financial indicators are critical in determining the feasibility, adoption, and success of PdM algorithms, artificial intelligence digital twin technologies, and the IoRT in big data-driven Industry 4.0 manufacturing systems. A company's size, industry type, technological readiness, and workforce skills shape how well PdM can be integrated, while financial performance metrics influence investment decisions and expected benefits. For PdM algorithms to be effective, enterprises should assess both their technical capabilities and financial health to ensure a well-planned, data-driven maintenance strategy that improves efficiency, reduces costs, and enhances long-term competitiveness.

Deep learning-assisted smart process planning configures Internet of Manufacturing Things-based enterprise digital maturity in the service industry by use of geospatial big data management algorithms [68–70]. Computer vision algorithms articulate sustainable supply chain practices and online knowledge networks through remote sensing data fusion techniques [71–73]. Enterprise generative artificial intelligence technologies and Internet of Things sensing networks can be harnessed in green logistics practices across deep learning-assisted smart process and supply chain management for cyber–physical production systems in the digital twin industrial metaverse [74–76]. Geospatial simulation and sensor fusion tools can be deployed for Internet of Robotic Things-based knowledge co-creation by use of big data management algorithms [77–79]. The digital twin industrial metaverse develops on deep learning and virtual simulation algorithms, generative artificial intelligence of things and Industry 4.0-based manufacturing systems, multisensory immersive extended reality and autonomous robotic technologies, big data simulation and modelling tools, and cyber–physical production networks [80–82]. The sustainable development of artificial intelligence-based Internet of Manufacturing Things systems integrates remote big data management tools, virtual machine and cognitive computing algorithms, and sensing and computing technologies for digital twin simulation modelling-based environment mapping [83–85].

6. Conclusions

The incorporation of PdM algorithms, artificial intelligence digital twin technologies, and the IoRT in big data-driven Industry 4.0 manufacturing systems signifies a substantial change in the manner in which manufacturing enterprises approach operational efficiency and maintenance. As traditional reactive and preventive maintenance practices are increasingly determined to be inefficient, PdM has emerged as a more sophisticated solution, utilising advancements in machine learning, the IoRT, and big data analytics. These technologies facilitate continuous monitoring and data-driven insights that anticipate potential failures prior to their occurrence, thereby reducing downtime, improving asset utilisation, and increasing operational profitability. This investigation underscores the substantial economic advantages of PdM algorithms, particularly for enterprises located in V4 countries. The statistical analysis confirms that organisations that implement PdM experience substantial increases in operating profit and decreases in the cost of goods sold.

The results showed that PdM significantly influences operating profit in the Czech Republic, Hungary, and Poland. For Slovakia, these differences were statistically insignificant. Additionally, the results showed that the differences in COGS in the Czech Republic, Poland, and Slovakia were statistically significant, but in Hungary, statistical significance was not confirmed. The highest median difference in operating profit was recorded in the Czech Republic (EUR +2198.17 thousand), while the impact of PdM was smallest in

Slovakia. On the other hand, for COGS, the most significant effect of PdM was identified in the Czech Republic, and the smallest effect was found in Hungary. These findings significantly correlate with the previous Kruskal–Wallis test results. The regression model's findings verified the immense importance of PdM algorithms, artificial intelligence digital twin technologies, and the IoRT in big data-driven Industry 4.0 manufacturing systems for companies. Companies that do not implement PdM generate lower operating profits than those that do. The impact of PdM on economic performance was consistent, even after accounting for other factors. The model also demonstrated that the final impact of PdM on economic performance is significantly influenced by national and organisational characteristics, as evidenced by the differences between the analysed countries and company types. The results of the decision tree reveal that the main predictor of PdM implementation is net sales. Companies with net sales less than EUR 11,947.07 thousand had a higher probability of implementing PdM than companies with higher values. The cost of goods sold was the second most critical factor. Companies with lower COGS (<EUR 8004.15 thousand) had a higher likelihood of implementing PdM. This outcome may indicate that organisations that had already implemented effective cost management strategies prioritised maintenance optimisation as a means of achieving additional cost savings. Conversely, organisations with greater COGS probably refrained from implementing PdM due to other constraints or priorities. The size of the company was the third critical factor. In comparison to large companies, those that were very large demonstrated an increased likelihood of implementing PdM. This discrepancy may be attributable to the increased complexity of production processes and the greater availability of financial resources, which PdM algorithms, artificial intelligence digital twin technologies, and IoRT in big data-driven Industry 4.0 manufacturing systems can effectively optimise. The results of our research are in accordance with the existing literature, which underscores the fact that PdM contributes to cost savings, extended equipment lifespan, and improved product quality, thereby resulting in increased customer satisfaction. Additionally, the regression model emphasises the financial benefits of PdM algorithms by demonstrating that companies that do not implement PdM have lower operating profits than those that do. The adoption of PdM is impeded by substantial challenges, such as the necessity for specialised expertise, the high initial implementation costs, and apprehensions about the reliability of predictive models, despite the apparent advantages. Organisations that are considering the adoption of PdM must exercise caution and implement strategic planning in order to address these obstacles. The decision-making process is further complicated by economic factors, such as the substantial upfront investment necessary for sensor infrastructure and system integration.

The empirical findings of this study make a valuable contribution to the current body of research on PdM algorithms, artificial intelligence digital twin technologies, and the IoRT in big data-driven Industry 4.0 manufacturing systems by providing quantitative evidence of their impact on company performance. Our research also serves as a foundation for future academic investigations, particularly in relation to the factors that influence the adoption of PdM. It is important to note that the analysis indicates that companies with lower net sales were more likely to implement PdM during the examined period, which implies that financial considerations may be a factor in the adoption of PdM.

7. Limitations, Implications, and Further Directions of Research

Nevertheless, this investigation is also a subject to certain limitations. The generalisability of the findings to other regions or industries is restricted by the geographical focus on the V4 countries. The outcomes of PdM adoption may be influenced by the variability in economic conditions, regulatory frameworks, and technological infrastructures. The scope of future research could be broadened to encompass other countries and industries, thereby facilitating a more comprehensive understanding of the applicability of PdM algorithms, artificial intelligence digital twin technologies, and the IoRT in big data-driven Industry 4.0 manufacturing systems. Furthermore, investigating organisational factors, including leadership attitudes, corporate culture, and decision-making processes, could offer a more comprehensive understanding of the internal factors that motivate the adoption of PdM.

This study offers significant insights into the economic advantages of PdM but contains some shortcomings that must be addressed in the future to enhance transparency and trustworthiness. A key concern is the possibility for sample bias, as this research exclusively examines large and very large firms within the Visegrad Group nations. This may not comprehensively reflect the experiences of small and medium-sized enterprises or sectors beyond manufacturing. Despite the stratified sample selection process, inherent biases may persist regarding industry representation, financial health, or technical maturity of the firms.

This research also faces regional and industry-specific constraints. While it offers comparative insights among the V4 countries, it fails to consider regulatory variations, governmental incentives, or changes in digital infrastructure, all of which could substantially influence PdM efficacy. Moreover, the results may not be applicable to sectors with reduced automation, where the implementation of predictive maintenance is less prevalent. Moreover, technological and organisational impediments pose hurdles to the implementation of PdM. This research recognises challenges including elevated initial expenses, complications in data management, and workforce preparedness, although it fails to quantify their precise influence on predictive maintenance adoption rates. Another constraint is the absence of longitudinal analysis. This research study provides an overview of PdM's economic impact but fails to evaluate long-term financial performance patterns or the possibility of declining returns from PdM over time. Mitigating these constraints in subsequent research would strengthen the findings, boost the validity of conclusions, and yield more useful insights for enterprises contemplating PdM adoption.

Nonetheless, it is evident that PdM provides substantial advantages for manufacturing enterprises in terms of operational efficiency, cost reduction, and economic performance. The long-term benefits of enhanced profitability, reduced downtime, and improved asset reliability are evident, despite the challenges associated with the implementation of PdM, including the necessity for technical expertise, high initial or infrastructure costs, cybersecurity risks, and data quality issues. These challenges can be perceived as points of interests in the future research. Furthermore, a broader industry-wide shift towards data-driven optimised maintenance practices is anticipated as big data-driven Industry 4.0 manufacturing systems continue to advance, resulting in an increasing number of companies across various sectors adopting PdM. The current challenges will likely be addressed by future research and technological advancements, which will facilitate a wider adoption of PdM.

Author Contributions: Conceptualisation, M.N. and G.L.; methodology, M.N., M.F. and K.V.; validation, K.V. and G.L.; formal analysis, M.N.; investigation, M.N. and M.F.; resources, K.V. and G.L.; data curation, G.L. and M.N.; writing—original draft preparation, K.V., G.L., M.N. and M.F.; writing—review and editing, K.V. and G.L.; visualisation, M.N. and M.F.; supervision, G.L. All authors have read and agreed to the published version of the manuscript.

Funding: This research was financially supported by the Slovak Research and Development Agency Grant VEGA 1/0494/24: Metamorphoses and causalities of indebtedness, liquidity and solvency of companies in the context of the global environment and faculty institutional project 4/FPEDAS/2024.

Data Availability Statement: The data presented in this study are available on request from the corresponding author.

Conflicts of Interest: The authors declare no conflicts of interest.

Appendix A

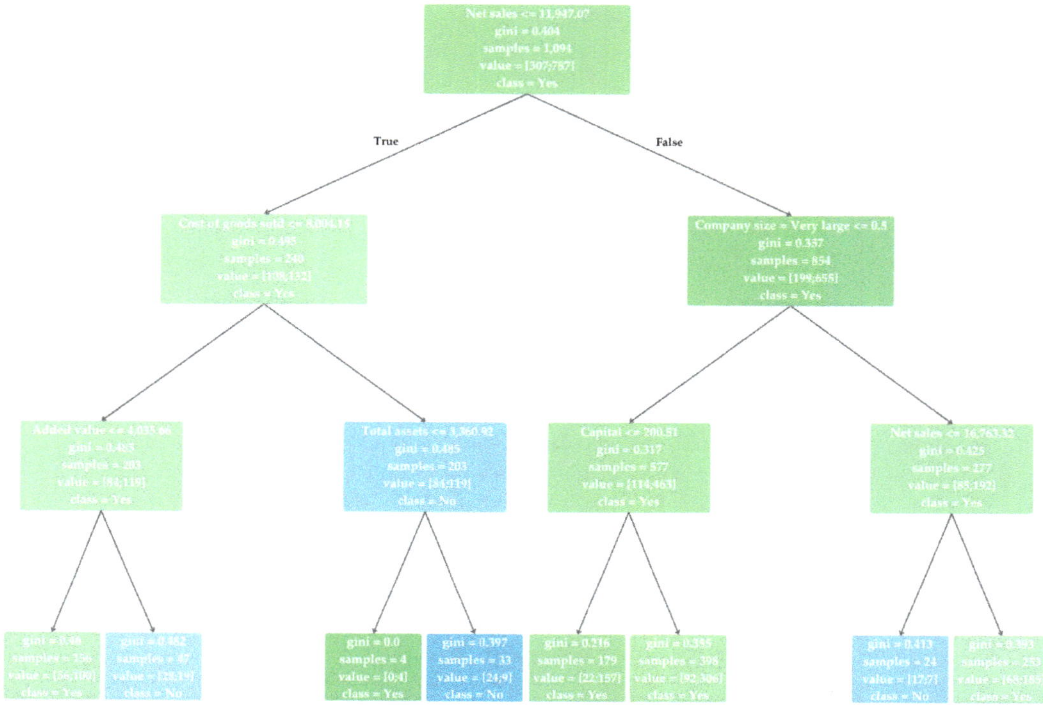

Figure A1. Decision tree of economic factors influencing companies to implement PdM. Source: authors' compilation.

References

1. Sajid, S.; Haleem, A.; Bahl, S.; Javaid, M.; Goyal, T.; Mittal, M. Data Science Applications for Predictive Maintenance and Materials Science in Context to Industry 4.0. *Mater. Today Proc.* **2021**, *45*, 4898–4905. [CrossRef]
2. Çınar, Z.M.; Abdussalam Nuhu, A.; Zeeshan, Q.; Korhan, O.; Asmael, M.; Safaei, B. Machine Learning in Predictive Maintenance towards Sustainable Smart Manufacturing in Industry 4.0. *Sustainability* **2020**, *12*, 8211. [CrossRef]
3. Baur, M.; Albertelli, P.; Monno, M. A Review of Prognostics and Health Management of Machine Tools. *Int. J. Adv. Manuf. Technol.* **2020**, *107*, 2843–2863. [CrossRef]
4. Ruiz-Sarmiento, J.-R.; Monroy, J.; Moreno, F.-A.; Galindo, C.; Bonelo, J.-M.; Gonzalez-Jimenez, J. A Predictive Model for the Maintenance of Industrial Machinery in the Context of Industry 4.0. *Eng. Appl. Artif. Intell.* **2020**, *87*, 103289. [CrossRef]
5. Meddaoui, A.; Hain, M.; Hachmoud, A. The Benefits of Predictive Maintenance in Manufacturing Excellence: A Case Study to Establish Reliable Methods for Predicting Failures. *Int. J. Adv. Manuf. Technol.* **2023**, *128*, 3685–3690. [CrossRef]
6. Carvalho, T.P.; Soares, F.A.A.M.N.; Vita, R.; Francisco, R.d.P.; Basto, J.P.; Alcalá, S.G.S. A Systematic Literature Review of Machine Learning Methods Applied to Predictive Maintenance. *Comput. Ind. Eng.* **2019**, *137*, 106024. [CrossRef]
7. Ayvaz, S.; Alpay, K. Predictive Maintenance System for Production Lines in Manufacturing: A Machine Learning Approach Using IoT Data in Real-Time. *Expert Syst. Appl.* **2021**, *173*, 114598. [CrossRef]
8. Benešová, A.; Hirman, M.; Steiner, F.; Tupa, J. Digital Predictive Maintenance: Case Study. In Proceedings of the 2024 International Conference on Diagnostics in Electrical Engineering (Diagnostika), Pilsen, Czech Republic, 3–5 September 2024; pp. 1–6. [CrossRef]
9. Maletič, D.; Maletič, M. An Exploration of Organisational Readiness for Industry 4.0: A Predictive Maintenance Perspective. *Qual. Innov. Prosper.* **2024**, *28*, 26–46. [CrossRef]

10. Gu, C.; He, Y.; Han, X.; Chen, Z. Product Quality Oriented Predictive Maintenance Strategy for Manufacturing Systems. In Proceedings of the 2017 Prognostics and System Health Management Conference (PHM-Harbin), Harbin, China, 9–12 July 2017; pp. 1–7. [CrossRef]
11. Cummins, L.; Sommers, A.; Ramezani, S.B.; Mittal, S.; Jabour, J.; Seale, M.; Rahimi, S. Explainable Predictive Maintenance: A Survey of Current Methods, Challenges and Opportunities. *IEEE Access* **2024**, *12*, 57574–57602 [CrossRef]
12. Dalzochio, J.; Kunst, R.; Pignaton, E.; Binotto, A.; Sanyal, S.; Favilla, J.; Barbosa, J. Machine Learning and Reasoning for Predictive Maintenance in Industry 4.0: Current Status and Challenges. *Comput. Ind.* **2020**, *123*, 103298. [CrossRef]
13. Serradilla, O.; Zugasti, E.; Rodriguez, J.; Zurutuza, U. Deep Learning Models for Predictive Maintenance: A Survey, Comparison, Challenges and Prospects. *Appl. Intell.* **2022**, *52*, 10934–10964. [CrossRef]
14. Passlick, J.; Dreyer, S.; Olivotti, D.; Grützner, L.; Eilers, D.; Breitner, M.H. Predictive Maintenance as an Internet of Things Enabled Business Model: A Taxonomy. *Electron. Mark.* **2021**, *31*, 67–87. [CrossRef]
15. Karki, B.R.; Porras, J. Digitalization for Sustainable Maintenance Services: A Systematic Literature Review. *Digit. Bus.* **2021**, *1*, 100011. [CrossRef]
16. de Jonge, B.; Scarf, P.A. A Review on Maintenance Optimization. *Eur. J. Oper. Res.* **2020**, *285*, 805–824. [CrossRef]
17. Juracka, D.; Nagy, M.; Zvarikova, K.; Popescu, G.H. Eco-Innovation: Analysis of EU Member States Performance in the EU Eco-Innovation Index 2013–2022. *Systems* **2024**, *12*, 452. [CrossRef]
18. Zheng, M.; Li, T.; Ye, J. The Confluence of AI and Big Data Analytics in Industry 4.0: Fostering Sustainable Strategic Development. *J. Knowl. Econ.* **2024**. [CrossRef]
19. Achouch, M.; Dimitrova, M.; Ziane, K.; Karganroudi, S.S.; Dhouib, R.; Ibrahim, H.; Adda, M. On Predictive Maintenance in Industry 4.0: Overview, Models, and Challenges. *Appl. Sci.* **2022**, *12*, 8081. [CrossRef]
20. Warren, D. Reactive to Proactive Maintenance Journey. *SAE Tech. Pap.* **2016**, *2016-01-0271*, 1–10. [CrossRef]
21. de Faria, H.; Costa, J.G.S.; Olivas, J.L.M. A Review of Monitoring Methods for Predictive Maintenance of Electric Power Transformers Based on Dissolved Gas Analysis. *Renew. Sustain. Energy Rev.* **2015**, *46*, 201–209. [CrossRef]
22. Ulansky, V.; Raza, A. A Historical Survey of Corrective and Preventive Maintenance Models with Imperfect Inspections: Cases of Constant and Non-Constant Probabilities of Decision Making. *Aerospace* **2024**, *11*, 92. [CrossRef]
23. Cheng, J.C.P.; Chen, W.; Chen, K.; Wang, Q. Data-Driven Predictive Maintenance Planning Framework for MEP Components Based on BIM and IoT Using Machine Learning Algorithms. *Autom. Constr.* **2020**, *112*, 103087. [CrossRef]
24. Kliestik, T.; Nica, E.; Durana, P.; Popescu, G.H. Artificial Intelligence-Based Predictive Maintenance, Time-Sensitive Networking, and Big Data-Driven Algorithmic Decision-Making in the Economics of Industrial Internet of Things. *Oeconomia Copernic.* **2023**, *14*, 1097–1138. [CrossRef]
25. Maryska, M.; Doucek, P.; Nedomova, L.; Sladek, P. The Energy Industry in the Czech Republic: On the Way to the Internet of Things. *Economies* **2018**, *6*, 36. [CrossRef]
26. Bui, H.N.; Duong, C.D. ChatGPT adoption in entrepreneurship and digital entrepreneurial intention: A moderated mediation model of technostress and digital entrepreneurial self-efficacy. *Equilib. Q. J. Econ. Econ. Policy* **2024**, *19*, 391–428. [CrossRef]
27. Vlasov, A.I.; Grigoriev, P.V.; Krivoshein, A.I.; Shakhnov, V.A.; Filin, S.S.; Migalin, V.S. Smart Management of Technologies: Predictive Maintenance of Industrial Equipment Using Wireless Sensor Networks. *Entrep. Sustain. Issues* **2018**, *6*, 489–502. [CrossRef]
28. Coandă, P.; Avram, M.; Constantin, V. A state of the art of predictive maintenance techniques. *IOP Conf. Ser. Mater. Sci. Eng.* **2020**, *997*, 012039. [CrossRef]
29. Er-Ratby, M.; Kobi, A.; Sadraoui, Y.; Kadiri, M. The Impact of Predictive Maintenance on the Performance of Industrial Enterprises. *SN Comput. Sci.* **2025**, *6*, 73. [CrossRef]
30. Lazaroiu, G.; Androniceanu, A.; Grecu, I.; Grecu, G.; Neguriță, O. Artificial Intelligence-Based Decision-Making Algorithms, Internet of Things Sensing Networks, and Sustainable Cyber-Physical Management Systems in Big Data-Driven Cognitive Manufacturing. *Oeconomia Copernic.* **2022**, *13*, 1047–1080. [CrossRef]
31. Jiang, Y.; Yin, S.; Li, K.; Luo, H.; Kaynak, O. Industrial Applications of Digital Twins. *Phil. Trans. R. Soc. A* **2021**, *379*, 20200360. [CrossRef]
32. Lazaroiu, G.; Negurita, O.; Grecu, I.; Grecu, G.; Mitran, P. Consumers' Decision-Making Process on Social Commerce Platforms: Online Trust, Perceived Risk, and Purchase Intentions. *Front. Psychol.* **2020**, *11*, 890. [CrossRef]
33. Cachada, A.; Moreira, P.M.; Romero, L.; Barbosa, J.; Leitno, P.; Gcraldcs, C.A.S.; Deusdado, L.; Costa, J.; Teixeira, C.; Teixeira, J.; et al. Maintenance 4.0: Intelligent and Predictive Maintenance System Architecture. In Proceedings of the 2018 IEEE 23rd International Conference on Emerging Technologies and Factory Automation (ETFA), Turin, Italy, 4–7 September 2018; IEEE: Piscataway, NJ, USA, 2018; pp. 139–146. [CrossRef]
34. Motaghare, O.; Pillai, A.S.; Ramachandran, K.I. Predictive Maintenance Architecture. In Proceedings of the 2018 IEEE International Conference on Computational Intelligence and Computing Research (ICCIC), Madurai, India, 13–15 December 2018; IEEE: Piscataway, NJ, USA, 2018. [CrossRef]

35. Franco, I.T.; de Figueiredo, R.M. Predictive Maintenance: An Embedded System Approach. *J. Control Autom. Electr. Syst.* **2023**, *34*, 60–72. [CrossRef]
36. Scaife, A.D. Improve Predictive Maintenance Through the Application of Artificial Intelligence: A Systematic Review. *Results Eng.* **2024**, *21*, 101645. [CrossRef]
37. Ren, Y. Optimizing Predictive Maintenance with Machine Learning for Reliability Improvement. *ASME J. Risk Uncertain. Part B* **2021**, *7*, 030801. [CrossRef]
38. Dabija, D.C.; Bejan, B.M.; Pușcaș, C. A qualitative approach to the sustainable orientation of generation z in retail: The case of Romania. *J. Risk Financ. Manag.* **2020**, *13*, 152. [CrossRef]
39. Deloitte. Predictive Maintenance: Transforming Maintenance Operations with Data and Analytics. Deloitte Insights. Available online: https://www2.deloitte.com/content/dam/Deloitte/us/Documents/process-and-operations/us-predictive-maintenance.pdf (accessed on 2 February 2025).
40. Arena, F.; Collotta, M.; Luca, L.; Ruggieri, M.; Termine, F.G. Predictive Maintenance in the Automotive Sector: A Literature Review. *Math. Comput. Appl.* **2022**, *27*, 2. [CrossRef]
41. Chinta, V.S.; Reddi, S.K.; Yarramsetty, N. Optimal Feature Selection on Serial Cascaded Deep Learning for Predictive Maintenance System in Automotive Industry with Fused Optimization Algorithm. *Adv. Eng. Inform.* **2023**, *57*, 102105. [CrossRef]
42. Golightly, D.; Kefalidou, G.; Sharples, S. A Cross-Sector Analysis of Human and Organisational Factors in the Deployment of Data-Driven Predictive Maintenance. *Inf. Syst. E-Bus. Manag.* **2018**, *16*, 627–648. [CrossRef]
43. Zonta, T.; da Costa, C.A.; Righi, R.d.R.; de Lima, M.J.; da Trindade, E.S.; Li, G.P. Predictive Maintenance in the Industry 4.0: A Systematic Literature Review. *Comput. Ind. Eng.* **2020**, *150*, 106889. [CrossRef]
44. Nunes, P.; Santos, J.; Rocha, E. Challenges in Predictive Maintenance—A Review. *CIRP J. Manuf. Sci. Technol.* **2023**, *40*, 53–67. [CrossRef]
45. Hurtado, J.; Salvati, D.; Semola, R.; Bosio, M.; Lomonaco, V. Continual Learning for Predictive Maintenance: Overview and Challenges. *Intell. Syst. Appl.* **2023**, *19*, 200251. [CrossRef]
46. Rüb, M.; Tuchel, P.; Sikora, A.; Mueller-Gritschneder, D. A Continual and Incremental Learning Approach for TinyML On-Device Training Using Dataset Distillation and Model Size Adaption. In Proceedings of the 2024 IEEE 7th International Conference on Industrial Cyber-Physical Systems (ICPS), St. Louis, MO, USA, 12–15 May 2024; pp. 1–8. [CrossRef]
47. Juracka, D.; Nagy, M.; Valaskova, K.; Nica, E. A Meta-Analysis of Innovation Management in Scientific Research: Unveiling the Frontier. *Systems* **2024**, *12*, 130. [CrossRef]
48. Giada, C.V.; Rossella, P. Barriers to Predictive Maintenance Implementation in the Italian Machinery Industry. *IFAC-PapersOnLine* **2021**, *54*, 1266–1271. [CrossRef]
49. Grand View Research. Predictive Maintenance Market Size, Share & Trends Analysis Report by Component (Software, Services), by Deployment (On-Premises, Cloud), by End-Use (Manufacturing, Automotive), by Region, and Segment Forecasts, 2023–2030. Grand View Research. 2023. Available online: https://www.grandviewresearch.com/industry-analysis/predictive-maintenance-market (accessed on 2 February 2025).
50. Elkateb, S.; Metwalli, A.; Shendy, A.; Abu-Elanien, A. Machine learning and IoT—Based predictive maintenance approach for industrial applications. *Alex. Eng. J.* **2024**, *88*, 298–308. [CrossRef]
51. Doran, N.M.; Badareu, G.; Puiu, S. Automation Systems Implications on Economic Performance of Industrial Sectors in Selected European Union Countries. *Systems* **2025**, *13*, 26. [CrossRef]
52. Piotrowska, K.; Wanicki, P. The Czech Republic as a country with the highest development rate in research and development among the v4 countries. *Zesz. Nauk.* **2023**, *9*, 125–135. [CrossRef]
53. Stefanini, R.; Tancredi, G.P.C.; Vignali, G.; Monica, L. Industry 4.0 and intelligent predictive maintenance: A survey about the advantages and constraints in the Italian context. *J. Qual. Maint. Eng.* **2022**, *29*, 37–49. [CrossRef]
54. Skoumpopoulou, D.; Toliyat, S.M.H.; Ojra, A.; Shokri, A.; Hu, S. Challenges of achieving digital transformation in manufacturing firms: The case of predictive maintenance and spare part inventory management. *J. Manuf. Technol. Manag.* **2025**, *36*, 159–178. [CrossRef]
55. Hojnik, B.B.; Huđek, I. Small and Medium-Sized Enterprises in the Digital Age: Understanding Characteristics and Essential Demands. *Information* **2023**, *14*, 606. [CrossRef]
56. Elahu, M.; Afolaranmi, S.O.; Lastra, J.L.M.; Garcia, J.A.P. A comprehensive literature review of the applications of AI techniques through the lifecycle of industrial equipment. *Discov. Artif. Intell.* **2023**, *3*, 43. [CrossRef]
57. Nyoni, R. Harnessing Data Analytics for Predictive Insights: Advancing Decision-Making with Big Data Innovations. *Int. J. Res. Publ. Rev.* **2025**, *6*, 2915–2936. [CrossRef]
58. Wan, Y.; Liu, Y.; Chen, Z.; Chen, C.; Li, X.; Hu, F.; Packianather, M. Making knowledge graphs work for smart manufacturing: Research topics, applications and prospects. *J. Manuf. Syst.* **2024**, *76*, 103–132. [CrossRef]
59. Ledro, C.; Nosella, A.; Pozza, I.D. Integration of AI in CRM: Challenges and guidelines. *J. Open Innov. Technol. Mark. Complex.* **2023**, *9*, 100151. [CrossRef]

60. Florian, E.; Sgarbossa, F.; Zennaro, I. Machine learning-based predictive maintenance: A cost-oriented model for implementation. *Int. J. Prod. Econ.* **2021**, *236*, 108114. [CrossRef]
61. Hoffmann, M.A.; Lasch, R. Unlocking the Potential of Predictive Maintenance for Intelligent Manufacturing: A Case Study On Potentials, Barriers, and Critical Success Factors. *Schmalenbach J. Bus. Res.* 2025; *early access*. [CrossRef]
62. Rahman, S.; Ghosh, T.; Aurna, N.F.; Kaiser, M.S.; Anannya, M.; Hosen, A.S. Machine learning and internet of things in industry 4.0: A review. *Meas. Sens.* **2023**, *28*, 100822. [CrossRef]
63. Lazaroiu, G.; Gedeon, T.; Rogalska, E.; Valaskova, K.; Nagy, M.; Musa, H.; Zvarikova, K.; Poliak, M.; Horak, J.; Crețoiu, R.I.; et al. Digital twin-based cyber-physical manufacturing systems, extended reality metaverse enterprise and production management algorithms, and Internet of Things financial and labor market technologies in generative artificial intelligence economics. *Oeconomia Copernic.* **2024**, *15*, 837–870. [CrossRef]
64. Alahi, M.E.E.; Sukkuea, A.; Tina, F.W.; Nag, A.; Kurdthongmee, W.; Suwannarat, K.; Mukhopadhyay, S.C. Integration of IoT-Enabled Technologies and Artificial Intelligence (AI) for Smart City Scenario: Recent Advancements and Future Trends. *Sensors* **2023**, *23*, 5206. [CrossRef]
65. Faizanbasha, A.; Rizwan, U. Optimizing burn-in and predictive maintenance for enhanced reliability in manufacturing systems: A two-unit series system approach. *J. Manuf. Syst.* **2025**, *78*, 244–270. [CrossRef]
66. Rojek, I.; Jasiulewicz-Kaczmarek, M.; Piechowski, M.; Mikołajewski, D. An Artificial Intelligence Approach for Improving Maintenance to Supervise Machine Failures and Support Their Repair. *Appl. Sci.* **2023**, *13*, 4971. [CrossRef]
67. Lei, X.; Sandborn, P.; Bakhshi, R.; Kashani-our, A.; Goudarzi, N. PHM based predictive maintenance optimization for offshore wind farms. In Proceedings of the 2015 IEEE Conference on Prognostics and Health Management (PHM), Austin, TX, USA, 22–25 June 2015; IEEE Xplore: New York, NY, USA, 2015; pp. 22–25. [CrossRef]
68. Zvolánková, S.S.; Krajčík, V. Digital maturity of Czech SMEs concerning the demographic characteristics of entrepreneurs and enterprises. *Equilib. Q. J. Econ. Econ. Policy* **2024**, *19*, 1363–1404. [CrossRef]
69. Pelau, C.; Dabija, D.-C.; Ene, I. What makes an AI device human-like? The role of interaction quality, empathy and perceived psychological anthropomorphic characteristics in the acceptance of artificial intelligence in the service industry. *Comput. Hum. Behav.* **2021**, *122*, 106855. [CrossRef]
70. Lăzăroiu, G.; Andronie, M.; Iatagan, M.; Geamănu, M.; Ștefănescu, R.; Dijmărescu, I. Deep Learning-Assisted Smart Process Planning, Robotic Wireless Sensor Networks, and Geospatial Big Data Management Algorithms in the Internet of Manufacturing Things. *ISPRS Int. J. Geo-Inf.* **2022**, *11*, 277. [CrossRef]
71. Gariba, M.I.; Rehman, F.U.; Prokop, V.; Giglio, C. Be digital to be sustainable! The mediating role of sustainable supply chain practices in triggering the effects of digitalisation on Sustainable Development Goals in the European Union. *Oeconomia Copernic.* **2024**, *15*, 1383–1425. [CrossRef]
72. Vătămănescu, E.-M.; Bratianu, C.; Dabija, D.-C.; Popa, S. Capitalizing online knowledge networks: From individual knowledge acquisition towards organizational achievements. *J. Knowl. Manag.* **2023**, *27*, 1366–1389. [CrossRef]
73. Nagy, M.; Lăzăroiu, G. Computer Vision Algorithms, Remote Sensing Data Fusion Techniques, and Mapping and Navigation Tools in the Industry 4.0-Based Slovak Automotive Sector. *Mathematics* **2022**, *10*, 3543. [CrossRef]
74. Chatzoudes, D.; Kadłubek, M.; Maditinos, D. Green logistics practices: The antecedents and effects for supply chain management in the modern era. *Equilib. Q. J. Econ. Econ. Policy* **2024**, *19*, 991–1034. [CrossRef]
75. Kliestik, T.; Dragomir, R.; Băluță, A.V.; Grecu, I.; Durana, P.; Karabolevski, O.L.; Kral, P.; Balica, R.; Suler, P.; Bușu, O.V.; et al. Enterprise generative artificial intelligence technologies, Internet of Things and blockchain-based fintech management, and digital twin industrial metaverse in the cognitive algorithmic economy. *Oeconomia Copernic.* **2024**, *15*, 1183–1221. [CrossRef]
76. Andronie, M.; Lăzăroiu, G.; Iatagan, M.; Uță, C.; Ștefănescu, R.; Cocoșatu, M. Artificial Intelligence-Based Decision-Making Algorithms, Internet of Things Sensing Networks, and Deep Learning-Assisted Smart Process Management in Cyber-Physical Production Systems. *Electronics* **2021**, *10*, 2497. [CrossRef]
77. Lizińska, W.; Sobol, D.; Rūtelionė, A. Internationalisation of science and technology parks and the stage of their life: The Central European perspective. *Equilib. Q. J. Econ. Econ. Policy* **2024**, *19*, 549–590. [CrossRef]
78. Pocol, C.B.; Stanca, L.; Dabija, D.-C.; Pop, I.D.; Mișcoiu, S. Knowledge Co-creation and Sustainable Education in the Labor Market-Driven University–Business Environment. *Front. Environ. Sci.* **2022**, *10*, 781075. [CrossRef]
79. Andronie, M.; Lăzăroiu, G.; Iatagan, M.; Hurloiu, I.; Ștefănescu, R.; Dijmărescu, A.; Dijmărescu, I. Big Data Management Algorithms, Deep Learning-Based Object Detection Technologies, and Geospatial Simulation and Sensor Fusion Tools in the Internet of Robotic Things. *ISPRS Int. J. Geo-Inf.* **2023**, *12*, 35. [CrossRef]
80. Kliestik, T.; Kral, P.; Bugaj, M.; Durana, P. Generative artificial intelligence of things systems, multisensory immersive extended reality technologies, and algorithmic big data simulation and modelling tools in digital twin industrial metaverse. *Equilib. Q. J. Econ. Econ. Policy* **2024**, *19*, 429–461. [CrossRef]

81. Gavkalova, N.; Gładysz, B.; Quadrini, W.; Sassanelli, C.; Asplund, F.; Ramli, M.R.; Detzner, P.; Deville, J.; Dragic, M.; Van Erp, T.; et al. Digital Innovation Hubs and portfolio of their services across European economies. *Oeconomia Copernic.* **2024**, *15*, 59–94. [CrossRef]
82. Nagy, M.; Lăzăroiu, G.; Valaskova, K. Machine Intelligence and Autonomous Robotic Technologies in the Corporate Context of SMEs: Deep Learning and Virtual Simulation Algorithms, Cyber-Physical Production Networks, and Industry 4.0-Based Manufacturing Systems. *Appl. Sci.* **2023**, *13*, 1681. [CrossRef]
83. Andronie, M.; Lăzăroiu, G.; Karabolevski, O.L.; Ştefănescu, R.; Hurloiu, I.; Dijmărescu, A.; Dijmărescu, I. Remote Big Data Management Tools, Sensing and Computing Technologies, and Visual Perception and Environment Mapping Algorithms in the Internet of Robotic Things. *Electronics* **2023**, *12*, 22. [CrossRef]
84. Valaskova, K.; Nagy, M.; Grecu, G. Digital twin simulation modeling, artificial intelligence-based Internet of Manufacturing Things systems, and virtual machine and cognitive computing algorithms in the Industry 4.0-based Slovak labor market. *Oeconomia Copernic.* **2024**, *15*, 95–143. [CrossRef]
85. Dadkhah, M.; Nedungadi, P.; Raman, R.; Dávid, L.D. Emerging and disruptive technologies and the sustainable development goals: A state of art and research agenda. *Equilib. Q. J. Econ. Econ. Policy* **2024**, *19*, 13–21. [CrossRef]

mathematics

MDPI

Article

Towards Efficient and Trustworthy Pandemic Diagnosis in Smart Cities: A Blockchain-Based Federated Learning Approach

Mohamed Abdel-Basset [1], Ibrahim Alrashdi [2], Hossam Hawash [1], Karam Sallam [3,*] and Ibrahim A. Hameed [4,*]

[1] Faculty of Computers and Informatics, Zagazig University, Zagazig, Sharqiyah 44519, Egypt; mohamedbasset@zu.edu.eg (M.A.-B.); hossamreda@zu.edu.eg (H.H.)
[2] Department of Computer Science, College of Computer and Information Sciences, Jouf University, Sakaka 2014, Saudi Arabia; irrashdi@ju.edu.sa
[3] School of IT and Systems, Faculty of Science and Technology, University of Canberra, Canberra 2601, Australia
[4] Department of ICT and Natural Sciences, Norwegian University of Science and Technology (NTNU), 7491 Ålesund, Norway
* Correspondence: karam.sallam@canberra.edu.au (K.S.); ibib@ntnu.no (I.A.H.)

Abstract: In the aftermath of the COVID-19 pandemic, the need for efficient and reliable disease diagnosis in smart cities has become increasingly serious. In this study, we introduce a novel blockchain-based federated learning framework tailored specifically for the diagnosis of pandemic diseases in smart cities, called BFLPD, with a focus on COVID-19 as a case study. The proposed BFLPD takes advantage of the decentralized nature of blockchain technology to design collaborative intelligence for automated diagnosis without violating trustworthiness metrics, such as privacy, security, and data sharing, which are encountered in healthcare systems of smart cities. Cheon–Kim–Kim–Song (CKKS) encryption is intelligently redesigned in BFLPD to ensure the secure sharing of learning updates during the training process. The proposed BFLPD presents a decentralized secure aggregation method that safeguards the integrity of the global model against adversarial attacks, thereby improving the overall efficiency and trustworthiness of our system. Extensive experiments and evaluations using a case study of COVID-19 ultrasound data demonstrate that BFLPD can reliably improve diagnostic accuracy while preserving data privacy, making it a promising tool with which smart cities can enhance their pandemic disease diagnosis capabilities.

Keywords: blockchain; federated learning; pandemic disease; smart cities; COVID-19; privacy-preserving; decentralized computing; Internet of Things (IoT); timely response to outbreaks

MSC: 68-04; 68Uxx; 68Vxx; 68Wxx

1. Introduction

The swift pace of urbanization and the rise of smart cities have transformed urban landscapes internationally. With the advent of advanced technologies and the Internet of Things (IoT), cities have become more interconnected, integrating various digital infrastructures to improve efficiency, sustainability, and quality of life [1]. This, in turn, has led to the propagation of data sources generating vast quantities of valued information [2]. Furthermore, the interconnected nature of smart cities enables seamless sharing and collaboration among different entities, fostering an environment conducive to the development of innovative solutions to societal challenges [3]. In this context, the challenge of pandemic disease diagnosis in smart cities encompasses several key aspects that need to be addressed in order to effectively alleviate the outbreak of infectious diseases [2–4]. These challenges stem from the unique characteristics of smart cities, such as their high population density, complex mobility patterns, and diverse sources of data. Timely identification and containment of outbreaks are crucial for preventing the rapid spread of infectious diseases and minimizing their impact on public health. Precise diagnosis not only supports the operative treatment and management of diseased individuals, but also enables proactive measures to be taken, such as resource allocation, targeted interventions, and public awareness campaigns [5]. Smart cities possess the capacity to transform pandemic disease diagnosis through the utilization of IoT and AI technologies, facilitating timely and evidence-based decision-making that can have life-saving implications and safeguard communities [3–5].

Unfortunately, the existing centralized approaches to pandemic disease diagnosis in smart cities face several challenges and limitations. First, they rely on centralized data repositories where sensitive health information is collected and saved in a single location, which poses noteworthy concerns concerning data privacy and security. Additionally, the sheer volume and velocity of data generated in smart cities can overwhelm centralized systems, causing delays and inefficiencies in diagnosis and replies to efforts [4,5]. Moreover, centralized approaches may encounter regulatory and governance matters, such as data ownership and access rights, impeding operative collaboration and information exchange among diverse stakeholders. Accordingly, smart cities are facing a pressing need for innovative solutions that leverage decentralized and privacy-preserving approaches to enhance the accuracy, efficiency, and security of pandemic disease diagnosis in smart cities [6,7].

Federated learning (FL) is a distributed AI mechanism that enables collaborative model training across multiple devices or entities without the need for centralized data aggregation [8,9]. In FL, instead of sending raw data to a central server, local models are trained on individual devices or entities, and only the model updates or aggregated parameters are shared. This decentralized approach ensures data privacy and security, as sensitive data remain on the local devices. Furthermore, FL leverages the collective intelligence of diverse data sources, allowing models to be trained on a more representative and inclusive dataset [9–11]. By involving various entities, such as healthcare providers, research institutions, and government agencies, in the collaborative training process, FL encourages knowledge sharing, cross-institutional cooperation, and the integration of diverse perspectives. This collaborative nature enhances the robustness and generalizability of the trained models, enabling more accurate and comprehensive pandemic disease diagnosis in smart cities. Moreover, FL empowers resource-constrained IoT devices to participate in the model training process, democratizing access to advanced AI techniques and fostering inclusivity in the healthcare and research domains [12].

The integration of federated learning and blockchain technology holds tremendous significance in the context of smart cities. The inherent properties of blockchains, such as decentralization, immutability, and transparency, offer a compelling solution for enhancing the security and trustworthiness of data sharing and storage in smart city environments [13]. By leveraging blockchains, smart cities can establish robust and tamper-resistant infrastructure that ensures the integrity and provenance of shared data, thereby addressing concerns related to data privacy, trust, and accountability. Furthermore, the collaborative nature of federated learning aligns perfectly with the distributed and interconnected nature of smart cities [14]. The significance of blockchains and federated learning in smart cities goes beyond data privacy and accuracy to promote inclusivity and equity, ensuring that all stakeholders have equal opportunities to participate in the diagnosis process [15–18]. This inclusiveness empowers communities, healthcare providers, and researchers, enabling them to collectively contribute to the identification, tracking, and containment of pandemic diseases.

The objectives of this study are twofold: first, to propose a new blockchain-based federated learning model that is personalized for the diagnosis of pandemic diseases in smart cities, hence named BFLPD; and second, to introduce an all-inclusive case study focusing on COVID-19 as an exemplar of our framework's applicability. Our proposed BFLPD aims to address the challenges of trustworthiness and scalability in the context of pandemic disease diagnosis while leveraging cooperative intelligence among distributed entities in smart cities. In this view, the contributions of this study can be summarized as follows:

- We introduce a pioneering framework that combines the decentralized and immutable nature of blockchain technology with the collaborative and privacy-preserving capabilities of FL, providing a secure and transparent environment for pandemic disease diagnosis in smart cities.
- A decentralized and secure aggregation method is presented to empower the global model to be unaffected by malicious updates from unknown clients in smart cities, which is aimed to improve the overall efficiency and trustworthiness of our system.
- We demonstrate the practicality and effectiveness of our framework through a rigorous case study on COVID-19, showcasing its potential to improve diagnostic accuracy and efficiency compared to traditional centralized approaches.
- We address critical considerations such as data privacy, security, and regulatory compliance within the smart city context, proposing mechanisms such as consensus algorithms, data aggregation techniques, and privacy-preserving mechanisms to overcome associated challenges.

2. Related Works

In this section, we review the existing literature and research labors that have investigated several aspects of blockchain, FL, and their applications in the context of smart city healthcare and pandemic management.

2.1. Pandemic Management in Smart Cities

In this section, we review and analyze a selection of literature studies that have examined the function of smart cities in managing and dealing with pandemics. These studies shed light on the innovative approaches, technological interventions, and governance strategies employed by smart cities worldwide to effectively tackle the challenges posed by public health emergencies. For instance, Hassankhani et al. [1] studied the crucial role of smart city technologies and their potential to effectively respond to and manage crisis situations. They explored the key lessons and insights gained from the COVID-19 pandemic, emphasizing the value of data-driven decision-making, real-time monitoring, and the integration of various technological solutions within a smart city framework. They offered valuable recommendations and guidelines for policymakers, urban planners, and stakeholders to use to enhance crisis management strategies and to build resilient smart cities capable of effectively addressing future pandemics or similar crises. In addition, Strielkowski et al. [2] contributed to the understanding of how to manage smart and sustainable cities in the aftermath of the COVID-19 pandemic. They examined the lessons learned from the pandemic and explored its implications for the management of cities striving to be smart and sustainable. They shed light on the importance of integrating technological advancements, such as IoT, big data analytics, and artificial intelligence, with sustainable urban planning and governance practices. The study emphasized the need for resilient infrastructure, citizen engagement, and effective decision-making processes to foster sustainable development in cities.

Das and Zhang [3] investigated the functions of technology and society in managing the COVID-19 pandemic in Singapore, a renowned smart city. They focused on Singapore's innovative approach to pandemic management, leveraging technology-driven solutions and societal cooperation. The authors also detailed the key technological interventions implemented by the government, including contact tracing apps, data analytics, and digital tools for remote work and learning. The study emphasized the critical role of societal factors, such as public trust, citizen compliance, and community engagement, in the successful deployment of these technological solutions. Kim et al. [4] explored the necessary adaptations to the structures of smart cities to effectively predict and overcome pandemics. They examined the roles of smart city technologies and infrastructure in enhancing preparedness and response to pandemic outbreaks. Through an in-depth analysis, they identified key areas where structural adjustments are needed, such as data collection and analysis, communication networks, healthcare systems, and urban planning. Furthermore, Costa et al. [5] provided a comprehensive review of smart city initiatives aimed at addressing the challenges posed by the COVID-19 pandemic. The research examined different technological solutions and strategies implemented by smart cities worldwide to combat the spread of the virus and mitigate its impacts. They studied key initiatives such as contact tracing applications, data analytics, remote healthcare services, and intelligent urban management systems, and highlighted the importance of leveraging smart city infrastructure and capabilities to enhance surveillance, early detection, and responses to new outbreaks.

2.2. Federated Learning in Smart Cities

In this subsection, we delve into the existing literature studies that focus on the application of FL in the context of smart cities, which is a collaborative and privacy-preserving ML paradigm, making it particularly relevant for the data-rich ecosystems found within smart cities. Zheng et al. [9] provided a comprehensive overview of the taxonomy of and the recent advances and open challenges in the application of federated learning in smart cities. They explored the potential of FL, as a distributed ML approach, to address privacy concerns and data limitations while enabling collaborative and efficient data analysis in smart city environments. They also classified and categorized the various applications of FL in smart cities, including traffic management, energy optimization, healthcare, and public safety. They emphasized the advantages and challenges associated with FL in smart city contexts and recognized significant research directions for future improvements. In [10], the authors surveyed the FL approaches that could leverage data from distributed sensors in a smart city environment while addressing data privacy and security concerns. They identified and discussed the key challenges associated with FL in smart city sensing, such as communication and coordination among sensors, data heterogeneity, and ensuring privacy-preserving mechanisms. The authors of [11] explored the concept of integrating FL techniques with digital twin technology, which designs virtual representations of physical resources and systems in smart cities. They discussed the ideas and theories related to FL-enabled digital twins, highlighting their potential applications in various domains of smart cities, including urban planning, infrastructure management, and resource optimization. They also discussed the advances in the field, such as FL algorithms, communication protocols, and privacy-preserving mechanisms. In [12], Rasha et al. studied the challenges and potential solutions for preserving privacy and ensuring data security when implementing federated learning in the context of smart cities. They investigated different aspects related to privacy and security, including data protection, secure communication protocols, privacy-preserving machine learning algorithms, and secure model aggregation. They debated the existing privacy and security issues in FL for smart cities and presented an overview of the state-of-the-art privacy and security mechanisms and techniques that have been proposed and implemented. Pandya et al. [13] surveyed the applications of FL that could address the unique challenges and requirements of smart city environments, such as data privacy, scalability, and resource-constrained devices. They explored various aspects of FL in smart cities, incorporating the architecture, algorithms, communication protocols, and their practical implementations in different sectors of smart cities, such as energy management, transportation, and healthcare.

2.3. Blockchain in Smart Cities

This section provides an overview of the existing research and literature related to the application of blockchain technology in the context of smart cities. This section aims to highlight the advancements, challenges, and potential opportunities offered by blockchains in various domains of smart city development. In [14], the authors offered a comprehensive review of the applications of blockchain technology in smart cities through the analysis of blockchain architectures, integration trends, and potential future research directions in the context of smart city development. They explored various aspects related to blockchain technology, such as consensus mechanisms, scalability, interoperability, and security. They studied the existing architectures and integration approaches of blockchains in smart cities and discussed the benefits and challenges associated with their implementation. In [15], the authors surveyed the application of blockchain technology in smart cities by analyzing the research issues and challenges associated with the implementation of blockchains in the context of smart city development. The authors explored various aspects related to blockchain technology, including its architecture, consensus mechanisms, privacy, security, scalability, and interoperability. Additionally, they discussed the challenges and open research issues that need to be addressed for the successful implementation of blockchain technology in smart cities. In [16], the authors focused on the application of blockchain technology in IoT-based smart cities though investigating the recent advances in blockchain technology and exploring its requirements and challenges in the context of smart cities. They examined the integration of Internet of Things (IoT) systems and blockchains and discussed the benefits and potential use cases of this combination in smart city environments. They also addressed the key requirements for implementing blockchains in IoT-based smart cities, such as scalability, interoperability, privacy, and security. Hakak et al. [17] studied the application of blockchain technology for enhancing the security of smart cities by examining the architecture, requirements, and challenges associated with securing smart cities using blockchain technology. They explored the potential of blockchain to address security vulnerabilities and risks in various domains of smart cities, including transportation, energy, healthcare, and public safety. They discussed the essential architectural components and requirements for implementing blockchains into smart cities, such as decentralized consensus, data integrity, privacy, and identity management. Furthermore, Kim et al. [18] presented a system of blockchain-based personal health records, which not only used an on-chain, off-chain system to regulate the consensus data of the patients, but also investigated the definite user experience based on a conducted questionnaire. Furthermore, Fiore et al. [19] reviewed the application of blockchains to tackle the supply chain challenges in healthcare, precisely for drugs, medical devices, lifeblood, and tissues. They conducted the review based on the PRISMA guidelines and an investigation of English studies in the PubMed and Proquest databases.

The prevailing literature on blockchain application in smart cities has made notable contributions in different parts. Nevertheless, a critical analysis reveals certain gaps that motivate the development of our proposed solution. Firstly, previous studies have primarily focused on the application of blockchains in smart cities without fully addressing the specific challenges and requirements of FL for pandemic diagnosis. Secondly, while some works have explored the integration of blockchains and FL, they regularly oversee the crucial facet of secure aggregation to alleviate the risks of malicious updates and preserve the integrity of the global model. These gaps in the present literature underline the requirement for our proposed framework, which not only fills these research voids, but also offers a healthy and protected solution for the cooperative analysis of distributed healthcare data in smart cities.

3. Methodology

In this section, we present a detailed discussion of the methodology of designing our proposed BFLPD framework for pandemic disease diagnosis in smart cities. Through the delineation of BFLPD methodology, we offer transparency and reproducibility, enabling researchers and practitioners to comprehend and replicate our framework in their own smart city healthcare systems.

3.1. System Model

Herein, we present the system model intended for the proposed BFLPD, providing inclusive details on each system component and its role in the diagnosis of pandemic diseases in smart cities. Figure 1 visualizes the architecture of the proposed system model, which is composed of several fundamental components that work together to enable the effective and secure federated diagnosis of pandemic diseases in a smart city environment. These components include the following.

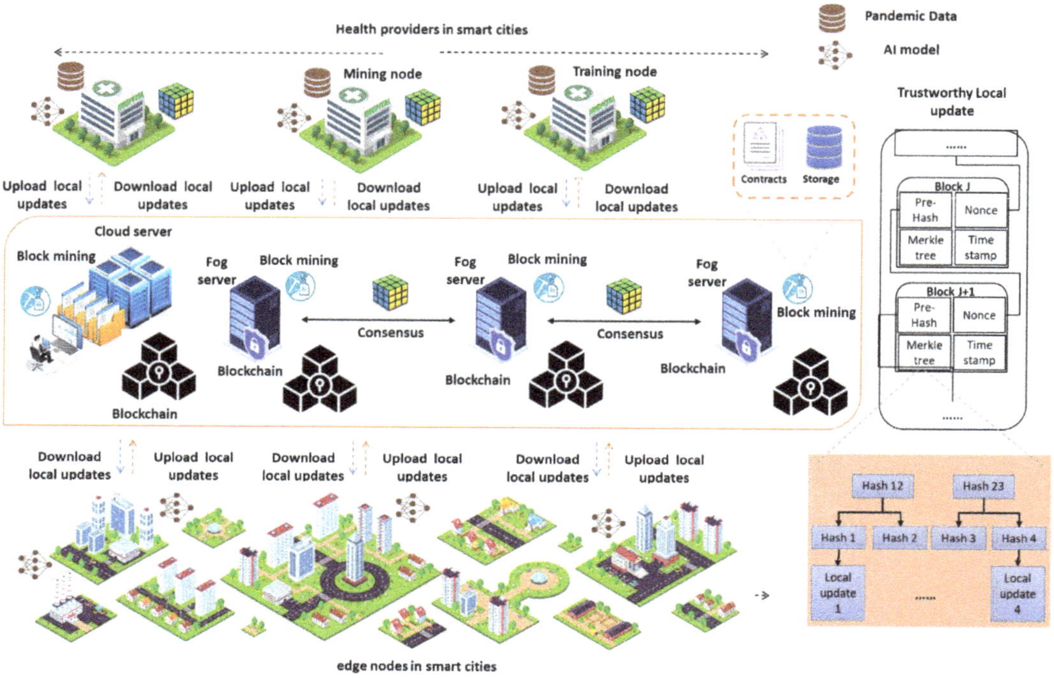

Figure 1. Visualization of our BFLPD system design for pandemic disease diagnosis in smart cities.

Data Sources: The BFLPD acknowledges the diverse nature of pandemic data and the requirement to gather information from diverse devices and sources. Pandemic data can take different formats, incorporating medical records, diagnostic test results, symptom reports, medical images, contact tracing data, health records, and environmental data. These data can originate from a variety of devices and systems, such as healthcare databases, wearable devices, IoT sensors, mobile applications, and social media platforms. The BFLPD highlights the integration and harmonization of different data types, guaranteeing interoperability among diverse systems and devices.

Edge Devices: These devices are distributed throughout the city and are equipped with sensors and data collection capabilities. They are in charge of collecting real-time data related to the health statuses of individuals, such as vital signs, symptoms, and potential exposure to the disease. They act as the frontline data sources, capturing valuable information directly from the individuals within the smart city environment. The distributed nature of the edge devices guarantees the obtainability of varied and up-to-date data, enhancing the accuracy and reliability of the pandemic disease's diagnosis and control. Furthermore, the inclusion of edge devices in BFLPD helps to minimize latency in data transmission and enables real-time monitoring, prompt replies, and control measures in the event of a pandemic outbreak.

Blockchain Network: This is a crucial component of our system model in which we incorporate the use of the hyperledger fabric blockchain network to ensure secure and transparent data management for pandemic diagnosis in smart cities. It is a kind of permissioned blockchain framework that provides a robust infrastructure for enterprise-grade applications. It consists of several key conditions and modules, which include the following. (1) We have the Peer Nodes, which are responsible for maintaining the distributed ledger and executing chain codes. These peer nodes are permitted, meaning that they require authentication and authorization to participate in the network. (2) The ordering service guarantees the consensus and order of transactions within the network. (3) The channel enables the segregation of network participants into distinct subgroups, allowing for privacy and confidentiality between different stakeholders through the promotion of selective sharing of data and transactions, thus guaranteeing that only authorized participants can access and authenticate specific information related to the pandemic diagnosis. The implementation of channels in hyperledger fabric includes the following common steps. First, we set up a rudimentary network by installing orderers, certificate authorities, and peers for each participant. Then, a consortium was created and participants were added to it. Then, the channel was defined and configured by specifying its name, the participation, and other parameters in the configuration file. Next, the genesis block was generated, and each organization joined the channel by signing and sending the channel configuration to the ordering service. Anchor peers are designated for cross-organization communication. Finally, participants can invoke transactions on the channel, which are endorsed, ordered, and committed to update the ledger's state.

Given that we are targeting pandemic disease diagnosis, the choice of the consensus method is crucial to guaranteeing the reliability, scalability, and security of the network. To accommodate these requirements in the BFLPD framework, the practical Byzantine fault tolerance (PBFT) is used for a high level of fault tolerance and Byzantine fault resistance, making it appropriate for serious and sensitive applications such as pandemic disease diagnosis. Unlike Proof of Work (PoW), which necessitates extensive computing power and the ingestion of energy for the purpose of resolving intricate puzzles, PBFT does not depend on resource-intensive mining actions. The PBFT algorithm's consensus mechanism is dependent on voting and a replication scheme, through which nodes can reach agreement on the validity of transactions. Unlike Proof of Stake (PoS), where block creation is limited to a select group of validators based on their stake, PBFT allows all participating nodes to contribute to the consensus procedure. This makes PBFT well-suited for a smart city environment with a large population and numerous nodes. It ensures that the consensus protocol can accommodate the computational demands of federated learning in a pandemic diagnosis scenario.

PBFT is derived according to a network of nodes, a predefined number of nodes of which are labeled as validators accountable for realizing a consensus on the order of transactions. In our case, the validators were designated as trusted nodes from various smart cities or healthcare institutions participating in the framework. The PBFT consensus method confirms that transactions are agreed upon by the utmost of the validators prior to being committed to the blockchain. PBFT provides a high degree of finality and reduces the risk of fraudulent or malicious activities in the network. Hence, it offers a fast consensus process, acknowledging high transaction throughput, which is decisive in scenarios involving large-scale pandemic diagnosis data and regular updates to the model, and has low energy consumption, which is beneficial in resource-constrained environments.

3.2. Methodological Design

The section outlines the step-by-step process followed in our study to implement the BFLPD framework for pandemic diagnosis. This section provides a thorough overview of the key stages involved in our methodology, including system initialization, local model updates, model aggregation, etc.

Initialization: In this phase of the BFLPD framework, the Key Management Center (KMC) was used to configure a pair of public/private keys (pk_x, sk_x) for each client node in the system. Additionally, another pair of public/private keys (pk_v, sk_v) was configured for the validator.

$$\begin{aligned} \text{client} &\overset{send}{\Leftarrow} (pk_x, sk_x) \leftarrow F_{keysgern}(\,) \\ \text{validator} &\overset{send}{\Leftarrow} (pk_v, sk_v) \leftarrow F_{keysgern}(\,) \end{aligned} \tag{1}$$

Once the key pairs had been adopted to perform homomorphic encryption computations, the original global model was uploaded to the blockchain, which was started by setting up an exclusive channel for each federated pandemic diagnosis within the network. This was followed by the addition of organization members, peer nodes, and ordered nodes to their respective channels, as well as the deployment of smart contracts.

Meanwhile, keys and certificates $(Pk_i, Sk_i, Cert_i)$ were issued by the Hyperledger Fabric Certificate Authority (CA) for each approved node joining the channel.

$$\text{node}_\text{i} \overset{send}{\Leftarrow} (Pk_i, Sk_i, Cert_i) \leftarrow F^{CA}_{keys-secgern}(\,) \quad (2)$$

Finally, a genesis block was created for each channel, and the block that contained the original global model was stored on the blockchain. This arrangement enabled the execution of each FL-based diagnosis model through the participating nodes on their respective blockchain channels.

Local update: During this phase of the BFLPD framework, a client c_i who wished to participate in an FL-based diagnosis was required to initiate a cycle of local training. In order to accomplish this, the client downloaded the most recent round's global model $[\![w_g^{r-1}]\!]_{pk_x}$ from the blockchain via the peer nodes. The global model was then decrypted by the client using their private key, sk_x, resulting in the local model w_i^{r-1}, which served as the starting point for the new round of training. The client then conducted the local model training utilizing a differentially private stochastic gradient descent (DP-SGD) procedure that optimized the loss function $\mathcal{L}$, which was defined on a batch size of b using sample X. X was a subset of the client's local dataset $\mathcal{D}_i$.

$$\mathcal{L}(w^{t-1}, X) = \frac{1}{b}\sum\nolimits_{x \in X} \ell\left(w^{t-1}, x\right) \quad (3)$$

where x denotes an instance of local pandemic data, and $\ell\left(w^{t-1}, x\right)$ denotes the relevant loss function. Thus, we obtained the local gradient, as follows:

$$g^t = \nabla\mathcal{L}\left(w^{t-1}, X\right) \quad (4)$$

Then, we injected Laplace noise according to the following:

$$\overset{\sim t}{g} = g^t \leftarrow \text{Lap}\left(\frac{\Delta f}{\varepsilon}\right)^d \quad (5)$$

where Δf is the sensitivity of the model, which represents the maximum change in the gradient due to the inclusion or exclusion of any single data point, and ε is the privacy budget or privacy parameter that determines the level of privacy protection. This was then used to implement local upgrades, as follows:

$$w^t = w^{t-1} - \eta \cdot g^t \quad (6)$$

By following this local update procedure, each participating client downloaded the newest global model, decoded it using their private key, performed local model training using their local dataset, and communicated the updated local model back to the local edge device.

To achieve this, the proposed BFLPD framework utilized the fully homomorphic CKKS (Cheon–Kim–Kim–Song) encryption scheme. This encryption scheme enabled the establishment of both floating-point addition and multiplication functions for complex or real numbers. During the encryption process and at each layer, the local updates were transformed into vectors, which were encrypted layer by layer. If a vector surpassed the predefined maximum length limit, it was divided into many vectors and encrypted separately.

The local model $\boldsymbol{w}_i^r$ was encrypted using the client's public key pk_x and the validator's public key pk_v, resulting in the encrypted versions, $[\![\boldsymbol{w}_i^r]\!]_{pk_x}$ and $[\![\boldsymbol{w}_i^r]\!]_{pk_v}$, respectively.

$$\begin{aligned} Blockchain &\overset{upload}{\Leftarrow} [\![w_i^r]\!]_{pk_x} \leftarrow F_{Enc}^{CKKS}(w_i^r, pk_x)\ \forall\, c_i \in \mathcal{C} \\ Blockchain &\overset{upload}{\Leftarrow} [\![w_i^r]\!]_{pk_v} \leftarrow F_{Enc}^{CKKS}(w_i^r, pk_v)\ \forall\, c_i \in \mathcal{C} \end{aligned} \tag{7}$$

Subsequently, the client c_i uploaded the encrypted editions, $[\![w_i^r]\!]_{pk_x}$ and $[\![w_i^r]\!]_{pk_v}$, to the blockchain like a transaction. At the same time, $[\![w_i^r]\!]_{pk_v}$ was dispatched to the resolver to guarantee that the updated model was securely saved in the blockchain and could be accessed by other participants for FL-based diagnosis.

Secure aggregation: Upon receiving the local updates from all clients, the validator utilized a model-averaging approach to aggregate these local updates:

$$w_g = \frac{1}{N} \cdot \sum_{i=1}^{N} w_i \tag{8}$$

However, it is crucial to address potential malicious activities where certain clients intentionally transmit incorrect local updates to disrupt the aggregation process, leading to an invalid aggregation result and failure of the entire training process. Furthermore, there is concern regarding the privacy of client data, as the "sincere" but "probing" server may attempt to infer sensitive information from the model parameters. Therefore, it becomes essential to mitigate the opposing impacts of poisoning attacks and safeguard the privacy of client data. Consequently, we must consider incorporating mechanisms to ensure secure aggregation, which can be expressed as follows:

$$[\![w_g]\!]_{pk_x} = \frac{1}{N} \cdot \sum_{i=1}^{N} [\![w_i]\!]_{pk_x} \tag{9}$$

To this end, we present an updated version of the Multi-Krum aggregation algorithm to carry out global model aggregation in the BFLPD framework, which allows for robust aggregation of the local updates from the participating clients to form a more accurate and trustworthy global model. Similarly to the Multi-Krum algorithm, our aggregation scheme is based on selecting a subset of clients' models that are deemed reliable and consistent based on their resemblances. It considers the divergence between the models and their gradients to identify the most agreeable models. The algorithm aims to exclude potential malicious or faulty models while aggregating the contributions from trustworthy clients. This, in turn, is advantageous for pandemic diagnosis models that have undergone the local training process, since it guarantees that the final global model represents a consensus among the participating clients, providing a dependable and accurate diagnosis of a pandemic disease. Rather than using Euclidean distance, we used total variation distance to estimate the similarity between local updates of two clients.

$$TV([\![w_i]\!]_{pk_x}, [\![w_j]\!]_{pk_x}) = \frac{1}{2} \int_{R^d} |p_{[\![w_i]\!]_{pk_x}}(u) - q_{[\![w_j]\!]_{pk_x}}(u)|\, du \tag{10}$$

Thus, for each client c_i, the Krum distance was computed as the sum of distances TV_{ij} between c_i and each of its $k-1$ next neighbors S_i,

$$d_{krum}(c_i) = \Sigma_{j \in S_i} TV_{ij} \tag{11}$$

where k is the number of models to be chosen for aggregation. Then, the threshold value T was calculated as the sum of the $k-1$ smallest Krum distances ($d_{krum}(c_i)$) among all of the clients' models. It denotes the maximum acceptable distance for a model to be considered reliable.

$$T = \Sigma_{c_i}\, min(d_{krum}(c_i)) \tag{12}$$

At this point, we selected the reliable local updates for aggregation, as follows:

$$S = \{c_i \mid d_{krum}(c_i) \le T\} \tag{13}$$

$$\begin{aligned} [\![w_g^r]\!]_{pk_v} &= [\![\tfrac{1}{n}\Sigma_{i=1}^{|S|}\, w_i^r]\!]_{pk_v} = \tfrac{1}{|S|}[\![\Sigma_{i=1}^{|S|}\, w_i^r]\!]_{pk_v} = \tfrac{1}{|S|}[\![w_1^r + w_2^r + \cdots + w_{|S|}^r]\!]_{pk_v} \\ &= \tfrac{1}{|S|}\left([\![w_1^r]\!]_{pk_v} + [\![w_2^r]\!]_{pk_v} + \cdots + [\![w_{|S|}^r]\!]_{pk_v}\right) = \tfrac{1}{|S|}\Sigma_{i=1}^{|S|}\, [\![w]\!]_{pk_v} \end{aligned} \tag{14}$$

The final global model $[\![w_g^r]\!]_{pk_v}$ was obtained by aggregating the reliable models contributed by the clients in set S. $|S|$ denotes the number of clients in set S.

4. Case Study and Setup

In this section, we describe the experimental setup used to evaluate the performance and effectiveness of the proposed BFLPD for pandemic disease diagnosis in smart cities. We outline the dataset which was utilized, including its sources and preprocessing steps, to ensure the reliability and representativeness of our experimental results. Moreover, we debate the hardware and software infrastructure which we deployed, providing details on the computing resources and tools leveraged for model training and evaluation.

For our experimental setup, we built a multi-source dataset based in three publicly available ultrasound datasets, namely, the POCUS [6], ICLUS-DB [7], and COVIDx-US datasets [8]. The combined dataset was built to train and evaluate our proposed method for pandemic disease diagnosis in smart cities.

The POCUS dataset encompassed a total of 261 recordings, comprising 59 images and 202 videos. These recordings were captured from 216 individuals using either linear or convex probes. It is worth noting that linear probes offer a higher frequency, resulting in superior resolution for examining irregularities near the pleural line. However, they may present challenges in distinguishing between B-line artifacts and hidden tissue due to their reduced penetration of lung tissue. This aspect emphasizes the importance of accurate diagnosis in distinguishing between different lung conditions. For curating the POCUS dataset, the images and videos were aggregated from diverse sources, including academic scientific publications, hospitals, ultrasound courses, health-tech corporations, and public medical repositories. Further details regarding the specific sources of data can be found in [20]. The COVID-19 cases included in the dataset were confirmed using RT-PCR testing, ensuring the inclusion of reliable and accurate samples. The videos in the dataset exhibited varying lengths (ranging from 160 to 144 frames) and were recorded at frame rates between 25 and 10 Hz, which added diversity to the dataset and simulated real-world scenarios. To ensure the quality and accuracy of the dataset, all samples were carefully reviewed and confirmed by a single medical expert possessing over 10 years of clinical experience and expertise as an academic instructor.

Second, ICLUS-DB [2] was composed of 277 lung ultrasound (LUS) videos from 35 patients, conforming to 58,924 frames. The data were aggregated from clinical centers using a diverse range of ultrasound scanners. Both linear and convex probes were utilized, depending on which was needed. The data were collected from a total of 35 patients, comprising 17 individuals with confirmed cases of COVID-19, 4 individuals suspected of having COVID-19, and 14 healthy individuals. Furthermore, COVIDx-US [3] consisted of 242 lung ultrasound videos comprising 29,651 processed images obtained from patients diagnosed with COVID-19 infection or pneumonia, healthy individuals, and patients with other lung diseases/conditions. It also encompasses a standardized and integrated lung ultrasound score per video file, allowing for better analysis while supporting other research opportunities, such as severity judgment. The dataset was systematically managed and authorized explicitly for the intent of creating and assessing AI models. Table 1 summarizes the class distribution in the combined dataset. In Figure 2, some instances of 2D ultrasound frames of healthy, pneumonia, and COVID-19 patients are displayed in the datasets.

Table 1. The class distribution of our combined dataset.

	Probe		COVID-19	Pneumonia	Healthy	Total
POCUS	CONVEX	Videos	64	52	66	182
		Images	18	20	15	53
	LINEAR	Videos	6	5	9	20
		Images	4	2	/	/
COVIDx-US	Convex	Videos	63	40	19	122
		Images	/	/	/	/
	Linear	Videos	8	9	9	26
		Images	/	/	/	/
ICLUS-DB	Convex/Linear	Videos	42	9	9	60

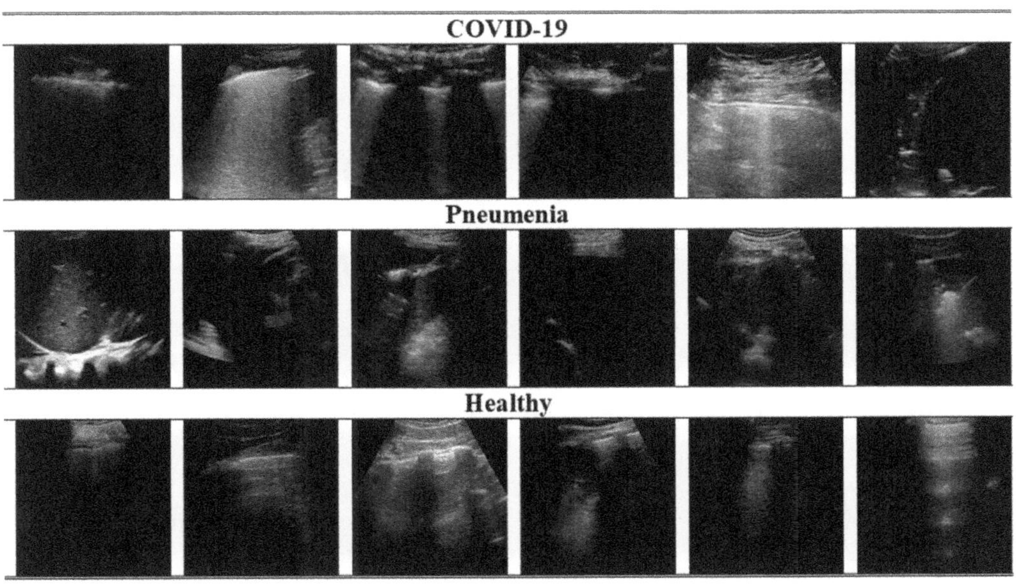

Figure 2. Visualization of randomly selected ultrasound frames from the combined dataset used in this study.

As a crucial step in the process of developing an ML solution, the data underwent pre-processing before being utilized for training purposes. During the experiments, convex ultrasound probes were employed to train the models. Nevertheless, the viral pneumonia class had a limited number of samples available (only three convex videos), which led us to pre-process the data by removing this class from the training set. Consequently, the training process solely utilized the data from the remaining three classes: healthy, pneumonia, and COVID-19 individuals. To facilitate the training process, all convex ultrasound samples were subjected to physical pre-processing. This involved dividing the videos into individual images at a frame rate of 3 Hz, resulting in a maximum of 30 frames per video. Subsequently, the generated image samples underwent cropping to remove artifacts, ratio bars, and text, ensuring that the focus was solely on the relevant content. The images were then resized to dimensions of 224 × 224 pixels, facilitating uniformity and compatibility during the training process. In this work, the reported results were obtained by applying a five-fold cross-validation technique which was bedded according to the number of examples in each class. The image samples were separated at the patient level to ensure that frames from a single video were present in only one of the five folds. Furthermore, the distribution of videos across classes was approximately equal for all folds, ensuring a balanced representation during model training and evaluation. All models were trained to classify images into three categories: healthy, pneumonia, and/or COVID-19. To enhance the robustness of the models and mitigate overfitting, data augmentation techniques were applied. Following the common augmentation practices, image flipping, rotations within the range of $-45°$ to $+45°$, and translations of up to 10% were applied to introduce variations in the data, differentiating the training samples and preventing the models from solely relying on specific patterns or features [19,21]. By undertaking these data preparation steps, we aimed to optimize the quality and suitability of the dataset for training our models, ensuring their robustness, generalization capability, and ability to accurately classify pandemic disease-related images in smart city environments.

In the process of evaluating the performance of our proposed BFLPD for pandemic disease diagnosis in smart cities, we employed several key evaluation metrics to assess its effectiveness and efficiency. These metrics included diagnostic *Accuracy*, *Precision*, *Recall*, and *F1-score*:

$$Accuracy = \frac{TP + TN}{TP + TN + FP + FN} \times 100, \tag{15}$$

$$Recall = \frac{TP}{TP + FN} \times 100, \tag{16}$$

$$Precision = \frac{TN}{TP + FP} \times 100, \tag{17}$$

$$F1\text{-}score(F1) = \frac{2TP}{2TP + FP + FN} \times 100 \tag{18}$$

To conduct simulations and evaluate the performance of our proposed BFLPD for pandemic disease diagnosis in smart cities, we implemented the framework using a combination of software tools, programming languages, and hardware resources. The blockchain infrastructure was developed using Ethereum, a widely adopted platform for creating decentralized applications. Smart contracts were written in Solidity, a programming language specifically designed for Ethereum smart contract development. We leveraged PyTorch to code the FL algorithms and train the models, which were run on Python 3.8.0 environment. The simulations were conducted in a controlled experimental environment (equipped with a powerful NVIDIA RTX GPU, multiple processors, and 128 GB RAM) with standardized settings and parameters. The specific configuration details, including the network architecture, hyperparameters, learning rate, batch size, and optimization algorithms, were carefully selected based on empirical studies and best practices in the field. By stating constant experimental settings across simulations, we guaranteed fair comparisons and consistent evaluations of our proposed framework's performance.

Since we focused on COVID-19 detection as a case study for our work, we chose common computer vision models (Shufflenet v2 [20], Mobilevit [22], Levit [23]) to be deployed as AI models in our BFLPD.

5. Results and Discussion

In this section, we present the results obtained from the evaluation of our proposed BFLPD, with a specific focus on the case study of COVID-19. We analyze and interpret the quantitative and qualitative outcomes of our experiments, shedding light on the performance indicators and the privacy preservation achieved by our framework. Furthermore, we discuss the implications of these results in the context of smart city healthcare and pandemic management, addressing the potential aids, confines, and future directions. Also, we provide visions into the efficacy and practicality of BFLPD, providing valuable perspectives for researchers, policymakers, and stakeholders involved in improving pandemic disease diagnosis in smart cities.

In order to assess the classification performance of the BFLPD framework for pandemic disease diagnosis in smart cities, we compared it against state-of-the-art models [24–29]. The results of this comparative analysis are presented in Table 2, showcasing the performance metrics obtained from the evaluation. Our framework exhibited highly promising results, outperforming existing state-of-the-art models in terms of diagnostic accuracy, precision, recall, and F1 score. The classification accuracy achieved by BFLPD (95.14%) surpassed the benchmark set by the state-of-the-art distributed models, demonstrating its greater competence in accurately classifying diverse classes of infection, including in COVID-19, pneumonia, and healthy individuals. Furthermore, the BFLPD framework demonstrated notable precision and recall values (95.26%, 95.77%) across all classes, reflecting its capacity to reduce false positives and false negatives. The high F1 score (95.52%) further indicated the balanced performance and effectiveness of our BFLPD framework in terms of pandemic disease diagnosis.

Table 2. Comparison of the performance of BFLPD against cutting-edge FL models.

Method	Accuracy (%)	Precision (%)	Recall (%)	F1-Score (%)
[24]	92.72 ± 5.43	93.57 ± 4.54	93.53 ± 5.97	93.55 ± 4.81
[25]	93.16 ± 5.89	95.08 ± 4.42	93.53 ± 2.89	93.8 ± 4.67
[26]	93.53 ± 5.11	94.99 ± 2.98	93.41 ± 1.96	93.7 ± 2.31
[27]	93.93 ± 1.54	92.39 ± 4.91	95.73 ± 4.44	94.03 ± 1.54
[28]	91.51 ± 0.17	92.33 ± 4.54	94.33 ± 0.88	93.32 ± 5.31
[29]	93.19 ± 4.56	92.36 ± 5.35	95.11 ± 1.21	93.72 ± 4.89
BFLPD	95.14 ± 1.76	95.26 ± 5.11	95.77 ± 5.66	95.52 ± 2.22

To determine the statistical significance of the comparative results presented in Table 2, we conducted appropriate statistical tests to validate the observed differences in performance between the BFLPD framework and the state-of-the-art models. The statistical significance test we employed depended on the nature of the data and the evaluation metrics being compared. To demonstrate the classification performance, we utilized the Wilcoxon signed-rank test, depending on the distributional assumptions and the size of the samples. This test examines whether the medians of two paired samples differ significantly. By examining the p-values in Table 3, we can determine the statistical significance of the results. It could be noted that the reported p-value was less than the specified significance level (typically 0.05), which signifies that the perceived differences are statistically significant; thus, we can reject the null hypothesis. This, in turn, enables us to raw confident conclusions about the superiority of our BFLPD framework and its statistically substantial improvements to pandemic disease diagnosis within smart city environments. The superiority of the BFLPD framework can be attributed to the innovative combination of blockchain technology and FL, which supports privacy-preserving and cooperative model training using distributed data sources. By leveraging the strengths of these two technologies, our framework effectively overcomes the limitations of centralized approaches, such as data privacy concerns and scalability issues, while achieving exceptional classification performance. With superior classification performance, along with its ability to preserve privacy and leverage distributed data sources, the BFLPD framework can be regarded as a compelling solution for accurate and efficient disease diagnosis in real-world scenarios.

Table 3. The *p*-values resulting from the Wilcoxon signed-rank test.

	Accuracy	F1-Score
[24]	8.600×10^{-9}	5.493×10^{-5}
[25]	7.088×10^{-4}	6.167×10^{-3}
[26]	3.169×10^{-3}	9.788×10^{-8}
[27]	1.670×10^{-8}	1.777×10^{-11}
[28]	7.072×10^{-7}	1.728×10^{-8}
[29]	9.242×10^{-5}	5.354×10^{-6}

As shown in Figure 3, we evaluated the performance of the BFLPD framework using different classifiers to assess its effectiveness in pandemic diagnosis, and we presented the confusion matrix of each experiment to provide a visual representation of the classifier. Our results consistently demonstrated high detection performance across all classifiers. The confusion matrix illustrated the ability of the framework to correctly classify instances of both COVID-19 and non-COVID-19 cases. The high detection performance indicated the robustness and accuracy of BFLPD framework in identifying potential cases of the pandemic, which is crucial for effective diagnosis and mitigation strategies. This further highlights the potential BFLPD framework for real-world applications in pandemic management.

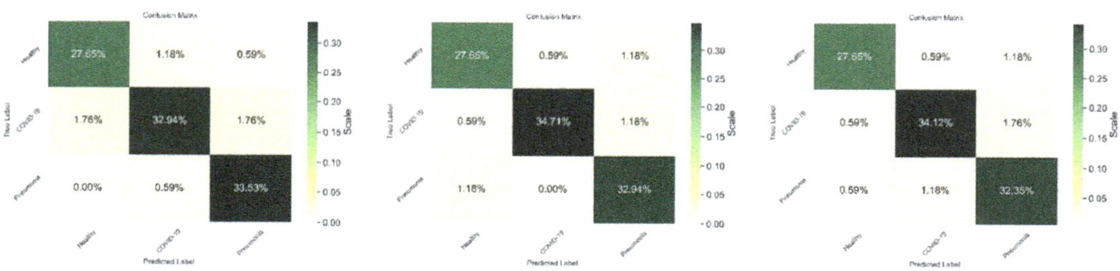

(A) (B) (C)

Figure 3. Comparison between the performance of the BFLPD framework under different classifiers. (**A**) Shufflenet v2 [20], (**B**) Mobilevit [22], (**C**) Levit [23].

In Figure 4, a heatmap visualization of the predictions of BFLPD framework is portrayed on randomly chosen samples from the test set. This heatmap specifies valuable insights into the regions of attention and relevance that contributed to the model's classification decisions. By overlapping the heat map with the ultrasound images, we were able to visually identify the regions that played significant roles in the classification process. Intense and bright colors suggest higher activation and importance, while cooler colors represent regions of lower importance. This visualization technique allowed us to understand which specific features or patterns captured by the ultrasound images influenced the prediction of the BFLPD framework. It can be observed that the BFLPD focused on certain areas that were revealing of COVID-19 or pneumonia conditions. These areas may include specific lung regions, lesions, or characteristic patterns associated with distinctive diseases. This interpretability enhances the trustworthiness of the model's predictions, granting medical specialists and researchers the chance to gain deeper insights into the diagnostic process.

Furthermore, we compared the detection performance of our BFLPD framework under different aggregation methods, e.g., model averaging, Krum, and Multi-Krum (see Figure 5). In the case of the averaging approach, we observed that this method yielded a satisfactory detection performance, but it was susceptible to the presence of malicious clients who intentionally sent incorrect models, compromising the aggregation process and potentially leading to inaccurate results. On the other hand, we can observe that the Krum and Multi-Krum aggregation algorithms were able to mitigate the impact of malicious clients. This algorithm considered the consensus among a subset of clients, discarding outliers and focusing on the models that demonstrated the highest agreement. Interestingly, our modified aggregation algorithm significantly improved the robustness of the aggregation process and enhanced the detection performance, overcoming the influence of malicious clients and becoming a more accurate and reliable global model than Multi-Krum aggregation.

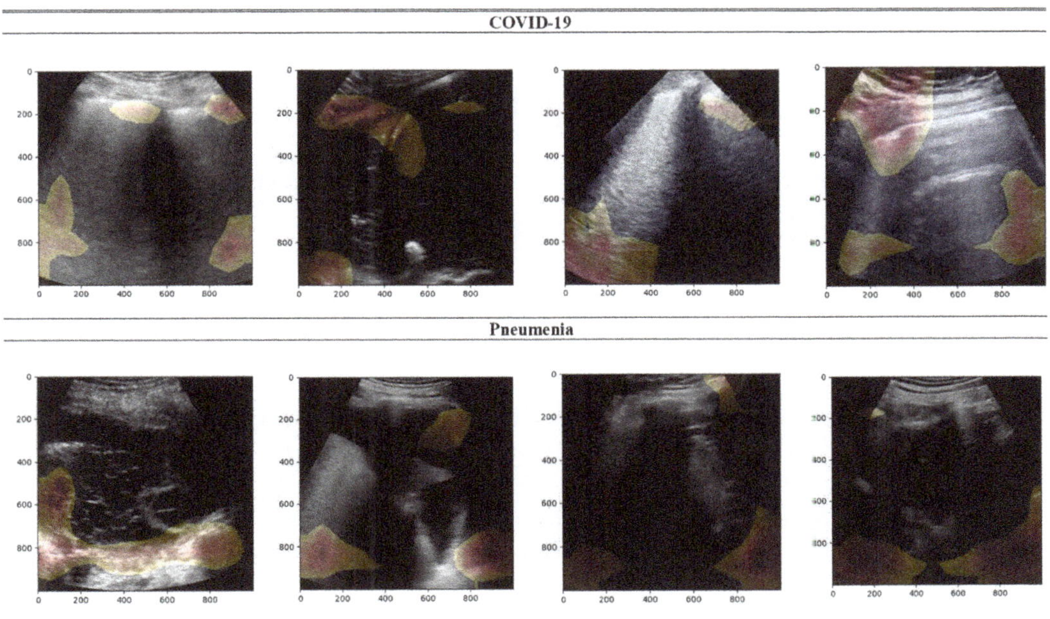

Figure 4. Visualization of gradient heatmaps for explaining the predictions of the BFLPD framework on samples from the test set.

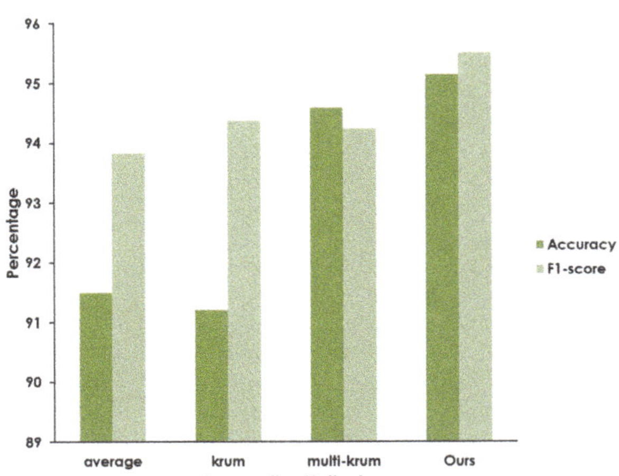

Figure 5. Comparison of the detection performance of the BFLPD framework according to different aggregation methods.

6. Conclusions

This study presents a robust blockchain-based FL framework that integrates blockchain and federated intelligence to improve pandemic disease diagnosis in smart cities while retaining the privacy and security of sensitive user data. Our framework demonstrates high detection performance amid a secure aggregation scheme, ensuring the accurate and reliable identification of pandemic diseases. The experimental results of a case study of COVID-19 validated the effectiveness of our framework in achieving an accurate and reliable pandemic diagnosis. The high detection performance achieved consistently across different classifiers underscores the robustness and generalizability of our approach. The thriving functioning of our framework holds great promise for improving the efficiency and effectiveness of pandemic management systems and participating in improved public health outcomes.

Author Contributions: Conceptualization, M.A.-B., I.A., H.H., K.S. and I.A.H.; Methodology, M.A.-B., H.H., I.A. and K.S.; Software, M.A.-B., H.H. and K.S.; Validation, M.A.-B., I.A., H.H., K.S. and I.A.H.; Formal analysis, M.A.-B., I.A., H.H., K.S. and I.A.H.; Investigation, M.A.-B.; Resources, M.A.-B. and I.A.H.; Writing—original draft, M.A.-B., I.A., H.H. and K.S.; Writing—review & editing, M.A.-B., I.A., H.H., K.S. and I.A.H.; Supervision, I.A. and K.S.; Funding acquisition, I.A.H. All authors have read and agreed to the published version of the manuscript.

Funding: The authors extend their appreciation to the Deputyship for Research & Innovation, Ministry of Education Saudi Arabia, for funding this research work through the project number 223202.

Data Availability Statement: The dataset used in this paper is available upon request from the authors.

Conflicts of Interest: The authors declare that there is no conflict of interest to declare in the research.

References

1. Hassankhani, M.; Alidadi, M.; Sharifi, A.; Azhdari, A. Smart city and crisis management: Lessons for the COVID-19 pandemic. *Int. J. Environ. Res. Public Health* **2021**, *18*, 7736. [CrossRef] [PubMed]
2. Strielkowski, W.; Zenchenko, S.; Tarasova, A.; Radyukova, Y. Management of Smart and Sustainable Cities in the Post-COVID-19 Era: Lessons and Implications. *Sustainability* **2022**, *14*, 7267. [CrossRef]
3. Das; Zhang, J.J. Pandemic in a smart city: Singapore's COVID-19 management through technology & society. *Urban Geogr.* **2020**, *42*, 408–416. [CrossRef]
4. Kim, J.H.; Kim, J.Y. How Should the Structure of Smart Cities Change to Predict and Overcome a Pandemic? *Sustainability* **2022**, *14*, 2981. [CrossRef]
5. Costa, D.G.; Peixoto, J.P.J. COVID-19 pandemic: A review of smart cities initiatives to face new outbreaks. *ET Smart Cities* **2020**, *2*, 64–73. [CrossRef]
6. Born, J.; Wiedemann, N.; Cossio, M.; Buhre, C.; Brändle, G.; Leidermann, K.; Goulet, J.; Aujayeb, A.; Moor, M.; Rieck, B.; et al. Accelerating detection of lung pathologies with explainable ultrasound image analysis. *Appl. Sci.* **2021**, *11*, 672. [CrossRef]
7. Roy, S.; Menapace, W.; Oei, S.; Luijten, B.; Fini, E.; Saltori, C.; Huijben, I.; Chennakeshava, N.; Mento, F.; Sentelli, A.; et al. Deep Learning for Classification and Localization of COVID-19 Markers in Point-of-Care Lung Ultrasound. *IEEE Trans. Med. Imaging* **2020**, *39*, 2676–2687. [CrossRef]
8. Ebadi, A.; Xi, P.; MacLean, A.; Florea, A.; Tremblay, S.; Kohli, S.; Wong, A. COVIDx-US: An Open-Access Benchmark Dataset of Ultrasound Imaging Data for AI-Driven COVID-19 Analytics. *Front. Biosci.* **2022**, *27*, 198. [CrossRef]
9. Zheng, Z.; Zhou, Y.; Sun, Y.; Wang, Z.; Liu, B.; Li, K. Applications of federated learning in smart cities: Recent advances, taxonomy, and open challenges. *Connect. Sci.* **2022**, *34*, 1–28. [CrossRef]
10. Jiang, J.C.; Kantarci, B.; Oktug, S.; Soyata, T. Federated learning in smart city sensing: Challenges and opportunities. *Sensors* **2020**, *20*, 6230. [CrossRef]
11. Ramu, S.P.; Boopalan, P.; Pham, Q.-V.; Maddikunta, P.K.R.; Huynh-The, T.; Alazab, M.; Nguyen, T.T.; Gadekallu, T.R. Federated learning enabled digital twins for smart cities: Concepts, recent advances, and future directions. *Sustain. Cities Soc.* **2022**, *79*, 103663. [CrossRef]
12. Rasha, A.H.; Li, T.; Huang, W.; Gu, J.; Li, C. Federated learning in smart cities: Privacy and security survey. *Inf. Sci.* **2023**, *632*, 833–857.
13. Pandya, S.; Srivastava, G.; Jhaveri, R.; Babu, M.R.; Bhattacharya, S.; Maddikunta, P.K.R.; Mastorakis, S.; Piran, J.; Gadekallu, T.R. Federated learning for smart cities: A comprehensive survey. *Sustain. Energy Technol. Assess.* **2023**, *55*, 102987. [CrossRef]
14. Bhushan, B.; Khamparia, A.; Sagayam, K.M.; Sharma, S.K.; Ahad, M.A.; Debnath, N.C. Blockchain for smart cities: A review of architectures, integration trends and future research directions. *Sustain. Cities Soc.* **2020**, *61*, 102360. [CrossRef]

15. Xie, J.; Tang, H.; Huang, T.; Yu, F.R.; Xie, R.; Liu, J.; Liu, Y. A survey of blockchain technology applied to smart cities: Research issues and challenges. *IEEE Commun. Surv. Tutor.* **2019**, *21*, 2794–2830. [CrossRef]
16. Majeed, U.; Khan, L.U.; Yaqoob, I.; Kazmi, S.A.; Salah, K.; Hong, C.S. Blockchain for IoT-based smart cities: Recent advances, requirements, and future challenges. *J. Netw. Comput. Appl.* **2021**, *181*, 103007. [CrossRef]
17. Hakak, S.; Khan, W.Z.; Gilkar, G.A.; Imran, M.; Guizani, N. Securing smart cities through blockchain technology: Architecture, requirements, and challenges. *IEEE Netw.* **2020**, *34*, 8–14. [CrossRef]
18. Kim, J.W.; Kim, S.J.; Cha, W.C.; Kim, T. A blockchain-applied personal health record application: Development and user experience. *Appl. Sci.* **2022**, *12*, 1847. [CrossRef]
19. Fiore, M.; Capodici, A.; Rucci, P.; Bianconi, A.; Longo, G.; Ricci, M.; Sanmarchi, F.; Golinelli, D. Blockchain for the healthcare supply chain: A systematic literature review. *Appl. Sci.* **2023**, *13*, 686. [CrossRef]
20. Ma, N.; Zhang, X.; Zheng, H.T.; Sun, J. Shufflenet v2: Practical guidelines for efficient cnn architecture design. In Proceedings of the European Conference on Computer Vision (ECCV), Munich, Germany, 8–14 September 2018; pp. 116–131.
21. Hejazi, H.D.; Khamees, A.A. Employees Motivational Factors toward Knowledge Sharing: A Systematic Review. *Int. J. Adv. Appl. Comput. Intell.* **2022**, *1*, 45–55.
22. Mehta, S.; Rastegari, M. Mobilevit: Light-weight, general-purpose, and mobile-friendly vision transformer. *arXiv* **2021**, arXiv:2110.02178.
23. Graham, B.; El-Nouby, A.; Touvron, H.; Stock, P.; Joulin, A.; Jégou, H.; Douze, M. Levit: A vision transformer in convnet's clothing for faster inference. In Proceedings of the IEEE/CVF International Conference on Computer Vision, Montreal, BC, Canada, 11–17 October 2021; pp. 12259–12269.
24. Zhang, W.; Zhou, T.; Lu, Q.; Wang, X.; Zhu, C.; Sun, H.; Wang, Z.; Lo, S.K.; Wang, F.-Y. Dynamic-fusion-based federated learning for COVID-19 detection. *IEEE Internet Things J.* **2021**, *8*, 15884–15891. [CrossRef] [PubMed]
25. Feki, I.; Ammar, S.; Kessentini, Y.; Muhammad, K. Federated learning for COVID-19 screening from Chest X-ray images. *Appl. Soft Comput.* **2021**, *106*, 107330. [CrossRef]
26. Kumar, R.; Khan, A.A.; Kumar, J.; Zakria, A.; Golilarz, N.A.; Zhang, S.; Ting, Y.; Zheng, C.; Wang, W. Blockchain-federated-learning and deep learning models for COVID-19 detection using CT imaging. *IEEE Sens. J.* **2021**, *21*, 16301–16314. [CrossRef] [PubMed]
27. Malik, H.; Naeem, A.; Naqvi, R.A.; Loh, W.K. DMFL_Net: A Federated Learning-Based Framework for the Classification of COVID-19 from Multiple Chest Diseases Using X-rays. *Sensors* **2023**, *23*, 743. [CrossRef]
28. Bian, G.; Qu, W.; Shao, B. Blockchain-Based Trusted Federated Learning with Pre-Trained Models for COVID-19 Detection. *Electronics* **2023**, *12*, 2068. [CrossRef]
29. Wibawa, F.; Catak, F.O.; Kuzlu, M.; Sarp, S.; Cali, U. Homomorphic encryption and federated learning based privacy-preserving cnn training: COVID-19 detection use-case. In Proceedings of the 2022 European Interdisciplinary Cybersecurity Conference, Barcelona, Spain, 15–16 June 2022; pp. 85–90.

mathematics

MDPI

Article

Hybrid Fuzzy Rule Algorithm and Trust Planning Mechanism for Robust Trust Management in IoT-Embedded Systems Integration

Nagireddy Venkata Rajasekhar Reddy [1], Pydimarri Padmaja [2], Miroslav Mahdal [3,*], Selvaraj Seerangan [4], Vrince Vimal [5], Vamsidhar Talasila [6] and Lenka Cepova [7]

1 Department of IT, MLR Institute of Technology, Hyderabad 500043, India; drrajasekhar@mlrit.ac.in
2 Department of Electronics and Communication Engineering, Teegala Krishna Reddy Engineering College, Hyderabad 500097, India; padmajavattem@gmail.com
3 Department of Control Systems and Instrumentation, Faculty of Mechanical Engineering, VSB-Technical University of Ostrava, 17. Listopadu 2172/15, 70800 Ostrava, Czech Republic
4 Department of Computer Science and Design, Kongu Engineering College, Perundurai, Erode 638060, India; selvaraj.cse@kongu.edu
5 Department of Computer Science and Engineering, Graphic Era Hill University, Dehradun 248002, India; vvi-mal@ec.iitr.ac.in
6 Department of Computer Science and Engineering, Koneru Lakshmaiah Education Foundation, Guntur 522302, India; talasila.vamsi@kluniversity.in
7 Department of Machining, Assembly and Engineering Metrology, Faculty of Mechanical Engineering, VSB-Technical University of Ostrava, 17. Listopadu 2172/15, 70800 Ostrava, Czech Republic; lenka.cepova@vsb.cz
* Correspondence: miroslav.mahdal@vsb.cz

Abstract: The Internet of Things (IoT) is rapidly expanding and becoming an integral part of daily life, increasing the potential for security threats such as malware or cyberattacks. Many embedded systems (ESs), responsible for handling sensitive data or facilitating secure online activities, must adhere to stringent security standards. For instance, payment processors employ security-critical components as distinct chips, maintaining physical separation from other network components to prevent the leakage of sensitive information such as cryptographic keys. Establishing a trusted environment in IoT and ESs, where interactions are based on the trust model of communication nodes, is a viable approach to enhance security in IoT and ESs. Although trust management (TM) has been extensively studied in distributed networks, IoT, and ESs, significant challenges remain for real-world implementation. In response, we propose a hybrid fuzzy rule algorithm (FRA) and trust planning mechanism (TPM), denoted FRA + TPM, for effective trust management and to bolster IoT and ESs reliability. The proposed system was evaluated against several conventional methods, yielding promising results: trust prediction accuracy (99%), energy consumption (53%), malicious node detection (98%), computation time (61 s), latency (1.7 ms), and throughput (9 Mbps).

Keywords: Internet of Things (IoT); embedded systems (ESs); trust management (TM); cyber-attacks; security

MSC: 94A16; 68T07; 68T20

1. Introduction

The Internet of Things (IoT) is a new technology that has quickly acquired prominence. Smart buildings, smart hospitals, smart transit, environmental control, smart devices, online atmosphere, and other IoT applications have a significant impact on how people live their everyday lives [1]. IoT is a system in which networks of computer devices are linked to each other. Due to the diversity, resource limitations, and low computing energy of IoT devices, there are several security and privacy concerns. These gadgets exchange data with minimal to no human contact [2]. Unfortunately, there are several security and trust issues brought on by IoT's growth. IoT devices are often found in distant areas, and the data they produce may be readily manipulated, resulting in false information, inconsistent data, and other protection and stability risks [3]. IoT safety has a big influence on how well IoT applications work. A guarantee of object reliability is required when an IoT object has to link with other items for information and communication safety purposes [4]. The study of organizational behavior offered by trust management is based on both current and past conduct. The adoption of trusted management may address issues with credential management, better user confidentiality, and data protection. There are now many suggested trust management solutions accessible on the IoT network [5]. While trying to evaluate the nodes' trust levels within an IoT network, many researchers significantly experience difficulties [6]. IoT devices are susceptible to security assaults because they operate in distant places in possibly hostile situations. These resource-constrained systems, unfortunately, are unable to implement standard safety techniques since they need strong hardware and software [7]. One may assume that a sizable quantity of critical information is handled in IoT devices given the size of IoT and the sectors that use this innovation. As a result, safety is essential. Trust management is one technique utilized to evaluate the network's dependability [8]. By providing a trust value to each node and identifying its degree of trust, trust management tries to assure the dependability of the network. As a result, the data supplied by a node with a higher trust value are regarded as accurate. The trustor and the trustee are the two parties who should be engaged to establish a trust contract [9].

Several sophisticated technologies, such as portable industrial robotics and automobile technologies, are driven by contemporary ESs. The ES is vulnerable to disruptions or abnormalities, particularly in such natural situations. The effects change with time while functioning in a non-stationary setting. For instance, the reliability of a sensor relies on its environment. Moreover, interior states such as the assessment of a mobile robot's present location may become more unreliable and unpredictable. As a result, the system's functioning becomes riskier. Conversely, safe functioning is essential, particularly with human involvement. Hence, as requirements for embedded systems rise, so do the chances of errors and unusual behavior. The former is mostly caused by the growing volume of and dependence on sensory input, as well as by being immersed in an atmosphere that is becoming more complicated and even disruptive and unpredictable [10]. In addition to these hardware difficulties, the programs utilized within the data flow can also induce difficulty due to inherent ambiguity in their outputs. Much worse, these ambiguities change during the system's lifespan, for example, as a result of drifting and fading impacts. These inconsistencies make the fundamental knowledge less trustworthy and add ambiguity and confusion, which seriously impairs safe functioning. Since the activities of an embedded system are dictated by the sensed data and analysis, which in turn affect the general security of the device and its connection with the surroundings, trust or trustworthiness is the key idea here. Hence, appropriate methods must be developed to handle the ambiguities or reliability, as appropriate [11]. The new area of data collection and transmission without human involvement comprises IoT and ES. It is described as a network of interconnected items that have embedded sensors, computers, and control mechanisms. IoT technology evolved as a consequence of scientific advancements. Applications for IoT and ESs are becoming more prominent in practically every industry. IoT applications integrate smarter cities and households, improve transportation and logistics, support healthcare applications, and monitor the environment and public safety. Industrial robots, the aerospace industry, networks and communications, and the automobile industry are a few examples of embedded system applications. Figure 1 displays a few of the predominant IoT and ESs applications [12].

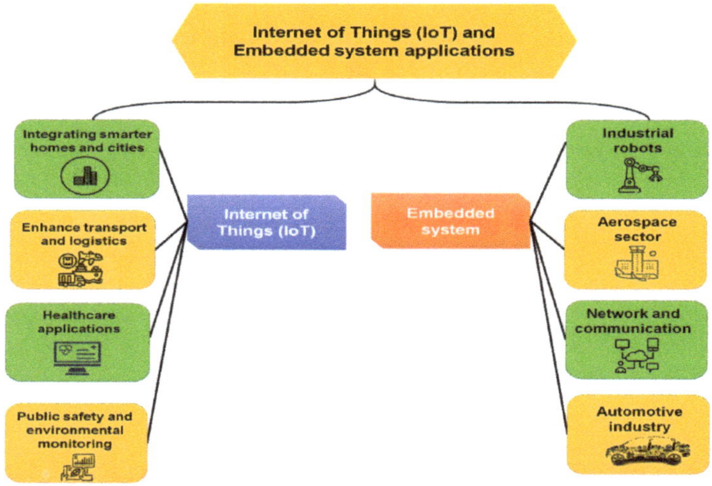

Figure 1. Applications of IoT and ES.

The IoT and ESs allow a variety of applications, yet they also present some security concerns. The idea of a conceptual approach for so-called trust management, which consists of infrastructure and certain procedures, will be proposed as a resolution. Its goal is to boost user confidence in framework efficiency to strengthen the framework's resilience and, as a result, enable efficiency optimization without compromising security.

The remaining parts are organized as follows: The problem statement and literature review are covered in Part II, followed by the suggested methodology in Part III, the proposed methodology's assessment in Part IV, and the conclusion in Part V.

2. Literature Survey

The suggested method was planned after referring to the relevant research works of various researchers on the aspects of Internet of Things, embedded systems, fuzzy rule algorithms, and trust planning mechanisms.

2.1. Review Based on Internet of Things

One of the key ideas in IoT is wireless sensor platforms, which include many network devices with limited resources. These asset-limited sensor nodes now face increased security risks as a result of IoT's widespread adoption. Praveena et al. [13] includes a comprehensive overview of the IoT, discussing its key features, enabling technologies, and potential applications in various domains such as healthcare, transportation, and smart cities. They highlighted the need for efficient communication protocols, scalable architectures, and security mechanisms to ensure the successful deployment and operation of IoT systems and suggested a blockchain-based trust management system (BBTM) to address the issue, wherein portable interface nodes would assess the reliability of IoT devices. For trust assessment, BBTM may design decentralized applications and validate the calculation procedure. The capacity and skill to recognize trustworthiness need more time. Trust is important in providing consumers of the IoT system with an efficient, dependable, expandable, and trusted setting where they may exchange personal data on a secured interaction network. Consequently, TM is a key component of a large-scale IoT platform's ability to transmit data successfully and securely across different nodes. For commercial IoT nodes that are unable to secure information, a lightweight strategy is suggested to address security concerns, delay, and the danger of malevolent actions [14].

2.2. Review Based on Embedded Systems

The evolution of and advancements in embedded systems have led to the proliferation of many embedded systems in various domains. Bitencourt et al. [15], combining embeddings and fuzzy time series for high-dimensional time series forecasting in (i), presented a comprehensive survey on many embedded systems, discussing their characteristics, challenges, and applications. They highlighted the key design considerations, including energy efficiency, real-time constraints, and communication protocols, to achieve many reliable and scalable embedded systems. In both controlled and distributed designs, the effect of rogue gadgets has been reduced through TM. Unfortunately, the majority of these conventional TM frameworks have difficulties with computing, memory, and transmission. The substantial number of factors that need to be modified is this system's primary shortcoming. For massive IoT operations, transferring reliable information across trusted parties is essential. However, the absence of trustworthy partnerships across IoT firms creates considerable obstacles to the aforementioned aim. The authors of [16] suggested a lightweight, secure, triple-trusting architecture (SLTA) that effectively utilizes a supportive system for blockchains. The design features a decentralized access control system that improves privacy and authority over digital records as well as an oracle-based information-collecting system that assures that the information gathered from IoT edge equipment is not altered. The power consumption for the computation is very high. Investigators require IoT-specific technologies, methodologies, and information to increase the degree of privacy for IoT.

2.3. Review Based on Fuzzy Rule Algorithms

Fahim et al. [17] offered an architecture for creating IoT context-aware safety mechanisms to identify malicious information in IoT application situations. To enhance the performance and efficiency of fuzzy rule algorithms, researchers have proposed various extensions and modifications. They introduced the concept of adaptive neuro-fuzzy inference systems (ANFIS), which combine fuzzy logic with neural networks to improve learning and adaptability. ANFIS has been applied in various applications, including system identification, time series prediction, and fault diagnosis. IoT-Flock is a recently developed, open source IoT information-generating program that makes the foundation. Investigators may create an IoT usage instance with both legitimate and malicious IoT systems using the IoT-Flock tool and create data. The suggested architecture also includes an open access application for transforming the traffic that IoT-Flock collects into an IoT database. The reliability of the Internet of Things (IoT) system depends on the detection of anomalous and malicious information. Diwan et al. [18] introduced a simple, limited-cost, greater privacy; lightweight, low-computational-time feature selection IoT intrusion detection approach feature entropy estimation (FEE). The link between all retrieved characteristics was then validated using FEE to detect fraudulent traffic in IoT systems. Due to the implementation of several commands, it is quite complicated. One of the major problems with the Android environment's cyber protection is the proliferation of malignant applications. It is almost impossible to manually identify malware applications in the Android environment due to the rapid growth of malware deployments for the platform. Machine learning is already a young method for detecting malware as a consequence. Abawajy et al. [19] analyzed how frequently utilized filter-based feature-selecting approaches function, with a focus on Android malware detection, and formulated the feature-selecting issue as a quadratic computing issue. The needed training time was longer [20,21]. The malicious program posed a danger to the safety of organizations and networking. Obfuscated malware that may avoid investigation and recognition by anti-malware programs has grown common, although anti-malware solutions are meant to safeguard devices and connections from malware assaults. As a result, a key worry now is how to identify and eliminate malware that has been hidden from the networks.

2.4. Reviews Based on Trust Planning Mechanism

In the context of IoT, trust planning mechanisms are crucial for ensuring the reliability and security of interconnected devices and systems. Mazumdar et al. [11] proposed a trust evaluation model for IoT environments based on fuzzy logic and multidimensional trust factors. They considered factors such as device reliability, communication quality, and security behavior to assess the trustworthiness of IoT devices, contributing to the effective management of trust in IoT ecosystems. In this study, a semi-supervised solution for obfuscated malware identification was proposed that combines deep learning, extraction of features, imagery modification, and computing approaches. With the advent of 5G and above systems, data protection and security have become key problems, giving rise to innovations such as blockchain and federated learning. They enable autonomous data to be used to train machine learning algorithms while maintaining anonymity. This research [22] addressed security concerns unique to this new framework for learning and explored the prospects for IoT malware identification made possible by federated learning (FL). In this regard, a methodology for detecting malware that affects IoT systems utilizing FL was provided. Because of the broad variety of sensitive apps that the Android system has embraced, including banking services, it is increasingly being targeted by malware that takes advantage of security system weaknesses. Methods for the identification of mobile malware have been suggested in a few studies. To obtain the highest level of effectiveness and performance, however, changes are necessary. To identify suspicious Android-directed assaults, we applied machine learning and deep learning technologies. The computation cost is very high [23]. Conventional network equipment has long encountered difficulties with security and reputation administration. Hence, this study provides software-defined networking (SDN), a unique scalable system for credential and trust management of IoT systems in IoT connectivity, a satisfactory proof of concept that shows the sustainability of the suggested system, using modeling that can hold the shared key of IoT systems on the blockchain and properly route networked congestion using SDN.

3. Problem Statement

Over the past several years, millions of IoT and ESs systems without adequate security features have been produced and deployed, and this number will continue to grow as breakthrough innovations become available. Notwithstanding the above information, IoT and ES's security problems are widespread. By taking advantage of these privacy holes, attackers may create a false network and carry out local or distant operations. Moreover, they can alter confidential data without authorization, stop the IoT from functioning normally, or even completely harm the IoT technology. IoT and ES's hardware and software elements are both potentially vulnerable. The demand for effective approaches to detect attacks in IoT and ESs systems within networks is driven by their susceptibility to assaults. To solve these security concerns, we thus introduced the TM-based FRA + TPM technique.

4. Research Methodology

IoT and ESs are effective technologies for many smart applications and have become a potent medium for sharing data and expressing/debating ideas. High-quality safety is a constant need to protect the sensitive data sent between ESs and IoT. One of the main issues to be addressed is safe routing to stop network impersonating attacks. Thus, we recommended the fuzzy rule algorithm and trust planning mechanism (FRA + TPM) for efficient trust management and to enhance IoT and ESs security.

4.1. Fuzzy Rule Algorithm (FRA)

We created FRA, an efficient trust management system for IoT and smarter environments. The FRA defines the mechanism's functioning and it explains how the proposed technique would operate. The IoT system enables the devices to join and terminates the connection at any time. IoT is a very volatile network as a result of this feature. The goal is to provide a technique that collects the network's adaptive properties to enhance the IoT platforms' overall architecture. Figure 2 shows the FRA procedure, where a server assesses the querying node to which it is sending the resources. When there are plenty of querying nodes, the FRA process may employ the First in First out (FIFO) activity to rank them in order of importance. After choosing a node and providing capabilities to it, the server changes its local storage to the transactions' results. The FRA is employed to assess trust in the IoT network ecosystem in the framework of trust in traditional cultures. We focused on the review process rather than the prioritization strategy for choosing the querying node.

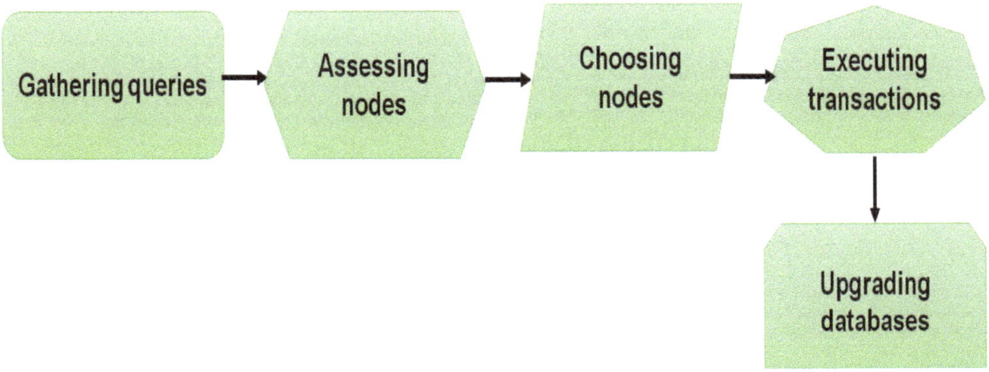

Figure 2. Process of FRA method.

Establishing and preserving trust between the IoT nodes of the intelligent system is the FRA's main goal. The FRA applies fuzzy logic computing to determine the reliability of IoT nodes. The server assesses each node's reliability after receiving a query from it before assigning a specific function to it. Utilizing trust qualities, such as each node's authenticity, suitability, and sensitivity to other nodes in the system, trust is assessed. When a node sends a demand to the server for any kind of operation, the system paradigm begins to function. The resource supplier (server) initially assesses the authorizing node using predefined trust characteristics, and only offers capabilities to trustworthy nodes if necessary. If not, the server considers the receiving node to be malignant or untrustworthy. The trust level is calculated utilizing fuzzy logic computing, where fuzzy rules are implemented to each querying node to maintain trust between the server and the querying node. Furthermore, the server receives the finalized trust value of the querying node after fuzzy logic analysis based on preset trust characteristics. The server chooses whether to serve the node or ignore it based on the ultimate trust value. To remove malignant nodes, a cutoff (CF) level is also specified. The FRA introduces a fuzzy filter that comprises two processes with predetermined thresholds.

The resulting trust level is re-evaluated by the fuzzy filter and compared to the predetermined cutoff value. The CF value could be flexible, changing in response to service requests. The threshold level is larger for extremely reactive equipment, while it is lower for less susceptible ones. The server offers resources to a node when its score is higher than or equal to the preset cutoff level, and once the operation is complete, the server changes its storage appropriately. The server refuses to send resources to the querying node and labels it as malignant or untrustworthy if the node does not reach the minimum cutoff value. The time-driven technique is utilized to increase the IoT platform's degree of confidence. A particular timestamp is assigned to the querying node, and the trust level of the querying node is saved in the storage at the moment the querying node is assessed by the server. There is no requirement to re-evaluate the node if it demands the resources a second time during the given period; instead, the server utilizes the node's previous trust level. The server must re-evaluate the querying node's trust level if the node demands resources after the specified time has passed. The period is fixed to 60 s for each node assessment. Moreover, every node is re-evaluated when the time has passed, and the storage is upgraded with the properly measured trust level for that specific node. Fuzzy filters are crucial for removing fraudulent nodes because they utilize methods to contrast calculated trust with a cutoff level. The server defines the asking node as trusted if the calculated trust is equivalent to or higher than the cutoff level; otherwise, the server identifies the querying node as malignant. The calculated trust of a node that the server flags as malignant is not kept in the server's internal storage. We developed an algorithm called Assessing Trustworthiness that analyzes the trustworthiness of arriving nodes in the system utilizing a predetermined cutoff level. The method in Algorithm 1 describes how the network operates for reliable nodes.

Algorithm 1: Trust node prediction

Configuration
- m_j-one Node for each query
- FC: Collection of fuzzy scores
- F = First trust score
- $L_T^{(T)}$ = Last trust score
- $P_C^{(T)}$ = Prior determined cutoff score
- n_T = Trusted node
- $\text{U}: (0.04 \rightarrow high(+), low(-))$
- Outcome: Trusted Node

Process
- Start
- $Q_j^{(New)} \rightarrow U$
- $n_i^q \rightarrow U$
- If U receive (q, n_i, M_T^T) then
  - $N_i^{(R)} \rightarrow n_i(N)$
  - If $\left(N_i^{(R)} \geq P_T\right) then$
    - $L_T^{(T)} \leftarrow P_T + U$
  - Else if $\left(N_i^{(R)} \geq P_T\right) then$
    - $L_T^{(T)} \leftarrow P_T - U$
  - End if
- End if
- $L_T^{(T)} \rightarrow$ Collection
- If $\left(L_T^{(T)} \geq H_T^{(T)}\right)$ then
  - $M_{n_{(i)}}^{(T)} \leftarrow n_i$
  - $M_n^{(T)} \leftarrow$ Solutions
  - $D_S \leftarrow M_{n_{(i)}}^{(T)}\left(L_T^{(T)}\right)$
- *else*
  - $U_n \leftarrow n_i$
  - $m_j \rightarrow V_t$
- End if
- End

Description of the symbols used in Algorithm 1:
Symbols in Configuration:
m_j: One node for each query.
FC: The collection of fuzzy scores.
F: The first trust score.
$L_T^{(T)}$: The last trust score.
$P_C^{(T)}$: The prior determined cutoff score.
n_T: The trusted node.

U: (0.04→high (+), low (−)): A membership function or fuzzy set representing the uncertainty level. Here, the value U ranges from 0.04 to high (+) or low (−).

Outcome: The trusted node as the outcome of the process.

Symbols in Process:

$Q_i^{(New)}$: The new query input.

n_i^q: A specific node associated with the query.

U receive (q, n_i, M_T^T): The condition where U receives a query, associated node, and a trust score.

$N_i^{(R)}$: The trust score received from node N.

P_T: A predetermined threshold or cutoff score.

$L_T^{(T)} \leftarrow P_T + U$: The update of the last trust score with the addition of U.

$L_T^{(T)} \leftarrow P_T - U$: The update of the last trust score with the subtraction of U.

$H_T^{(T)}$: A high trust score threshold.

$M_{n_{(i)}}^{(T)}$: The trusted node associated with query n_i.

$M_n^{(T)}$: The solutions or outcome based on the trusted node.

D_S: A collection of solutions associated with the trusted node $M_{n_{(i)}}^{(T)}\left(L_T^{(T)}\right)$ and the last trust score $L_T^{(T)}$.

U_n: A specific node denoted as uncertain.

$m_j \rightarrow V_t$: The transfer or assignment of query node m_j to a variable V_t.

This algorithm, Identifying Malicious Node, which functions as a filtration, was created for malignant node identification. As soon as a node has finished using the reasoning engine and the server has received the final trust level from the querying node, this method is once again evaluated on that specific node. This method labels a node as malignant or untrusted if it has lower trust than the cutoff level. The FRA is effective and clever in identifying malevolent or hacked nodes and taking appropriate action. Two trust properties, namely, a node's reliability and flexibility, are specified to calculate a node's trust level. Reliability, a trust characteristic, indicates the degree of certainty and trust for a querying node. Based on its scope, we split the attribute reliability into three groups, including (0, 1). Flexibility, a trust characteristic, denotes a node's interoperability with the server and if a node may legitimately demonstrate its trust in the trustworthiness node. The characteristic flexibility is also separated into three groups based on input data. We obtain a clear range of outcomes for the property total trust by employing these input data to the inference engine along with the fuzzy system. The detail of detecting malicious nodes is provided in Algorithm 2. The suggested mechanism's entire workflow is shown in Figure 3.

Description of the symbols used in Algorithm 2:

Symbols in Configuration:

$M_{n_{(i)}}^{(T)}$: The last trust value associated with node n_i.

FC: The collection of fuzzy scores.

$P_C^{(T)}$: The prior determined cutoff score.

T_n: The novel trust score used for malignant node detection.

U: Trust value up/down (0.05) per request: A membership function or fuzzy set representing the trust value adjustment per request. Here, the value U is 0.05, which can be added or subtracted to the trust score.

Outcome: The malignant node as the outcome of the procedure.

Symbols in Procedure:

$A_n^{(T)}$: The trust value associated with node n_i.

$U_{receive} A_n^{(T)}$: The condition where U receives the trust value $A_n^{(T)}$.

$T_{Pr}^{(T)}$: A temporary trust score.

$P_{req}^{(T)}$: A requested trust score.

T_n: The updated trust score for the node.

$T_n \rightarrow S$: The transfer or assignment of the trust score T_n to a variable S.

$P_C^{(T)}$: The cutoff score or threshold.

Malicious: A malignant node.
$n_i^{(B)}$: A specific node categorized as a malignant node.
Allowed: A node that is allowed or trusted.
$n_i^{(R)}$: A specific node categorized as a trusted node.

Algorithm 2: Malignant nodes detection

Configuration
$M_{n_{(i)}}^{(T)} =$ last trust value of n_i
FC = Collection of fuzzy scores
$P_C^{(T)}$ = Prior determined cutoff score
T_n= Novel trust score for malignant node detection
U: Trust value up/down (0.05) per request
Outcome: Malignant node
Procedure:
Initialize
$A_n^{(T)} \leftarrow M_{n_{(i)}}^{(T)}$
$if\ U_{receive} A_n^{(T)} \leftarrow M_{n_{(i)}}^{(T)}$ then
$T_{Pr}^{(T)} \leftarrow A_n^{(T)}$
$if\left(P_{req}^{(T)} \geq P_C^{(T)}\right) then$
$T_n \leftarrow A_n^{(T)} + U$
$else\ if\left(T_{Pr}^{(T)} \geq P_C^{(T)}\right)$ then
$T_n \leftarrow A_n^{(T)} - U$
$end\ if$
$end\ if$
$T_n \rightarrow S$
$if\ T_n \leftarrow P_C^{(T)}$ then
$Malicious \leftarrow n_i^{(B)}$
Else
$Allowed \leftarrow n_i^{(R)}$
$end\ if$
End

4.2. Trust Planning Mechanism (TPM)

The TPM is intended for significant platforms with a collection of predetermined functions as shown in Figure 4. It was designed specifically for embedded systems utilized in mission-critical, safety, and security-related activities. A dependable operator initializes and configures the system, and it is not anticipated that the user would make changes to it while using it. Basic embedded system elements such as the CPU, RAM, system bus, a peripheral bus, peripherals, and trusted planning additions such as the trusted zone, planner, application-aware MPU, time-slice and synchronization analyzers, and a peripheral bus administrator are all included in TPM. The planner controls how programs are performed and notifies the trusted zone when a novel service has to be run. Programs are started by the planner either by its predetermined plan or in reaction to outside circumstances. As a result, the planner is set up with the planning rules it has to follow and is notified of the active processes on the platform. The trusted zone preserves the program's present condition and passes a command to the novel program when a novel program is planned to run. The trusted zone deploys and enables the appropriate CPU time-slice tracker (for example, a timing interruption) and the synchronization analyzers that transmit responsibility back to it from the program before the program is executed. The trusted zone may always reclaim management of the CPU with the least amount of delay thanks to the CPU duration and synchronization analyzers working in combination. As the planner operates effectively on the CPU and is separated from the programs and various platform parts, it is constantly able to pause and resume the operation of programs.

Every program executing on the platform is assigned a unique specialized memory sector, and an assessment knowing storage security module (SSM) enforces application borders to guarantee activity separation. Lastly, we implement a peripheral bus administrator to guarantee that bus connection to the peripherals is safely managed. The CPU (the executing apps) and other good peripherals cannot be denied connection to the peripheral bus by misbehaving devices thanks to the peripheral bus controller, which manages access from the different peripherals. Moreover, it makes sure that every program is limited to using the peripherals permitted in the trusted zone.

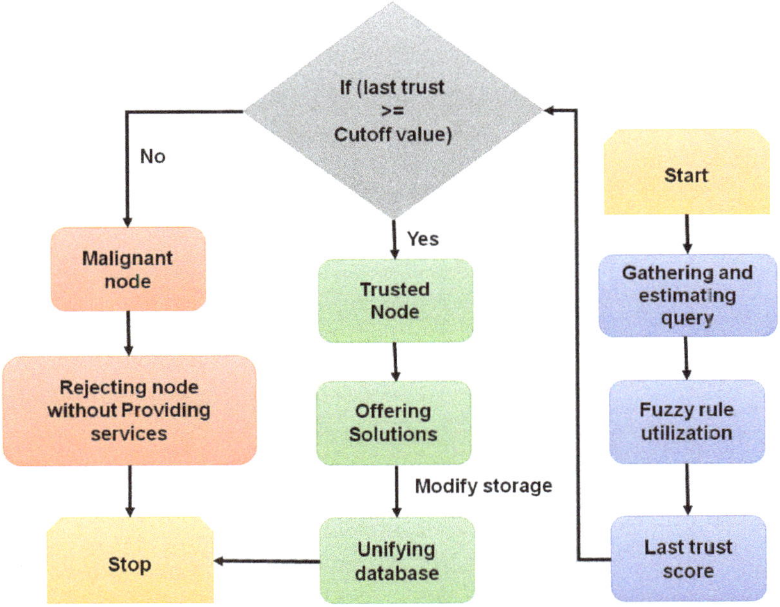

Figure 3. Workflow of proposed FRA.

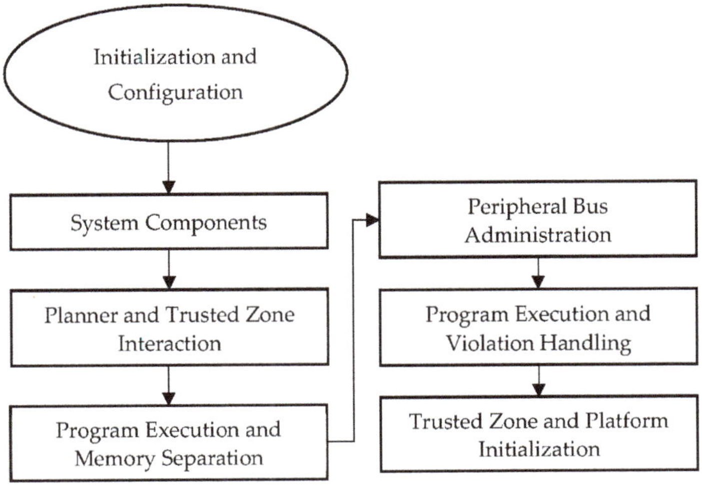

Figure 4. Flow of TM module.

Every program begins running at the first command it receives and keeps running until it crashes, breaks a security rule, or is interrupted. Whenever a program attempts to access or alter a program or content that does not correspond to it, a privacy violation is triggered. Moreover, if a program attempts to perform an atomic segment that exceeds the predetermined limit or if its allotted CPU period runs out, a safety violation is generated. Moreover, if any program attempts to use a service to which it has not been allowed access or if the duration of any bus operation is beyond a pre-maximum limit, the peripheral bus administrator generates a safety violation. As an alternative, the atomicity monitor may also be used to impose the maximum limit on the duration of bus operations. Responsibility is returned to the trusted zone (through the hardware planner) when an activity is interrupted or forcefully ended as a result of a safety violation. The device planner then continues the operation of any outstanding programs.

We presume that a trustworthy manager provides platform initiation parameters to the trusted zone in terms of platform activation. When the device is turned on, the trusted zone runs first and initiates system startup using the executive's commands. The number of apps, the peripherals they utilize, the storage architecture, the operation plan, and preemption rules (such as the maximal CPU time-slice for every program, etc.) are a few examples of these factors, although they are not restricted to them. The scheduler is set up by the trusted zone to run programs regularly or in response to events. Moreover, it sets up specific software and data divisions for every activity by configuring the asset allocator's CPU scheduling settings and the software SSM. The trusted zone also constructs the peripheral bus administrator with knowledge of the services' needs for peripheral connectivity after initializing the software storage of each unique activity.

5. Result and Discussion

In this part, the computation outcomes of the suggested scheme are compared to those of the conventional strategies in terms of trust level prediction, energy consumption, malignant node detection, computation time, latency, and throughput. The existing comparison systems include Trust2Vec, Trust-FTSR, RFA, and EMBTR.

5.1. Trust Prediction Accuracy (%)

Predicting the amount of trust among organizations and consumers who are not already linked is the procedure of forecasting a new trustworthy relationship. Clients must be able to estimate the trust level to have trustworthy communication in IoT and ESs applications. The trust prediction accuracy of the suggested and conventional systems is shown in Figure 5. The outcome of the trust prediction accuracy is depicted in Table 1. This demonstrates that the proposed system's indicated trust level forecast is accurate.

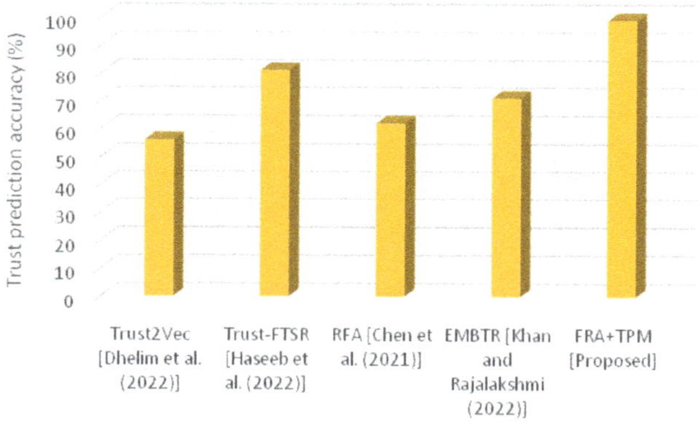

Figure 5. Trust prediction accuracy for suggested and conventional strategies i.e., Trust2Vec [24], Trust-FTSR [25], RFA [26] and EMBTR [27].

Table 1. Outcome of trust prediction accuracy.

Methods	Trust Prediction Accuracy (%)
Trust2Vec [24]	56
Trust-FTSR [25]	81
RFA [26]	62
EMBTR [27]	71
FRA + TPM [Proposed]	99

5.2. Throughput (Mbps)

The quantity of data efficiently sent between two locations in a certain length of the period is known as throughput. It may also be defined as the highest required capability that a method can handle. The throughput results using the suggested and conventional systems are shown in Figure 6. The outcome of the throughput is depicted in Table 2. It demonstrates that the recommended approach offers more throughput than the existing strategies.

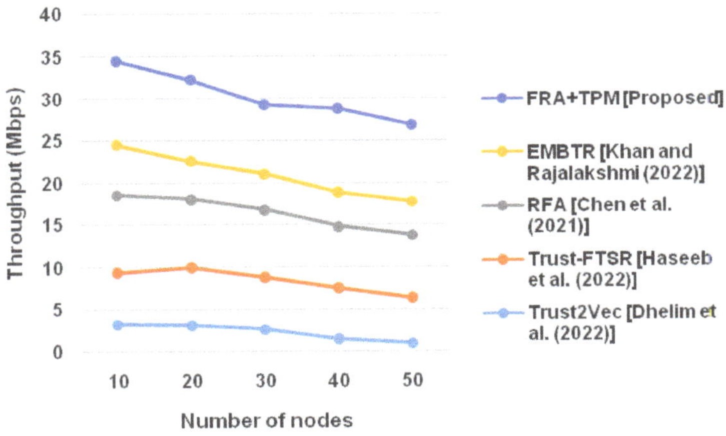

Figure 6. Throughput for suggested and conventional strategies i.e., Trust2Vec [24]. Trust-FTSR [25], RFA [26] and EMBTR [27].

Table 2. Outcome of throughput.

Number of Nodes	Throughput (Mbps)				
	Trust2Vec [24]	Trust-FTSR [25]	RFA [26]	EMBTR [27]	FRA + TPM [Proposed]
10	3.2	6.2	9.2	5.9	10
20	3.1	6.9	8.1	4.5	9.7
30	2.6	6.2	8	4.3	8.2
40	1.5	6	7.3	4.1	9.9
50	1	5.3	7.5	4	9

5.3. Latency (ms)

In information exchange, latency refers to lag. It displays the duration required for data to travel through the system. Networks with increased delays reduce performance. Less delay ensures excellent transmission dependability. The latency results using the suggested and conventional systems are shown in Figure 7. The outcome of the latency is depicted in Table 3. It demonstrates that the recommended technique has lower latency than the existing strategies, allowing data to be transferred more effectively in the IoT and ES.

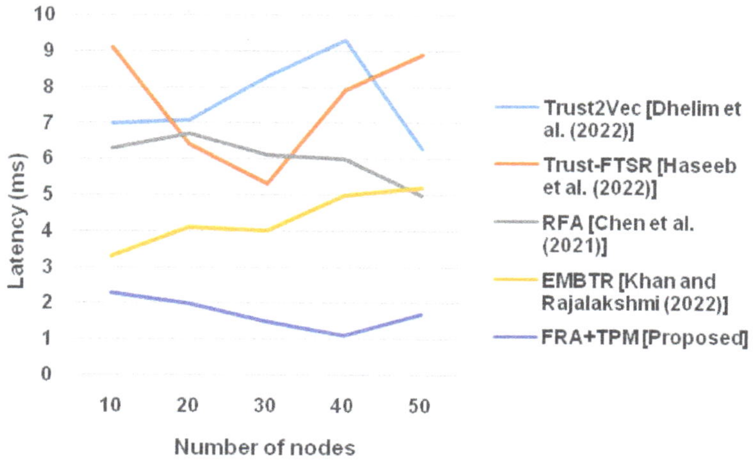

Figure 7. Latency for suggested and conventional strategies i.e., Trust2Vec [24], Trust-FTSR [25], RFA [26] and EMBTR [27].

Table 3. Outcome of latency.

Number of Nodes	Latency (ms)				
	Trust2Vec [24]	Trust-FTSR [25]	RFA [26]	EMBTR [27]	FRA + TPM [Proposed]
10	7	9.1	6.3	3.3	2.3
20	7.1	6.4	6.7	4.1	2
30	8.3	5.3	6.1	4	1.5
40	9.3	7.9	6	5	1.1
50	6.3	8.9	5	5.2	1.7

5.4. Energy Consumption (%)

The strategy's entire energy usage is referred to as energy consumption. It is the difference between the energy structure in its initial condition and the energy structure in its most recent state. The energy consumption results using the suggested and conventional systems are shown in Figure 8. The outcome of the energy consumption is depicted in Table 4. It is noticeable from Figure 7 that the recommended method requires less energy while computing. For interaction to be efficient, energy consumption must be maintained to a minimum.

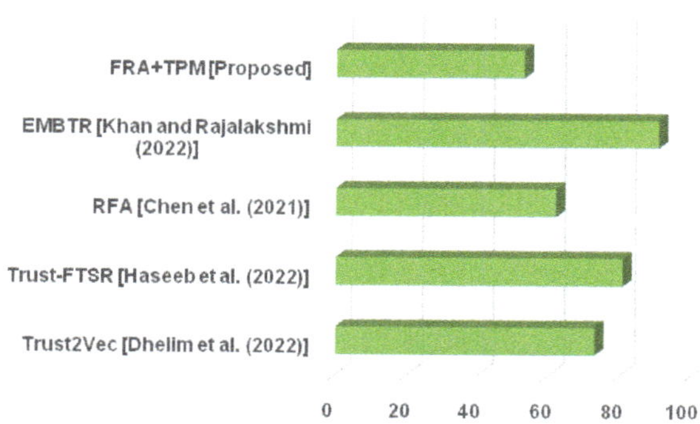

Figure 8. Energy consumption for suggested and conventional strategies i.e., Trust2Vec [24], Trust-FTSR [25], RFA [26] and EMBTR [27].

Table 4. Outcome of energy consumption.

Methods	Energy Consumption (%)
Trust2Vec [24]	73
Trust-FTSR [25]	81
RFA [26]	62
EMBTR [27]	91
FRA + TPM [Proposed]	53

5.5. Malignant Node Detection (%)

A malignant node attempts to prevent other nodes in the system from providing services to devices connected. The malignant node may reduce the system performance rate, which would reduce the network's service time. It is necessary to identify these nodes in wireless communication. The malignant node detection results using the suggested and conventional systems are shown in Figure 9. The outcome of the malignant node detection is depicted in Table 5. It shows that the provided approach effectively detects malicious nodes.

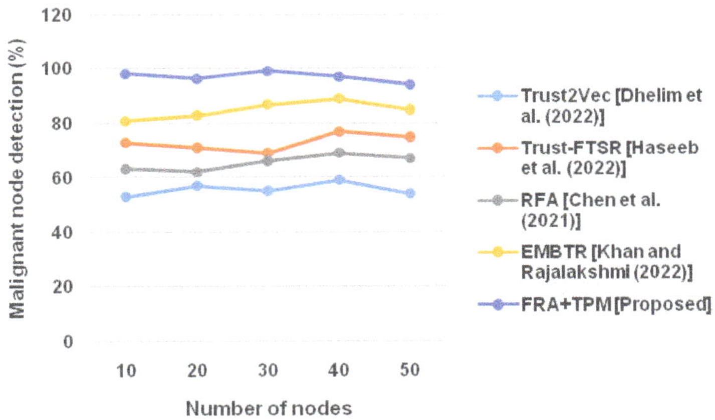

Figure 9. Malignant node detection for suggested and conventional strategies i.e., Trust2Vec [24], Trust-FTSR [25], RFA [26] and EMBTR [27].

Table 5. Outcome of malignant node detection.

Number of Nodes	Malignant Node Detection (%)				
	Trust2Vec [24]	Trust-FTSR [25]	RFA [26]	EMBTR [27]	FRA + TPM [Proposed]
10	53	73	63	81	98
20	57	71	62	83	96
30	55	69	66	87	99
40	59	77	69	89	97
50	54	75	67	85	94

5.6. Computation Time (s)

Computation time is the length of time required to accomplish a calculation (also known as execution time). This indicates how rapidly the modal can detect the result. The computation time results using the suggested and conventional systems are shown in Figure 10. The outcome of the computation time is depicted in Table 6. As a result, the recommended method takes less time to forecast trust and find malignant nodes.

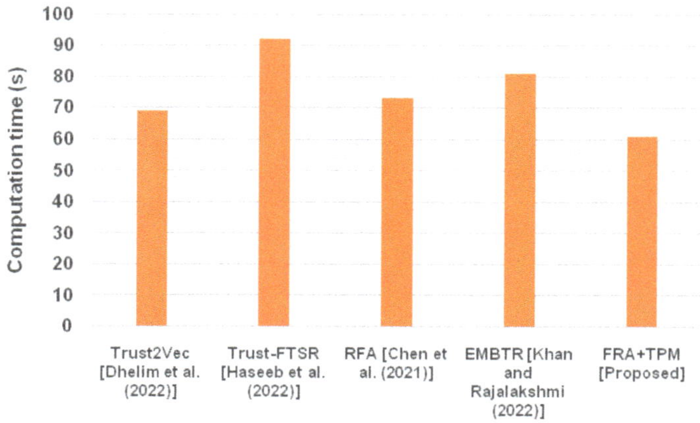

Figure 10. Computation time for suggested and conventional strategies i.e., Trust2Vec [24], Trust-FTSR [25], RFA [26] and EMBTR [27].

Table 6. Outcome of computation time.

Methods	Computation Time (s)
Trust2Vec [24]	69
Trust-FTSR [25]	92
RFA [26]	73
EMBTR [27]	81
FRA + TPM [Proposed]	61

6. Discussion

Dhelim et al. [24] presented the Trust2Vec TM technology, which could maintain connections in large-scale IoT networks and counter large-scale trust assaults carried out by many hostile units. They provided a network embeddings population identification technique that recognizes and inhibits communities of harmful nodes to identify assaults such as self-promotion and maligning. The accuracy of malicious detection is extremely low. To enhance trustworthy and cooperative connectivity in smart cities, the authors of [25] suggested the fault-tolerant supervised routing (Trust-FTSR) paradigm for TM in the IoT environment. For dependable and efficient system architecture, every node assesses the actions of its partners and develops mutual contact. A fault-tolerant transmitting mechanism was further provided by utilizing guided network learning without adding any extra overheads. The cost of computing is expensive. For computing immediate trust, using a gliding window and the time-decaying factor may significantly speed up the convergence rate. To efficiently screen out incorrect suggestions and reduce the impact of dangerous items, the authors of [26] designed a recommendation-filtering algorithm (RFA). As a means of adapting to the periodically dangerous situation, an adaptive weight was created to blend straight trust and suggestion trust into synthesis trust more effectively. The algorithm's complexity contributes to the computation's poor performance. The highest level of privacy is a constant need to protect the sensitive data sent between ESs and IoT. To complete the routing process safely based on nodes' trustworthy values, an enhanced multi-attribute-based attack resistance (EMBTR) method was suggested by the authors of [27]. By excluding the misbehaving nodes from the transmission channel based on the computed trusted measures for the network nodes, the method described in the study may offer a trustworthy route. The trust value forecasting process takes a long time to complete.

Owing to these shortcomings in the existing strategies, we offered an FRA+TPM for an effective TM in the IoT and ES.

7. Discussion of Analysis of the Presented Fuzzy Method with the Exact Method

One of the key advantages of the fuzzy method is its ability to handle uncertainty and imprecision. In IoT environments, where data and conditions can be uncertain or vague, the fuzzy approach allows for more flexible reasoning and decision making. In contrast, the exact method might struggle to handle uncertain or imprecise data effectively.

The fuzzy method provides a more granular and flexible approach to trust management. It allows for the definition of linguistic variables, membership functions, and fuzzy rules, enabling the system to reason and make decisions based on linguistic terms rather than rigid numerical values. This flexibility can be beneficial in capturing and expressing complex trust relationships. The exact method, on the other hand, may have more limited expressiveness and granularity due to its reliance on crisp values and strict rules.

The exact method may excel in situations where precision and deterministic decision making are crucial. However, in dynamic and uncertain IoT environments, the fuzzy method's ability to handle imprecision and uncertainty can contribute to increased robustness. The fuzzy method prioritizes adaptability and resilience to variations and changes in the environment, which may be advantageous in real-world IoT deployments.

It is worth considering the computational complexity of both methods. The exact method, depending on its mathematical or rule-based nature, may be computationally efficient but may lack the flexibility to adapt to changing conditions. On the other hand, the fuzzy method, particularly when involving complex fuzzy rule bases or large-scale systems, may require more computational resources for inference and decision making.

The fuzzy method offers unique advantages in handling uncertainty, providing flexibility, and achieving robustness in trust management within IoT-embedded systems. While the exact method may excel in certain scenarios that require precise decision making, the fuzzy method's ability to handle imprecision and uncertainty makes it particularly suitable for the complexities of IoT environments. The choice between the two methods depends on the specific requirements, context, and trade-offs desired in the trust management system.

8. Conclusions

IoT has the opportunity to become one of the best-known platforms in the age of online computation with the advancement in enabling capabilities and the invention of implementation methods. Any unit in the IoT platform that is capable of recognizing, detecting, connecting, and computing can interact with every other unit to achieve a specific goal using a variety of interaction styles. Embedded systems operate in more dynamic, unpredictable, and non-stationary ecosystems like those found in robot navigation and autonomous vehicle systems. As a result, there is an increase in requirement for their reliability, particularly when dealing with people. IoT device availability, performance efficiency, security, and sustainability issues have become greater due to the quick rise in IoT device utilization. To offer consumers improved services, these problems must be resolved. To manage trust effectively and increase IoT and ES's dependability, we proposed a hybrid fuzzy rule algorithm (FRA) and trust planning mechanism (TPM) (FRA+TPM). The proposed system was evaluated, and the findings are: trusted prediction (99%), energy consumption (53%), malignant node detection (98%), computation time (61 s), latency (1.7 ms), and throughput (9 Mbps). This proves that the proposed method is effective in TM in IoT and ES.

Based on the comprehensive work carried out in this study, the following management insights can be drawn:

- The integration of the FRA and trust planning mechanism offers enhanced trust management capabilities in IoT-embedded systems. By incorporating fuzzy logic, the approach effectively handles uncertainty, imprecision, and complex relationships among entities, enabling more accurate and adaptive trust assessment. This can provide decision makers with valuable insights into the trustworthiness of IoT devices and facilitate informed decision making processes.

- The proposed approach's ability to adapt to changing conditions and handle uncertainties contributes to the robustness of trust management in dynamic IoT environments. The fuzzy-logic-based reasoning allows for flexible adjustments to evolving trust relationships and varying levels of trustworthiness, ensuring the system's resilience against environmental changes and potential threats.

Despite the rigorousness of the study, the following limitations need to be addressed:

- It is important to consider the computational complexity associated with the integration of FRA and trust planning mechanism. The process of fuzzy inference, rule evaluation, and defuzzification can introduce computational overhead, especially for large-scale IoT deployments or real-time applications. Further research and optimization techniques may be needed to mitigate potential performance issues.
- The proposed approach's applicability may vary depending on the specific characteristics and requirements of the IoT-embedded systems integration. The effectiveness of the FRA and trust planning mechanism may depend on the availability and quality of input data, the complexity of trust relationships, and the nature of the IoT environment. Thorough evaluation and validation of the approach in various scenarios would help assess its suitability for different contexts.
- The integration of the hybrid fuzzy rule algorithm and trust planning mechanism offers valuable insights into trust management in IoT-embedded systems. The approach enhances trust assessment, provides robustness in dynamic environments, and enables decision makers to make informed choices. However, the computational complexity and context-specific limitations should be considered when applying the proposed approach, highlighting the need for further research and validation to fully realize its potential in practical IoT deployments.

In further studies, the communication and effectiveness of the suggested TM mechanism will be optimized by various innovative techniques.

Author Contributions: N.V.R.R.—conceptualization, methodology, formal analysis, investigation, writing—original draft, P.P.—conceptualization, methodology, formal analysis, investigation, writing—original draft, M.M.—conceptualization, methodology, resources, writing—review and editing, S.S.—methodology, validation, writing—original draft, V.V.—methodology, validation, writing—original draft, V.T.—methodology, validation, writing—review and editing, L.C.—conceptualization, methodology, validation, writing—review and editing. All authors have read and agreed to the published version of the manuscript.

Funding: This article was supported by the project SP2023/074 Application of Machine and Process Control Advanced Methods supported by the Ministry of Education, Youth and Sports, Czech Republic.

Data Availability Statement: Not applicable.

Conflicts of Interest: The authors declare no conflict of interest.

References

1. Lv, Z.; Song, H. Mobile Internet of Things Under Data Physical Fusion Technology. *IEEE Internet Things J.* **2019**, *7*, 4616–4624. [CrossRef]
2. Lv, Z.; Cheng, C.; Song, H. Digital Twins Based on Quantum Networking. *IEEE Netw.* **2022**, *36*, 88–93. [CrossRef]
3. Kumar, R.; Sharma, R. Leveraging blockchain for ensuring trust in IoT: A survey. *J. King Saud Univ. Comput. Inf. Sci.* **2022**, *34*, 8599–8622. [CrossRef]
4. Lv, Z.; Qiao, L.; Li, J.; Song, H. Deep-Learning-Enabled Security Issues in the Internet of Things. *IEEE Internet Things J.* **2020**, *8*, 9531–9538. [CrossRef]
5. Qu, Z.; Zhang, Z.; Liu, B.; Tiwari, P.; Ning, X.; Muhammad, K. Quantum detectable Byzantine agreement for distributed data trust management in blockchain. *Inf. Sci.* **2023**, *637*, 118909. [CrossRef]
6. Rana, K.; Singh, A.V.; Vijaya, P. Recent Trust Management Models for Secure IoT Ecosystem. *Int. J. Intell. Syst. Appl. Eng.* **2022**, *10*, 23–33.

7. Lv, Z.; Wu, J.; Li, Y.; Song, H. Cross-Layer Optimization for Industrial Internet of Things in Real Scene Digital Twins. *IEEE Internet Things J.* **2022**, *9*, 15618–15629. [CrossRef]
8. Dai, X.; Xiao, Z.; Jiang, H.; Alazab, M.; Lui, J.C.S.; Dustdar, S.; Liu, J. Task Co-Offloading for D2D-Assisted Mobile Edge Computing in Industrial Internet of Things. *IEEE Trans. Ind. Inform.* **2022**, *19*, 480–490. [CrossRef]
9. Liu, Y.; Yu, W.; Rahayu, W.; Dillon, T. An Evaluative Study on IoT ecosystem for Smart Predictive Maintenance (IoT-SPM) in Manufacturing: Multi-view Requirements and Data Quality. *IEEE Internet Things J.* **2023**, 1. [CrossRef]
10. Babaei, A.; Khedmati, M.; Jokar, M.R.A.; Tirkolaee, E.B. Designing an integrated blockchain-enabled supply chain network under uncertainty. *Sci. Rep.* **2023**, *13*, 1–19. [CrossRef]
11. Mazumdar, H.; Kaushik, A.; Gohel, H.A. To mitigate primary user emulation attack trajectory using cognitive single carrier frequency division multiple access approaches: Towards next generation green IoT. *Eng. Rep.* **2023**, e12672. [CrossRef]
12. Mishra, N.; Pandya, S. Internet of Things Applications, Security Challenges, Attacks, Intrusion Detection, and Future Visions: A Systematic Review. *IEEE Access* **2021**, *9*, 59353–59377. [CrossRef]
13. Praveena, J.; Arivazhagan; Reddy, P.V.P. Blockchain based sensor system design for embedded IoT. *J. Comput. Inf. Syst.* **2023**, 1–18. [CrossRef]
14. Din, I.U.; Bano, A.; Awan, K.A.; Almogren, A.; Altameem, A.; Guizani, M. LightTrust: Lightweight Trust Management for Edge Devices in Industrial Internet of Things. *IEEE Internet Things J.* **2021**, *10*, 2776–2783. [CrossRef]
15. Bitencourt, H.V.; de Souza, L.A.F.; Santos, M.C.D.; Silva, R.; de Lima e Silva, P.C.; Guimarães, F.G. Combining embeddings and fuzzy time series for high-dimensional time series forecasting in internet of energy applications. *Energy* **2023**, *271*, 127072. [CrossRef]
16. Shi, P.; Wang, H.; Yang, S.; Chen, C.; Yang, W. Blockchain-based trusted data sharing among trusted stakeholders in IoT. *Softw. Pract. Exp.* **2019**, *51*, 2051–2064. [CrossRef]
17. Fahim, M.; El Mhouti, A.; Boudaa, T.; Jakimi, A. Modeling and implementation of a low-cost IoT-smart weather monitoring station and air quality assessment based on fuzzy inference model and MQTT protocol. *Model. Earth Syst. Environ.* **2023**, 1–18. [CrossRef]
18. Diwan, T.D.; Choubey, S.; Hota, H.S.; Goyal, S.B.; Jamal, S.S.; Shukla, P.K.; Tiwari, B. Feature Entropy Estimation (FEE) for Malicious IoT Traffic and Detection Using Machine Learning. *Mob. Inf. Syst.* **2021**, *2021*, 1–13. [CrossRef]
19. Abawajy, J.; Darem, A.; Alhashmi, A.A. Feature Subset Selection for Malware Detection in Smart IoT Platforms. *Sensors* **2021**, *21*, 1374. [CrossRef]
20. Zheng, W.; Deng, P.; Gui, K.; Wu, X. An Abstract Syntax Tree based static fuzzing mutation for vulnerability evolution analysis. *Inf. Softw. Technol.* **2023**, *158*, 107194. [CrossRef]
21. Gong, J.; Rezaeipanah, A. A fuzzy delay-bandwidth guaranteed routing algorithm for video conferencing services over SDN networks. *Multimed. Tools Appl.* **2023**, 1–30. [CrossRef] [PubMed]
22. Rey, V.; Sánchez, P.M.S.; Celdrán, A.H.; Bovet, G. Federated learning for malware detection in IoT devices. *Comput. Netw.* **2022**, *204*, 108693. [CrossRef]
23. Alkahtani, H.; Aldhyani, T.H.H. Artificial Intelligence Algorithms for Malware Detection in Android-Operated Mobile Devices. *Sensors* **2022**, *22*, 2268. [CrossRef] [PubMed]
24. Dhelim, S.; Aung, N.; Kechadi, M.T.; Ning, H.; Chen, L.; Lakas, A. Trust2Vec: Large-Scale IoT Trust Management System Based on Signed Network Embeddings. *IEEE Internet Things J.* **2022**, *10*, 553–562. [CrossRef]
25. Haseeb, K.; Saba, T.; Rehman, A.; Ahmed, Z.; Song, H.H.; Wang, H.H. Trust Management With Fault-Tolerant Supervised Routing for Smart Cities Using Internet of Things. *IEEE Internet Things J.* **2022**, *9*, 22608–22617. [CrossRef]
26. Chen, G.; Zeng, F.; Zhang, J.; Lu, T.; Shen, J.; Shu, W. An adaptive trust model based on recommendation filtering algorithm for the Internet of Things systems. *Comput. Netw.* **2021**, *190*, 107952. [CrossRef]
27. Khan, A.F.; Rajalakshmi, C.N. A multi-attribute based trusted routing for embedded devices in MANET-IoT. *Microprocess. Microsyst.* **2022**, *89*, 104446. [CrossRef]

mathematics

MDPI

Article

Order-Preserving Pattern Matching with Partition

Joong Chae Na [1,†], Youngjoon Kim [2,†], Seokchul Kang [2] and Jeong Seop Sim [2,*]

1 Department of Computer Science and Engineering, Sejong University, Seoul 05006, Republic of Korea; jcna@sejong.ac.kr
2 Department of Computer Engineering, Inha University, Incheon 22212, Republic of Korea; yjkim12130@gmail.com (Y.K.); stoneiron02@inha.edu (S.K.)
* Correspondence: jssim@inha.ac.kr
† These authors contributed equally to this work.

Abstract: Order-preserving pattern matching, which considers the relative orders of strings, can be applied to time-series data analysis. To perform a more meaningful analysis of time-series data, approximate criteria for the order-isomorphism are necessary, considering diverse types of errors. In this paper, we introduce a novel approximation criterion for the order-isomorphism, called the partitioned order-isomorphism. We then propose an efficient $O(n + sort(m))$-time algorithm for the order-preserving pattern matching problem considering the criterion of partition. A comparative experiment demonstrates that the proposed algorithm is more effective than the exact order-preserving pattern matching algorithm.

Keywords: order-preserving pattern matching; approximate order-preserving pattern matching; pattern matching with partition; string matching

MSC: 68W32

1. Introduction

Two strings of the same length are *order-isomorphic* when they have the same relative orders. For example, strings $X = (5, 9, 7)$ and $Y = (23, 51, 47)$ are order-isomorphic because the relative order of individual characters in both strings is the same as $(1, 3, 2)$. Given a text T of length n and a pattern P of length m, the order-preserving pattern matching (OPPM) problem is finding all substrings of T which are order-isomorphic to P. For example, given a text $T = (10, 23, 5, \underline{3, 30, 8, 27, 15, 25, 12, 6}, 17, 11, 4)$ and $P = (1, 8, 3, 7, 5, 6, 4, 2)$, P occurs at the fourth position of T because the underlined substring $(3, 30, 8, 27, 15, 25, 12, 6)$ of T is order-isomorphic to P (see Figure 1). The OPPM problem can be solved in $O(n + sort(m))$ time [1–3] based on the KMP algorithm [4], where $sort(m)$ is time for sorting P. Practical algorithms for the OPPM problem using the Horspool approach [5] and a filtration method have been studied [6–9]. Additionally, various related studies have been conducted, including research on order-preserving string regularities [10,11], OPPM with multiple patterns [1,12], parallelism in OPPM [12,13], order-preserving suffix trees [14], and OPPM in a tree and directed acyclic graphs [15].

Since, in OPPM, the relative order of characters is more important than their absolute values, it can be effectively applied to analyze various types of time-series data, such as stock prices, energy consumption, annual temperature variance, and music melodies [7,16–18]. For example, when there are two stocks in the same industry but with different face values, the actual price fluctuations of the two stocks over the same period may differ, but the pattern of change can be similar. In such a case, using order-isomorphism allows for detecting the same or similar patterns. However, it is not always easy to apply OPPM algorithms directly due to various factors such as differences in the measurement intervals of time-series data, noise occurring in the measurement process, errors in the data transformation process, etc. Additionally, even if this error occurs at just one position, it can affect the overall relative order, which makes it difficult to detect in the reference data. For example, if the measurements of a certain time-series data are $(3,30,8,27,15,25,12,6)$, the relative order of each measurement is $(1,8,3,7,5,6,4,2)$. If the second value (character), 30, is changed to 24 due to a measurement or recording error, the relative order of each measurement changes significantly to $(1,6,3,8,5,7,4,2)$, making it difficult to detect in the reference data or possibly leading to the detection of an incorrect position. Therefore, to perform a more meaningful pattern search, prediction, and analysis of time-series data, it is necessary to consider diverse types of errors.

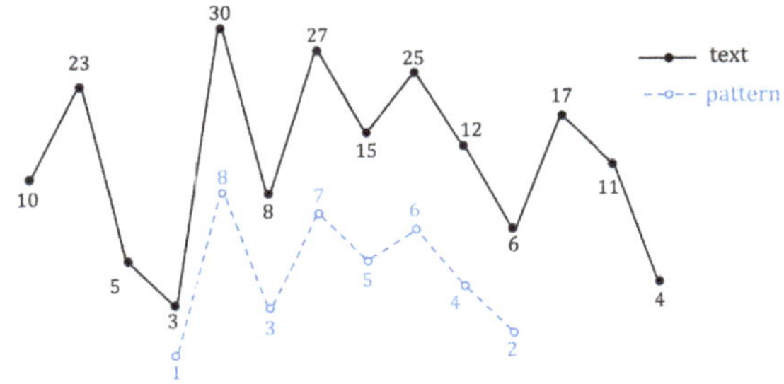

Figure 1. An OPPM example for $T = (10,23,5,3,30,8,27,15,25,12,6,17,11,4)$ and $P = (1,8,3,7,5,6,4,2)$.

Recently, approximate OPPM problems that allow errors in matching have been studied. Gawrychowski and Uznański [17] studied the OPPM problem with k-mismatches. Their algorithm finds occurrences of a pattern within a text where the relative order of elements is preserved, allowing up to k mismatches, with a time complexity of $O(n(\log\log m + k\log\log k))$. Russo et al. [18] studied OPPM problems on indeterminate strings, which considers the relative order of elements, even when some elements are not precisely determined. They presented an $O(rm\log r)$-time algorithm, where r is a bound on the number of uncertain characters per position. Kim et al. [19] defined a scaling approximation and presented algorithms for the OPPM problem with scaling, which considers the relative orders of cusps and the scale of lengths of strings between them. They presented an $O(n + m\log m)$-time algorithm to solve the OPPM problem with scaling, which is the same time bound as the best known exact OPPM algorithm.

In some cases in OPPM, it might be more effective to compare two strings part by part rather than as whole strings. Consider two strings $x = (1,8,3,7,5,6,4,2)$ and $y = (3,30,8,27,15,25,12,6)$ for example. The relative orders of these two strings are the same with $(1,8,3,7,5,6,4,2)$. Assume that the second character of y is changed to 24 due to some error, as in the previous example. That is, let us assume y is changed to $y' = (3,24,8,27,15,25,12,6)$. Then the relative orders of x and y' become quite different, with $(1,8,3,7,5,6,4,2)$ and $(1,6,3,8,5,7,4,2)$, respectively. Despite their overall differences in order, however, when x and y' are divided at the third position, the orders of the two divided parts of x and y' are the same with $(1,3,2)$ and $(5,3,4,2,1)$, respectively, i.e., the two respective parts match each other. See Figure 2.

The above approach can be applied to plagiarism detection in music melodies. OPPM can be used to detect parts of a full melody that exactly match a specific short melody to determine plagiarism. If one of the notes in the detected section is modified to a similar but different note, it may sound similar when played but might not be detected by OPPM. However, if the section is divided into two parts and OPPM is performed separately, the likelihood of detection increases. It can also be applied to stock price prediction. If the price movement of a particular stock over a certain period resembles the movement patterns of other stocks in the past, future changes can be predicted to some extent. However, as these patterns become longer, it becomes more challenging to find exact matches using OPPM. By dividing the pattern into two parts and performing OPPM separately, the chances of finding matching sections increase.

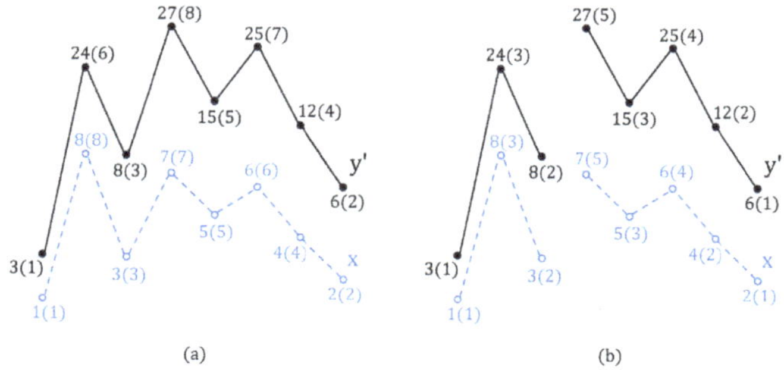

Figure 2. Illustration of the partitioned order-isomorphism (the numbers in parentheses represent the relative orders). For $x = (1, 8, 3, 7, 5, 6, 4, 2)$ and $y' = (3, 24, 8, 27, 15, 25, 12, 6)$, (**a**) x and y' are not order-isomorphic, but (**b**) when dividing them at the third position, two divided parts of x and y' match each other.

Our contributions are as follows:

- We introduce a novel approximation criterion for the order-isomorphism, called the partitioned order-isomorphism for the first time. In OPPM, unlike classical pattern matching, a change in just one character can significantly alter the overall ranking of the entire string. As a result, parts that could not be found when the two strings were compared directly may be detected when the strings are divided and compared in two parts.
- We present an efficient algorithm for the problem using the Z-function [10,20]. Even with the consideration of the approximation criterion, our algorithm achieves a time complexity of $O(n + sort(m))$, which is equivalent to the best-known exact OPPM algorithm.
- We demonstrate the effectiveness of our algorithm through comparative experiments using three types of real time-series data. The experiments are conducted by varying the lengths of patterns for each data set, which demonstrates that our algorithm is more effective than the OPPM algorithm.

This paper is organized as follows. In Section 2, we give some preliminary works. In Section 3, we first define the partitioned order-isomorphism and the OPPM problem with partition formally. And then we give an efficient algorithm for the OPPM problem with partition. In Section 4, we present experimental comparisons between the algorithm introduced in [1] and the algorithm proposed in this study. In Section 5, we conclude this research.

2. Related Works

Let Σ denote the set of characters where two characters can be compared in constant time. A string S is a sequence of characters drawn from Σ. We denote by $|S|$ the length of S and by $S[i]$ the ith character of S $(1 \le i \le |S|)$. The substring $S[i] \cdots S[j]$ $(1 \le i, j \le |S|)$ is denoted by $S[i..j]$, and $S[i..j]$ is an empty string when $i > j$. Substrings $S[1..i]$ and $S[i..|S|]$ $(1 \le i \le |S|)$ are called a prefix and a suffix of S, respectively. The concatenation of two strings X and Y is denoted by XY. For convenience, we assume that the characters in a string are all distinct as in [1].

The relative order (or rank) of a character $S[i]$ in a string S is defined as the number of characters in S that are smaller than or equal to $S[i]$, including itself. A natural way to represent the order relations of the characters in S is to enumerate their relative orders in S, which is called the *natural representation* of the order relations. For example, the natural representation of the string $(54, 12, 38, 69, 45, 22)$ is $(5, 1, 3, 6, 4, 2)$. Although the natural representation is intuitive and easy to understand, it is not suitable for pattern matching [1]. The *prefix representation* is an alternative way to represent order relations, where the relative order of a character $S[i]$ is considered within $S[1..i]$ rather than in the entire string S. For example, the prefix representation of the string $(54, 12, 38, 69, 45, 22)$ is $(1, 1, 2, 4, 3, 2)$. It has been shown that there is a one-to-one mapping between the natural representation and the prefix representation [1].

We formally define the order-isomorphism of two strings. Two strings X and Y of the same length are *order-isomorphic*, denoted by $X \approx Y$, when $X[i] < X[j] \Leftrightarrow Y[i] < Y[j]$ for all $1 \leq i, j \leq |X|$. Obviously, the following lemma and corollary hold by the definition of the order-isomorphism.

Lemma 1. *For two strings X and Y of the same length, if X and Y are order-isomorphic, then substrings $X[i..j]$ and $Y[i..j]$ (for all $1 \leq i \leq j \leq |X|$) are also order-isomorphic.*

Corollary 1. *For two strings X and Y of the same length, if there exist substrings $X[i..j]$ and $Y[i..j]$ ($1 \leq i \leq j \leq |X|$) that are not order-isomorphic, then X and Y are not order-isomorphic.*

The order-isomorphism can be efficiently determined using the nearest neighbor representation [1–3,6], which is represented by the location tables $LMax_X$ and $LMin_X$.

$$LMax_X[i] = \begin{cases} j & \text{if } X[j] = \max\{X[k] : X[k] < X[i],\ 1 \leq k \leq i-1\} \\ -1 & \text{if there is no such } j \end{cases}$$

$$LMin_X[i] = \begin{cases} j & \text{if } X[j] = \min\{X[k] : X[k] > X[i],\ 1 \leq k \leq i-1\} \\ -1 & \text{if there is no such } j \end{cases}$$

That is, $LMax_X[i]$ and $LMin_X[i]$ are the nearest neighbors of $X[i]$ in the sorted sequence of characters in $X[1..i]$. Table 1 shows $LMax_X$ and $LMin_X$ for $X = (54, 12, 38, 69, 45, 22)$. The location tables can be computed in $O(|X| \log |X|)$ time using a sorting algorithm or an order-statistic tree. Using the location tables for X, we can determine the order-isomorphism of X and Y in $O(|X|)$ time by checking the following inequality for every position $1 \leq i \leq |X|$:

$$Y[LMax_X[i]] < Y[i] < Y[LMin_X[i]] \tag{1}$$

(if $LMax_X[i]$ or $LMin_X[i]$ is equal to -1, we assume the respective inequality is true). Moreover, we can find the longest prefix β of Y that is order-isomorphic to a prefix of X in $O(|\beta|)$ time by checking the above inequality in increasing order of i until we reach a position where the inequality is false (Algorithm 1). We call β the longest order-isomorphic prefix (loip) of X and Y, which is called the longest common prefix (lcp) in classical matching.

Table 1. Location tables and Z-function for $X = (54, 12, 38, 69, 45, 22)$.

i	1	2	3	4	5	6
$X[i]$	54	12	38	69	45	22
$LMax_X[i]$	−1	−1	2	1	3	2
$LMin_X[i]$	−1	1	1	−1	1	3
Z_X	6	1	1	2	2	1

For a string S, the Z-function $Z_S(i)$ ($1 \leq i \leq |S|$) represents the length of the longest prefix of $S[i..|S|]$ which is order-isomorphic to a prefix of S, i.e., the loip of $S[i..|S|]$ and $S[1..|S| - i + 1]$ [10]. It is a modified version for order-isomorphism of the Z-function introduced in [20] to solve classical string matching problems. With the nearest neighbor representation of S, Z_S can be computed in $O(|S|)$ time.

Algorithm 1 Computing loip

Input: two strings $X[1..m]$ and $Y[1..m]$, and an integer $k \geq 0$ such that $X[1..k] \approx Y[1..k]$.
output: length of the loip of X and Y.

```
procedure COMPLOIP (X, Y, k)
    for i ← k + 1 to m do                        ▷ Assume LMax_X and LMin_X are given.
        if LMax_X[i] ≠ −1 and Y[LMax_X[i]] ≥ Y[i] then
            return i − 1
        end if
        if LMin_X[i] ≠ −1 and Y[i] ≥ Y[LMin_X[i]] then
            return i − 1
        end if
    end for
    return m
end procedure
```

3. Order-Preserving Pattern Matching (OPPM) with Partition

In this section, we define the OPPM problem with partition and present an efficient algorithm for solving the problem.

3.1. Problem Definition

We define the partitioned order-isomorphism. Let us consider two strings $X = (5, 11, 18, 7, 3, 9)$ and $Y = (1, 2, 3, 5, 4, 6)$, which are not order-isomorphic. When we partition X (resp. Y) into two substrings $X[1..3]$ and $X[4..6]$ (resp. $Y[1..3]$ and $Y[4..6]$), the partitioned substrings of X are order-isomorphic to those of Y, i.e., $X[1..3] \approx Y[1..3]$ and $X[4..6] \approx Y[4..6]$. Then, we say that X is partitioned order-isomorphic to Y. That is, the *partitioned order-isomorphism* is formally defined as follows.

Definition 1. *For two strings X and Y of length m and a position t* $(0 \leq t \leq m)$*, if* $X[1..t] \approx Y[1..t]$ *and* $X[t+1..m] \approx Y[t+1..m]$*, then X and Y are partitioned order-isomorphic at t, denoted by* $X \approx_{\vdash}^{t} Y$*. If there exists any position t such that* $X \approx_{\vdash}^{t} Y$*, then X and Y are partitioned order-isomorphic.*

In the example above, $X \approx_{\vdash}^{3} Y$ since $X[1..3] \approx Y[1..3]$ and $X[4..6] \approx Y[4..6]$. However, X is not partitioned order-isomorphic to $Z = (1, 2, 3, 4, 5, 6)$ because there exists no t such that $X \approx_{\vdash}^{t} Z$. We call position t a *partition position.* (Although we consider the cases when a string is partitioned into two substrings, Definition 1 can be extended to the cases of partitioning into more than two substrings.)

The following lemma shows that partition positions are consecutive.

Lemma 2. *For two strings X and Y of length m, let* l_{max} *be the largest l such that* $X[1..l] \approx Y[1..l]$ $(1 \leq l \leq m)$*. Similarly, let* r_{min} *be the smallest r such that* $X[r..m] \approx Y[r..m]$ $(1 \leq r \leq m)$*. Then,* $X \approx_{\vdash}^{t} Y$ *if and only if* $r_{min} - 1 \leq t \leq l_{max}$*.*

Proof. Note that l_{max} and r_{min} always exist since two strings of length 1 are order-isomorphic.
($\Rightarrow$) We show by contradiction that if $X \approx_{\vdash}^{t} Y$, then $r_{min} - 1 \leq t \leq l_{max}$. Suppose that $X \approx_{\vdash}^{t} Y$ for some t such that $t < r_{min} - 1$ or $t > l_{max}$. Consider the case when $t > l_{max}$. Since $X \approx_{\vdash}^{t} Y$, $X[1..t] \approx Y[1..t]$, which contradicts the definition of l_{max}. Similarly, a contradiction occurs when $t < r_{min} - 1$.
($\Leftarrow$) We show that $X \approx_{\vdash}^{t} Y$ if $r_{min} - 1 \leq t \leq l_{max}$. Since $X[1..l_{max}] \approx Y[1..l_{max}]$ and $t \leq l_{max}$, $X[1..t] \approx Y[1..t]$ by Lemma 1. Similarly, $X[t+1..m] \approx Y[t+1..m]$. Therefore, $X \approx_{\vdash}^{t} Y$. □

Corollary 2. *If* $r_{min} - 1 > l_{max}$*, then X and Y are not partitioned order-isomorphic.*

The *OPPM problem with partition* is to find all substrings of a text T that are partitioned order-isomorphic to a pattern P, formally defined as follows.

Problem 1. *The order-preserving pattern matching (OPPM) problem with partition.*
Input: A text T of length n and a pattern P of length m $(n \geq m)$.
Output: Every pair $(i, [a : b])$ where i is a position in T $(1 \leq i \leq n - m + 1)$ and $[a : b]$ is a position range in P $(0 \leq a \leq b \leq m)$ such that $T[i..i + m - 1] \approx^{t}_{\vdash} P$ if and only if $a \leq t \leq b$.

For example, given $T = (13, 92, 34, 88, 77, 63, 37, 40, 70, 54, 35, 24, 50)$ and $P = (54, 12, 38, 69, 45, 22)$, the output is $(2, [3 : 3])$ and $(6, [2 : 5])$. See Table 2, which shows l_{max} and r_{min} for P and $T[i..i + m - 1]$ $(1 \leq i \leq n - m + 1)$.

Table 2. Given $T[1..13] = (13, 92, 34, 88, 77, 63, 37, 40, 70, 54, 35, 24, 50)$ and $P[1..6] = (54, 12, 38, 69, 45, 22)$, the values of l_{max} and r_{min} for P and $T[i..i + 5]$ $(1 \leq i \leq 8)$.

i	1	2	3	4	5	6	7	8
l_{max} $(= L_i)$	1	3	1	2	2	5	1	1
r_{min} $(= 7 - R_i)$	4	4	6	6	5	3	4	6
occurrence	-	$(2, [3 : 3])$	-	-	-	$(6, [2 : 5])$	-	-

3.2. An Algorithm for the OPPM Problem with Partition

Let T_i denote the substring of T of length m starting at position i, i.e., $T_i = T[i..i + m - 1]$ $(1 \leq i \leq n - m + 1)$. Let L_i and R_i be the lengths of the longest prefix and the longest suffix of T_i that are order-isomorphic to a prefix and a suffix of P, respectively. When we are given L_i's and R_i's for all $1 \leq i \leq n - m + 1$, we can solve Problem 1 in $O(n)$ time using Lemma 2 and Corollary 2 (note that $l_{max} = L_i$ and $r_{min} - 1 = m - R_i$ for T_i and P). See Algorithm 2.

Now we focus on the problem of computing all L_i's. (We omit the details of computing R_i's since it can be computed using the same way by considering reversed strings of T and P.) Note that L_i is the length of the loip of T_i and P. So we can compute each L_i in $O(m)$ time using the nearest neighbor representation of P (Algorithm 1), resulting in a total time complexity of $O(nm + m \log m)$, where $O(m \log m)$ represents the time required for computing the nearest neighbor representation $LMax_P$ and $LMin_P$ of P. More efficiently, we can compute all L_i's in $O((n + m) \log(n + m))$ time by concatenating P and T (say $S = PT$) and computing Z_S.

Algorithm 2 OPPM with partition

```
Input: a text T[1..n] and a pattern P[1..m]
Output: all pairs (i, [a : b]) representing occurrences of Problem 1.
procedure OPPMwPartition(T, P)
    compute L_i and R_i for 1 ≤ i ≤ n − m + 1 using Algorithm 3
    for i ← 1 to n − m + 1 do
        if m − R_i ≤ L_i then
            report (i, [m − R_i : L_i]) as an occurrence
        end if
    end for
end procedure
```

However, we can improve the time complexity to $O(n + m \log m)$ time by computing Z_P and all L_i's separately. We first preprocess P for computing $LMax_P$, $LMin_P$, and Z_P. Then, we compute L_i in increasing order of i using Z_P and L_j $(1 \leq j < i)$, which is similar to the Z-algorithm in [20]. For any position k $(1 \leq k \leq n)$, the L-box starting at k is defined as the range $(k, k + L_k - 1)$ of length L_k. (Note that $L_k > 0$ because the first characters of two strings are trivially order-isomorphic.) For each i $(1 \leq i \leq n)$, let (l_i, r_i) be the L-box whose end position is the rightmost among the L-boxes whose start position is less than or equal to i. (If more than one L-box has the rightmost end position, we may choose any L-box.)

Algorithm 3 Computing L_i's

input: a text $T[1..n]$ and a pattern $P[1..m]$
Output: all pairs $(i, [a:b])$ representing occurrences of Problem 1.

```
procedure COMPUTEL(T, P)
    Compute LMax_P, LMin_P, and Z_P for P                 ▷ Preprocessing P
    (l, r) ← (0, 0)                                       ▷ Initialize L-box (l_0, r_0)
    for i ← 1 to n − m+1 do
        if i > r then                                     ▷ Case 1
            L_i ← COMPLOIP(P, T[i..i + m − 1], 0)
            (l, r) ← (i, i + L_i − 1)                     ▷ Set (l_i, r_i) as the new L-box
        else                                              ▷ Case 2
            z ← Z_P(i − l + 1)
            if z = |T[i..r]| then                         ▷ Case 2(a)
                L_i ← COMPLOIP(P, T[i..i + m − 1], z)
                (l, r) ← (i, i + L_i − 1)                 ▷ Set (l_i, r_i) as the new L-box
            else                                          ▷ Case 2(b) and 2(c)
                L_i = min(z, |T[i..r]|)
                                ▷ (l, r) remains unchanged, i.e., (l_i, r_i) = (l_{i−1}, r_{i−1})
            end if
        end if
    end for
end procedure
```

Algorithm 3 shows the pseudo-code of our algorithm for computing L_i's. At each iteration i $(1 \leq i \leq n-m+1)$, we compute L_i and (l_i, r_i) using (l_{i-1}, r_{i-1}). Initially, we set $(l_0, r_0) = (0, 0)$. The details of each iteration i are as follows:

1. If $i > r_{i-1}$, then compute L_i by explicitly checking the inequality (1) for T_i $(= T[i..i+m-1])$ and P character by character, starting from the first position and continuing until either the inequality becomes false or the check reaches the last position. Since $i > r_{i-1}$, we set (l_i, r_i) to $(i, i+L_i-1)$, i.e., the new L-box starting at position i.
2. If $i \leq r_{i-1}$, then $T[i]$ is contained in the L-box $(l_{i-1}..r_{i-1})$. Let $\beta = T[i..r_{i-1}]$ and $z = Z_P(i - l_{i-1} + 1)$. Then, $\beta[1..\ell] \approx P[1..\ell]$ where $\ell = \min(|\beta|, z)$, i.e., it is guaranteed that L_i is at least $\min(|\beta|, z)$, which is proved in Lemma 3.
    (a) If $z = |\beta|$, then L_i can be larger than $z = r_{i-1} - i + 1$. Thus, we compute L_i by explicitly checking the inequality (1) for T_i and P character by character, starting from the position $z+1$. Furthermore, we set (l_i, r_i) as $(i, i+L_i-1)$, i.e., the new L-box starting at position i.
    (b) If $z < |\beta|$, then $L_i = z$ and $(l_i, r_i) = (l_{i-1}, r_{i-1})$.
    (c) If $z > |\beta|$, then $L_i = |\beta|$ and $(l_i, r_i) = (l_{i-1}, r_{i-1})$.

We illustrate how our algorithm computes L_i's for $T = (13, 92, 34, 88, 77, 63, 37, 40, 70, 54, 35, 24, 50)$ and $P = (54, 12, 38, 69, 45, 22)$. We first preprocess P (see Table 1 for $LMax_P$, $LMin_P$, and Z_P). Initially, $(l_0, r_0) = 0$.

- Iteration $i = 1$ (Case 1). We obtain $L_1 = 1$ by explicitly checking the inequality (1) for T_1 and P. Note that the inequality is false at position 2 since $LMin_P[2] = 1$ and $T[2](= 92) \not< T[1](= 13)$, i.e., $T[1..2] \not\approx P[1..2]$. Thus, the L-box starting at position 1 is $(1, 1)$, and it is set as (l_1, r_1) since $r_0 < 1$.
- Iteration $i = 2$ (Case 1). We compute $L_2 = 3$ explicitly and set $(l_2, r_2) = (2, 4)$. Note that $T_2[1..3] = (92, 34, 88) \approx P[1..3] = (54, 12, 38)$ but $T_2[1..4] = (92, 34, 88, 77) \not\approx P[1..4] = (54, 12, 38, 69)$.

- Iteration $i = 3$ (Case 2(b)). We have $\beta = T[3..4]$, $z = Z_P(3-2+1) = Z_P(2) = 1$, and $\ell = 1$. Then, $T[3..3] = T_3[1..1] \approx P[1..1]$, but $T_3[1..2] = (34, 88) \not\approx P[1..2] = (54, 12)$, i.e., $L_3 = 1$, which we can know without checking the inequality (1) as proven in Lemma 3. The new L-box starting at position 3 is $(3, 3)$ and it is included in $(l_2, r_2) = (2, 4)$. Hence $(l_3, r_3) = (2, 4)$.
- Iteration $i = 4$ (Case 2(a)). We have $\beta = T[4..4]$ and $z = Z_P(4-2+1) = Z_P(3) = 1$. Since $|\beta| = z$, we obtain $L_4 = 2$ by explicitly checking the inequality (1) starting from position 2. Note that $T_4[1..2] = (88, 77) \approx P[1..2] = (54, 12)$, but $T_4[1..3] = (88, 77, 63) \not\approx P[1..3] = (54, 12, 38)$. Moreover, we set $(l_4, r_4) = (4, 5)$.
- Iteration $i = 5$ (Case 2(a)). We have $\beta = T[5..5]$ and $z = Z_P(5-4+1) = Z_P(2) = 1$. Similarly, we obtain $L_5 = 2$ and $(l_5, r_5) = (5, 6)$.
- Iteration $i = 6$ (Case 2(a)). We have $\beta = T[6..6]$ and $z = Z_P(6-5+1) = Z_P(2) = 1$. Similarly, we obtain $L_6 = 5$ and $(l_6, r_6) = (6, 10)$.
- Iteration $i = 7$ (Case 2(b)). We have $\beta = T[7..10]$ and $z = Z_P(7-6+1) = Z_P(2) = 1$. Similarly to iteration 3, we obtain $L_7 = 1$ and $(l_7, r_7) = (6, 10)$.
- Iteration $i = 8$ (Case 2(b)). We have $\beta = T[8..10]$ and $z = Z_P(8-6+1) = Z_P(3) = 1$. Similarly to iteration 7, we obtain $L_8 = 1$ and $(l_8, r_8) = (6, 10)$.

Lemma 3. *Algorithm 3 correctly computes L_i and (l_i, r_i) for all $1 \le i \le n - m - 1$.*

Proof. It can be proved by induction. Initially (when $i = 1$), since $i > r_0 = 0$ (Case 1), we compute L_i explicitly and set (l_1, r_1) as $(1, L_1)$, which is the only L-box whose start position is less than or equal to 1.

Now we consider an iteration where $i > 1$. In Case 1, we obtain correct L_i since we compute it by explicitly checking inequality (1). Since $L_i > 0$ (note that the first characters of two strings are trivially order-isomorphic), the end position $i + L_i - 1$ of the new L-box starting at i is greater than or equal to i. Additionally, since $i > r_{i-1}$ (by the condition of Case 1), $i + L_i - 1 > r_{i-1}$. Hence, $(i, i + L_i - 1)$ is the L-box whose end position is the rightmost among the L-boxes whose start position is less than or equal to i. Therefore, we compute (l_i, r_i) correctly.

In Case 2, $T[i]$ is contained in substring $T[l_{i-1}..r_{i-1}]$ such that $T[l_{i-1}..r_{i-1}] \approx P[1..r_{i-1} - l_{i-1} + 1]$, which means $T[i..r_{i-1}] \approx P[i - l_{i-1} + 1..r_{i-1} - l_{i-1} + 1]$ by Lemma 1. Let $\beta = T[i..r_{i-1}]$, $i' = i - l_{i-1} + 1$, and $z = Z_P(i')$ (see Figure 3). Then, $\beta \approx P[i'..i' + |\beta| - 1]$ and $P[i'..i' + z - 1] \approx P[1..z]$ (by the definition of z). Thus, $\beta[1..\ell] \approx P[1..\ell]$ where $\ell = \min(|\beta|, z)$, i.e., $L_i \ge \min(|\beta|, z)$.

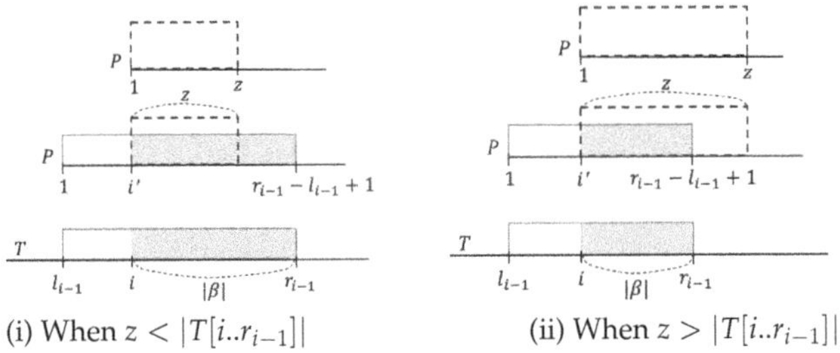

(i) When $z < |T[i..r_{i-1}]|$ (ii) When $z > |T[i..r_{i-1}]|$

Figure 3. Illustration of several cases under the condition $i \le r_{i-1}$.

In Case 2(a), we obtain correct L_i since we compute it by explicitly checking inequality (1) starting from position $z + 1$. Note that we do not need to check the inequality for the position k ($1 \le k \le z$) since $T_i[1..z] \approx P[1..z]$. Additionally, since $L_i \ge z = |\beta|$ ($= r_{i-1} - i + 1$), it follows that $i + L_i - 1 \ge r_{i-1}$. Therefore, (l_i, r_i) should be $(i, i + L_i - 1)$.

In Case 2(b), $T_i[z+1]$ is contained in β and $T_i[1..z+1] \approx P[i'..i' + z]$ (see Figure 3(i)). Since $P[i'..i' + z] \not\approx P[1..z+1]$ by the definition of $z = Z_P(i')$, it follows that $T_i[1..z+1] \not\approx$

$P[1..z+1]$ and thus $L_i = z$. Moreover, since the end position $i+z-1$ of the new L-box starting at position i is to the left of r_{i-1} $(= i+|\beta|-1)$, (l_i, r_i) is the same as (l_{i-1}, r_{i-1}).

In Case 2(c), $P[|\beta|+1]$ is contained in $P[1..z]$ and $P[1..|\beta|+1] \approx P[i'..i'+|\beta|]$ (see Figure 3(ii)). Since $P[i'..i'+|\beta|] \not\approx T_i[1..|\beta|+1]$ by the definition of β, it follows that $T_i[1..|\beta|+1] \not\approx P[1..|\beta|+1]$ and thus $L_i = |\beta|$. Additionally, the end position $i+|\beta|-1$ of the new L-box starting at position i is equal to r_{i-1}. Hence, (l_i, r_i) can be either (l_{i-1}, r_{i-1}) or $(i, i+L_i-1)$ (where (l_{i-1}, r_{i-1}) is chosen in our algorithm).

Therefore, Algorithm 3 computes L_i and (l_i, r_i) correctly in all cases. □

Lemma 4. *Given $LMax_P$, $LMin_P$, and Z_P, we can compute all L_i's $(1 \le i \le n-m+1)$ in $O(n)$ time.*

Proof. Each iteration takes constant time except for checking the inequality (1) in procedure COMPLOIP. The check ends when the inequality becomes false or the check reaches the last position. Let x_i be the number of times when the inequality is true during iteration i. If $x_i > 0$, r_i increases by x_i (and never decreases) from r_{i-1} (Cases 1 and 2(a)). Since $r_i \le n$, the total sum of x_i's for all iterations is at most n. Therefore, given $LMax_P$, $LMin_P$, and Z_P, all L_i's can be computed in $O(n)$ time. □

Theorem 1. *Given a text T of length n and a pattern P of length m, the OPPM problem with partition can be solved in $O(n + m\log m)$ time. If the characters of P can be sorted in linear time, the problem can be solved in $O(n+m)$ time.*

Proof. We can compute $LMax_P$ and $LMin_P$ in $O(m \log m)$ time. (If we can sort the characters of P in linear time, for example, such as when the characters are drawn from an integer alphabet, we can compute $LMax_P$ and $LMin_P$ in $O(m)$ time [2]). Then, Z_P and L_i's can be computed in $O(m)$ time and $O(n)$ time, respectively. Moreover, R_i's can be computed similarly using the reversed strings of P and T. For each position i $(1 \le i \le n-m+1)$ in Algorithm 2, given L_i and R_i, we can determine whether T_i is partitioned isomorphic to P and compute the range of partition positions in constant time. Therefore, we obtain the theorem. □

4. Experimental Results

The experimental environment was as follows: the operating system was Microsoft Windows 10 Education (64-bit), the CPU was an AMD Ryzen 9 5950X 16-Core, with 64 GB of RAM. The development tool was gcc 13.1.0, and the development language was C++. Three types of time-series data were used in the experiment: a power consumption index, particulate matter (PM2.5) levels, and the Dow Jones Index. The power consumption index consisted of data on the average household voltage per minute in Sceaux, France, from 00:00 on 16 December 2006, to 22:00 on 2 December 2008 [21]. The PM2.5 levels were recorded in Beijing at one-hour intervals from 00:00 on 2 January 2010, to 22:00 on 9 October 2014 [22]. The Dow Jones Index data included the daily closing prices of the Dow Jones Industrial Average from 2 May 1885, to 12 April 2019 [23]. For brevity, the power consumption index data are referred to as PCI, the particulate matter level data as PML, and the Dow Jones Index as DJI. We generated the experimental data as follows. First, the text was created by multiplying the real-valued time-series data by appropriate values to convert all values into integers. Since using a random string as the pattern could result in significantly fewer occurrences, a fixed-length substring starting from a random position in the converted text was designated as the pattern.

We compare the number of occurrences detected by our OPPM with partition algorithm (referred to as OPPM_PART) with those detected by the exact OPPM algorithm proposed by Kim et al. [1] (referred to as OPPM_EXACT). The lengths n of text T for PCI, PML, and DJI were 2,049,280, 41,757, and 38,104, respectively. Pattern P was generated by randomly extracting strings of lengths $6, 8, 10, 12$, and 14 from T. The reason for selecting the pattern lengths in this way is that, due to the nature of OPPM, the longer the pattern length, the fewer matching parts occur. In fact, when the pattern length exceeded 15, there were almost no occurrences.

Table 3 compares the number of occurrences for each algorithm, which are the averages obtained from executing the algorithm on 100 different patterns. As shown in the table, our algorithm, OPPM_PART, detects approximately 2.2 (for $m = 14$) to 51.0 (for $m = 8$) times more occurrences than OPPM_EXACT with PCI. With PML, OPPM_PART detects approximately 7.5 (for $m = 14$) to 35.3 (for $m = 10$) times more occurrences than OPPM_EXACT. With DJI, OPPM_PART detects approximately 1.9 (for $m = 14$) to 25.1 (for $m = 8$) times more occurrences than OPPM_EXACT. In all cases we experimented with, OPPM_PART detected significantly more meaningful parts than OPPM_EXACT.

Table 4 compares the execution times for each algorithm, which are the averages obtained from running the algorithm on 100 different patterns. As shown in Table 4, OPPM_PART runs approximately 2.3 (for $m = 10$) to 2.5 (for $m = 6$) times slower than OPPM_EXACT with PCI. With PML, OPPM_PART runs approximately 2.3 (for $m = 14$) to 2.8 (for $m = 6$) times slower than OPPM_EXACT. With DJI, OPPM_PART runs approximately 2.2 (for $m = 14$) to 2.8 (for $m = 6$) times slower than OPPM_EXACT. In all the cases we experimented with, OPPM_PART was approximately twice as slow as OPPM_EXACT. Although both algorithms, OPPM_EXACT and OPPM_PART, have the same time complexity of $O(n + m \log m)$, OPPM_PART was slower due to the additional $O(n)$ time required for calculating L_i and R_i, as confirmed by additional experiments (not presented in a table). However, considering that the actual execution time is measured in microseconds, this increase in runtime is not significant (see Figure 4).

Table 3. Comparison of the number of occurrences for PCI, PML, and DJI data (averages of 100 times of executions).

Data	Algorithm	Pattern Length (m)				
		$m = 6$	$m = 8$	$m = 10$	$m = 12$	$m = 14$
PCI (n = 2,049,280)	OPPM_EXACT	10,138.7	465.5	80.9	40.7	34.2
	OPPM_PART	176,917.0	23,751.0	2927.8	315.9	76.8
PML (n = 41,757)	OPPM_EXACT	475.5	83.4	3.6	1.2	1.1
	OPPM_PART	4608.4	1030.1	127.2	22.1	8.3
DJI (n = 38,104)	OPPM_EXACT	220.7	21.0	3.1	1.3	1.0
	OPPM_PART	3495.5	527.5	75.7	12.1	1.9

Table 4. Comparison of execution times (microseconds) for PCI, PML, and DJI data (averages of 100 times of executions).

Data	Algorithm	Pattern Length (m)				
		$m = 6$	$m = 8$	$m = 10$	$m = 12$	$m = 14$
PCI (n = 2,049,280)	OPPM_EXACT	12,154.5	12,060.2	11,997.9	12,043.4	12,134.3
	OPPM_PART	29,979.6	28,583.7	28,102.4	28,472.5	28,552.8
PML (n = 41,757)	OPPM_EXACT	241.5	243.9	246.8	242.2	243.6
	OPPM_PART	672.4	629.5	620.7	563.9	549.3
DJI (n = 38,104)	OPPM_EXACT	226.9	231.7	232.9	231.5	235.2
	OPPM_PART	624.6	592.6	580.2	565.2	519.2

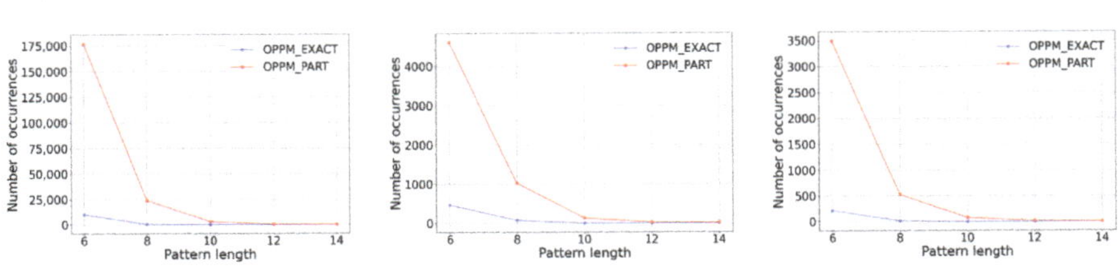

(**a**) PCI (**b**) PML (**c**) DJI

Figure 4. The number of occurrences according to pattern length.

5. Conclusions

We have introduced the concept of partitioned order-isomorphism as a novel criterion for approximate order-isomorphism. This approximation criterion is significant not only because it can be utilized in analyzing time-series data where errors may occur, but also because it provides a criterion for approximation that is challenging to address in classical pattern matching based solely on character value comparison. Using this criterion, we have defined the OPPM problem with partition and presented an efficient algorithm for solving it. One advantage of our criterion is that it allows for approximate matching without increasing the time complexity compared to the exact OPPM.

Our experimental results show that we can find more occurrences using our approximation criterion. As shown in the experimental results, our algorithm detected significantly more meaningful parts in all cases we tested. Therefore, we expect that our criterion will also enable more meaningful pattern search, prediction, and analysis in real-world applications. Moreover, as the pattern length increases, the difference of the number of occurrences between OPPM_EXACT and OPPM_PART decreases. This suggests that partitioning into more than two parts may be necessary for longer patterns.

Although our current focus is on partitioning a string into two parts, our definition can be extended to cases where the string is partitioned into three or more parts. However, developing an efficient algorithm for partitioning into even three parts is not straightforward. In the two-part case, we solved the problem efficiently by preprocessing T (i.e., computing L_i's and R_i's), since the start position of the left part and the end position of the right part are fixed. In contrast, when partitioning into three parts, neither the start nor the end position of the middle part is fixed, which can increase the complexity. We plan to explore this in future research.

Author Contributions: J.C.N. and Y.K. designed and analyzed the algorithm and wrote the draft of the paper. S.K. implemented and experimented with the algorithms. J.C.N. and J.S.S. reviewed and revised the paper. J.S.S. analyzed the algorithm, provided algorithmic support, and was the project manager. All authors read and approved the final manuscript.

Funding: This work was supported by the Institute of Information & Communications Technology Planning & Evaluation (IITP) grant funded by the Korean Government (MSIT) (No. RS-2022-00155915, Artificial Intelligence Convergence Innovation Human Resources Development (Inha University)) and by the INHA UNIVERSITY Research Grant.

Data Availability Statement: Data are contained within the article.

Conflicts of Interest: The authors declare they have no conflicts of interest.

Abbreviations

The following abbreviations are used in this manuscript:

OPPM	order-preserving pattern matching
KMP	Knuth–Morris–Pratt
loip	longest order-isomorphic prefix
lcp	longest common prefix
PCI	power consumption index data
PML	particulate matter level data
DJI	Dow Jones Index data
OPPM_EXACT	OPPM algorithm proposed in [1]
OPPM_PART	OPPM with partition algorithm proposed in this study

References

1. Kim, J.; Eades, P.; Fleischer, R.; Hong, S.; Iliopoulos, C.S.; Park, K.; Puglisi, S.J.; Tokuyama, T. Order-preserving matching. *Theor. Comput. Sci.* **2014**, *525*, 68–79. [CrossRef]
2. Kubica, M.; Kulczyński, T.; Radoszewski, J.; Rytter, W.; Waleń, T. A linear time algorithm for consecutive permutation pattern matching. *Inf. Process. Lett.* **2013**, *113*, 430–433. [CrossRef]

3. Kim, J.; Amir, A.; Na, J.C.; Park, K.; Sim, J.S. On Representations of Ternary Order Relations in Numeric Strings. *Math. Comput. Sci.* **2017**, *11*, 127–136. [CrossRef]
4. Knuth, D.E.; Morris, J.H.; Pratt, V.R. Fast Pattern Matching in Strings. *SIAM J. Comput.* **1977**, *6*, 323–350. 10.1137/0206024.
5. Horspool, R.N. Practical Fast Searching in Strings. *Softw. Pract. Exp.* **1980**, *10*, 501–506. [CrossRef]
6. Cho, S.; Na, J.C.; Park, K.; Sim, J.S. A fast algorithm for order-preserving pattern matching. *Inf Process. Lett.* **2015**, *115*, 397–402. [CrossRef]
7. Chhabra, T.; Tarhio, J. A filtration method for order-preserving matching. *Inf. Process. Lett.* **2016**, *116*, 71–74. [CrossRef]
8. Na, J.C.; Lee, I. A simple heuristic for order-preserving matching. *IEICE Trans. Inf. Syst.* **2019**, *102*, 502–504.
9. Kim, Y.; Kim, Y.; Sim, J.S. An improved order-preserving pattern matching algorithm using fingerprints. *Mathematics* **2022**, *10*, 1954. [CrossRef]
10. Hasan, M.M.; Islam, A.S.M.S.; Rahman, M.S.; Rahman, M.S. Order preserving pattern matching revisited. *Pattern Recognit. Lett.* **2015**, *55*, 15–21. [CrossRef]
11. Gourdel, G.; Kociumaka, T.; Radoszewski, J.; Rytter, W.; Shur, A.M.; Walen, T. String periods in the order-preserving model. *Inf. Comput.* **2020**, *270*, 104463. [CrossRef]
12. Park, S.; Park, J.; Kim, Y.; Sim, J.S. Order-Preserving Multiple Pattern Matching in Parallel. *Appl. Sci.* **2023**, *13*, 5142. [CrossRef]
13. Chhabra, T.; Faro, S.; Külekci, M.O.; Tarhio, J. Engineering order-preserving pattern matching with SIMD parallelism. *Softw. Pract. Exp.* **2017**, *47*, 731–739. [CrossRef]
14. Crochemore, M.; Iliopoulos, C.S.; Kociumaka, T.; Kubica, M.; Langiu, A.; Pissis, S.P.; Radoszewski, J.; Rytter, W.; Waleń, T. Order-preserving indexing. *Theor. Comput. Sci.* **2016**, *638*, 122–135. [CrossRef]
15. Nakamura, T.; Inenaga, S.; Bannai, H.; Takeda, M. Order preserving pattern matching on trees and dags. In *International Symposium on String Processing and Information Retrieval*; Springer: Berlin/Heidelberg, Germany, 2017; pp. 271–277.
16. Cantone, D.; Faro, S.; Külekci, M.O. The order-preserving pattern matching problem in practice. *Discret. Appl. Math.* **2020**, *274*, 11–25. [CrossRef]
17. Gawrychowski, P.; Uznański, P. Order-preserving pattern matching with k mismatches. *Theor. Comput. Sci.* **2016**, *638*, 136–144. [CrossRef]
18. Russo, L.M.S.; Costa, D.M.; Henriques, R.; Bannai, H.; Francisco, A.P. Order-preserving pattern matching indeterminate strings. *Inf. Comput.* **2022**, *289*, 104924. [CrossRef]
19. Kim, Y.; Kang, M.; Na, J.C.; Sim, J.S. Order-preserving pattern matching with scaling. *Inf. Process. Lett.* **2023**, *180*, 106333. [CrossRef]
20. Gusfield, D. *Algorithms on Strings, Trees, and Sequences—Computer Science and Computational Biology*; Cambridge University Press: Cambridge, UK, 1997. [CrossRef]
21. Hebrail, G.; Berard, A. Individual Household Electric Power Consumption Data Set. 2012. Available online: https://archive.ics.uci.edu/ml/datasets/individual+household+electric+power+consumption (accessed on 26 September 2024).
22. Chen, S.X. Beijing PM2.5 Data Set. 2017. Available online: https://archive.ics.uci.edu/dataset/381/beijing+pm2+5+data (accessed on 26 September 2024).
23. Williamson, S. Daily Closing Values of the DJA in the United States, 1885 to Present, Measuring Worth. 2021. Available online: https://www.measuringworth.com/datasets/DJA/index.php (accessed on 26 September 2024).

mathematics

MDPI

Article

Get Spatial from Non-Spatial Information: Inferring Spatial Information from Textual Descriptions by Conceptual Spaces

Omid Reza Abbasi [1], Ali Asghar Alesheikh [1,*] and Seyed Vahid Razavi-Termeh [2]

[1] Department of Geospatial Information Systems, K. N. Toosi University of Technology, Tehran 19697, Iran; oabbasi@mail.kntu.ac.ir
[2] Department of Computer Science & Engineering and Convergence Engineering for Intelligent Drone, XR Research Center, Sejong University, Seoul 05006, Republic of Korea; razavi@sejong.ac.kr
* Correspondence: alesheikh@kntu.ac.ir

Abstract: With the rapid growth of social media, textual content is increasingly growing. Unstructured texts are a rich source of latent spatial information. Extracting such information is useful in query processing, geographical information retrieval (GIR), and recommender systems. In this paper, we propose a novel approach to infer spatial information from salient features of non-spatial nature in text corpora. We propose two methods, namely DCS and RCS, to represent place-based concepts. In addition, two measures, namely the Shannon entropy and the Moran's I, are proposed to calculate the degree of geo-indicativeness of terms in texts. The methodology is compared with a Latent Dirichlet Allocation (LDA) approach to estimate the accuracy improvement. We evaluated the methods on a dataset of rental property advertisements in Iran and a dataset of Persian Wikipedia articles. The results show that our proposed approach enhances the relative accuracy of predictions by about 10% in case of the renting advertisements and by 13% in case of the Wikipedia articles. The average distance error is about 13.3 km for the advertisements and 10.3 km for the Wikipedia articles, making the method suitable to infer the general region of the city in which a property is located. The proposed methodology is promising for inferring spatial knowledge from textual content that lacks spatial terms.

Keywords: textual content; geo-indicativeness; conceptual spaces; topic modeling

MSC: 68T50; 68T30

1. Introduction

Geospatial information services are typically based on collected quantitative data about objects and phenomena. They can provide answers, limited by the accuracy of the underlying data, to spatial queries such as 'where is ... ?' The answer is extracted from the stored data, which is basically coordinate-based. These services have been extremely helpful in a broad spectrum of disciplines ranging from agriculture [1], ecology [2,3], mining [4,5], and archeology [6] to urban planning [7,8], public health [9], and water resources [10]. The role of the rapid growth of spatial data acquisition technologies in the success of GIS is undeniable. Mappers all over the world are using equipment such as surveying tools, GPS receivers, and drones to provide accurate data for organizations and corporations. Clearly, this is a costly and time-consuming procedure. In addition, it requires a great number of people to be involved in the procedure.

Due to the developments of the internet and the high penetration rate of social networks in people's daily lives, a vast resource of information has been enabled. Of this, a noticeable share is enriched with spatial information. This information is mainly latent in textual content in the form of online books, user reviews, and blog posts [11]. As users utilize a natural language to communicate through the internet, this information appears in terms of qualitative descriptions of objects and events [12]. Although these descriptions rarely contain the spatial location of objects (e.g., in terms of postal addresses [13]), they may contain place names, spatial relations, and directions. Research on the extraction of spatial relations from textual content and natural languages has been growing rapidly in recent years [14–16]. Such studies aim at extracting spatial information in three forms that correspondingly answer three types of queries. Specifically, they intend to find topological states, proximity relations, and spatial directions. Topological relations describe the way objects are connected. It is very useful in geographic information retrieval (GIR) and qualitative reasoning (QR) to know systematically how places noted in a text are located in relation to each other [17,18]. These relations often appear in natural language as prepositions [19]. While they inform us about the way objects are contacting each other, they do not compare the situation where the objects are in one topological form. For instance, if two objects are topologically disjoint, the topological models (e.g., RCC-8 [20]) do not provide information about how much the two objects are disjoint. Proximity queries aim to find objects near a reference object represented by a place name in text. As the exact locations of objects in textual corpora are not known, the interpretation of how much of a distance is considered as near is rather subjective [21]. Directions, whether cardinal or relative, and orientations are important spatial relations usually observable in textual content. Again, they may be subjectively interpreted and are rarely implemented in GIS [15].

While the focus of the existing literature on the subject has been on the extraction of the mentioned spatial information from spatial descriptors within texts, we argue that some spatial information can be inferred from non-spatial terms. As only a few sentences in textual resources contain spatial descriptors [22], the extraction of spatial information from non-spatial terms would be of high interest in applications that need to locate textual resources. Our approach is especially useful in applications where resources, rather than toponyms within resources, are to be located. To give an example, consider the description '*The house is furnished, has a roof garden, and a good view of the mountains is available*' for a house in the city of Tehran that has been advertised on the web. In this description, there is no sign of explicit spatial descriptors. However, when seeing the description, a resident of the city may learn that the house is located in a northern neighborhood. The resident knows that some aspects of housing in northern Tehran are different from those in other neighborhoods. In addition, there may be some terms in the description that do describe the house but imply the region in which it is located. In the example above, a roof garden is a feature of houses mainly located in northern Tehran. Also, the houses located in northern Tehran often have a clear view of the mountains.

Stock, et al. [19] categorized expressions within texts into three classes: geospatial, other-spatial, and non-spatial terms. The geospatial class includes those expressions that contain a geographical reference and a spatial relation term, hence referring to an absolute geographic coordinate. The other-spatial category refers to expressions that contain spatial relations but lack a geographic reference, leading to relative coordinates. Non-spatial expressions are those terms that do not fall within the previous categories. In this paper, we further extend Stock's non-spatial class into two distinct classes. In fact, we recognize another class of expressions that, while lacking spatial relations, contain terms implying geospatial information. We borrow the term geo-indicative from Adams and Janowicz [23] to refer to this class. The difference between the two classes is neither trivial nor explicit, as they both do not contain geospatial or other-spatial terms. However, geo-indicative expressions contain words that describe some characteristics belonging to a certain geographical region. This is in contrast with the non-spatial expressions, where terms describe features that do not belong to a specific region in the study area. Therefore, some terms that are identified as geo-indicative in some study areas may be non-spatial in other regions. While spatial relations appear in texts often as prepositional phrases, geo-indicative terms are latent in texts as nouns and adjectives. Table 1 shows real-world examples for each class of expressions.

Table 1. Different classes of expressions, their definitions, and real-world examples from our dataset.

Class of Expression	Definition	Example [1]
Geospatial expressions	Contain direct references to geographical features, such as place names	The house is located on the south side of **the Valiasr metro station**.
Other-spatial expressions	Contain terms that refer to the relative position of features, such as topological relations	A supermarket is available **across** the street.
Geo-indicative expressions	Contain terms that do not directly refer to geographical features but describe the characteristics of a certain geographical region	The house is furnished, has **a roof garden**, and has a good view of **the mountains**.
Non-spatial expressions	Contain terms that neither refer to nor describe any specific geographical region	The house has two rooms. The building has been recently renovated, and the walls are wallpapered.

[1] Words that can be used to infer spatial information from are bold.

In this paper, we aim to extract spatial information from the non-spatial terms that imply a space (i.e., geo-indicative terms) in large corpora with the help of the theory of conceptual spaces. The proposed methods prove more versatile than other approaches reliant on geospatial terms in texts, considering that few texts contain such terms. The method finds broad applicability in cases where a system is to be designed to georeference texts without explicit geospatial terms and conduct spatial analysis or recommend textual items to users. For instance, in the Divar application, from which we extracted one of our datasets, rental property advertisers have the option to provide exact location information by placing a pin on a map. However, if a user chooses not to specify the exact location of the property, the neighborhood in which the property is situated must be manually determined via a search box. In this real-world scenario, our proposed method could offer two valuable services: Firstly, the platform can utilize our approach to automatically predict and suggest the neighborhood name for the advertiser in advance, streamlining the process. Secondly, the platform's recommendation engine presently suggests items in other areas based on the adjacency of the neighborhoods to the current neighborhood. With our proposed method, the recommendation system gains greater flexibility since our estimation approach does not rely on the physical topology of neighborhoods. As a result, the recommendations can be more relevant and encompass a wider range of options, enhancing user experience and satisfaction. The contribution of our study is three-fold:

1. We propose two novel methods, specifically the regional conceptual space (RCS) and an adapted version of directional conceptual space (DCS) from [24]. These methods are founded on the conceptual spaces theory and are designed to represent terms in textual content within a high-dimensional space.
2. We suggest utilizing Moran's I as a measure for assessing the geo-indicativeness of terms in textual content. As far as we are aware, Moran's I has not been employed previously for the specific purpose of gauging the geo-indicativeness of terms in text.
3. We contrast the methods we propose with a previously suggested approach relying on Latent Dirichlet Allocation (LDA).

In Section 2, we review some studies conducted on the topic of spatial information extraction from textual content. In Section 3, the theory of conceptual spaces is introduced, and our adaptation of the theory is explained. In Section 4, the methodology of the work is presented. Then, we provide the results and discuss them in Section 5. Finally, we conclude the paper and highlight some suggestions for future research.

2. Related Work

In this section, we provide a review of research conducted on extracting spatial information from textual content. In this area of research, most studies have focused on identifying spatial relations, including topological relations, in texts. In other words, a significant share of such studies have chosen geospatial and other-spatial expressions to extract spatial information from corpora. Since the focus of our study is on the extraction of spatial information from geo-indicative expressions, we limit our review to this topic.

Most studies in the area of the extraction of spatial information from text have utilized both geospatial and place names to estimate the location of resources. In almost all of these efforts, no distinction is made between place names and geo-indicative expressions, and both types are included in the computations. For instance, Han, et al. [25] focused on the feature selection procedure to enhance the accuracy of georeferencing. Their idea was that many words in texts are not useful (i.e., they are non-spatial). Hence, they proposed three feature selection methods to distinguish between terms that highly imply spatial information and non-spatial terms. They developed term frequency-inverse corpus frequency (TF-ICF), analogous to TF-IDF, along with information gain ratio (IGR) and maximum entropy (ME) as criteria to select features. The methods were applied on large geographical scales. A grid of $0.5'$ in $0.5'$ divided the Earth, and the tweets far from cities were eliminated. They reported a 10% improvement in the accuracy of georeferencing using only geospatial and geo-indicative words in comparison with using full features. They also noted that place names played a noticeable role in the improvement.

Place name recognition is vital to the georeferencing and enrichment of geospatial gazetteers. However, it is a time-consuming task and needs massive labeled datasets [26]. In most cases, there is even no place name within the text. It is especially true for short texts such as messages on social media and advertisements across the web. A third group of researchers focused only on the geo-indicative terms. For instance, Hollenstein and Purves [27] explored the terms used by users to describe city centers. They used geo-tagged Flickr photos to examine how different city centers across the globe are described. Then, the terms were used to derive the boundaries of the city centers. Wang, et al. [28] employed content analysis to explore the perception of green spaces in Beijing, China. They developed a structured lexicon based on which landscape features were extracted and fed into machine learning techniques. Their approach relies on prior knowledge about the phenomenon under study. Chang, et al. [29] proposed a probabilistic framework to estimate the city-level location of Twitter users based on the content of their tweets. In their approach, a classification component is used to automatically identify words with a high probability of referring to a specific city (the word *rockets* probably refers to *Houston*). Their results showed that 51% of Twitter users can be placed within 100 miles of their actual location. McKenzie and Janowicz [30] explored the regional variation between Foursquare's POI types. They applied a statistical hypothesis test to the dataset and found that there is regional variation between POI types that is not the result of random fluctuations. Adams and Janowicz [23] chose georeferenced Wikipedia articles and travel blog posts to train a Latent Dirichlet Allocation (LDA) model. They estimated a probability surface over the Earth for each topic. A Kernel Density Estimation (KDE) analysis was used to represent the probability of correspondence of each topic to the regions. They applied the method to places in the United States and achieved an accuracy of about 500 km for 75% of the test Wikipedia articles. In [31], the same authors extended their work and viewed the subject from a *semantic signature* perspective. Analogous to spectral signatures, semantic signatures can be used to illustrate various thematic features of the space. They showed that their proposed approach automatically categorizes documents into place types in existing ontologies. The focus of the mentioned study has been on the enrichment of knowledge graphs and inferring the relationships between place types in ontologies. While the topic of their study is closely related to ours, we proposed the use of conceptual spaces, rather than LDA, to build a similarity space of unstructured text. Therefore, the proposed method is more suitable in applications such as recommender systems, where similarity is at the core of computations.

The aim of the latter studies is similar to ours. They use only geo-indicative terms to infer the location of textual resources. In this article, we seek to extract geo-indicative expressions. It should be noted that geo-indicative terms refer to a specific region of the study area. Hence, the spatial distribution of documents containing geo-indicative terms would be clustered. Therefore, geo-indicative terms inherently play the role of place names in text. Our study employs the theory of conceptual spaces, which, as a cognitive knowledge representation framework, is more suitable to the task of analyzing semantic information in textual corpora. In addition, as we construct a similarity space in which more semantically similar documents are placed closer together, our methodology becomes more interpretable and the properties of the space can be discussed. Another feature of our research is the size of the study area. In this study, we estimate the location of textual resources within the city of Tehran, the capital of Iran. According to the definition of geo-indicativeness, the different parts of the underlying study area must have distinguishing properties so that the language can describe them and discriminate from other regions. The previous studies mainly applied their method to large-scale areas such as the world and countries. Obviously, different regions of a country have specific properties, such as the accent, culture, geography, and climate, among many others, and can be described by diverse and distinguishing terms. We applied our method to the level of a city, which makes the prediction task more challenging. Unlike some other studies which seek to find the relationships between the location of a text and the user, our proposed method aims to infer the location of general textual resources, irrespective of the location of text's creator. Our approach does not inherently violate user privacy. Of course, if the method is to be applied in scenarios where privacy concerns matter, the users' consent is required. In the forthcoming sections, we present and scientifically compare our results with those of the above studies.

3. Materials and Methods

Figure 1 demonstrates the workflow of our study. The details of the methods shown in the figure are described in this section. First, the LDA method, as the reference method with which we compare our results, is explained. In order to provide the readers an insight into how the words and topics are represented over the study area, Kernel Density Estimation (KDE) is introduced. Then, the theory of conceptual spaces, as proposed by Gärdenfors [32] and formalized by Adams and Raubal [33], is briefly highlighted. After introducing the original theory, the modified versions that we utilized in the study is introduced. We propose two data-driven adaptations of conceptual spaces. The results of both methods are evaluated and compared against the LDA and a random estimator.

3.1. Semantic Signatures

Adams and Janowicz [23] proposed semantic signatures to address the problem of discovering latent spatial information in natural language. The method is based on Latent Dirichlet Allocation (LDA), an unsupervised probabilistic method over discrete data such as text corpora [34,35]. LDA is commonly used in the tasks of topic modeling [36] and recommendation [37]. It assumes that documents are probability distributions of topics, and each topic is a probability distribution of words within documents. Formally, it is assumed that a text corpus is comprised of k topics z_n and contains $|V|$ words. For a document D containing N words, the probability of assigning each topic to document D is computed as [35]:

$$p(\mathbf{w}|\alpha,\beta) = \int_\theta p(\theta|\alpha)\left(\prod_{n=1}^{N}\sum_{z_n=1}^{k} p(w_n|z_n,\beta)p(z_n|\theta)\right)d\theta \quad (1)$$

where $\mathbf{w}$ is the set of words in document D, and θ is a k-dimensional random variable sampled from a Dirichlet distribution, and α and β are the parameters of the model.

In order to represent the probabilities provided by LDA as surfaces over the study area, Adams and Janowicz [23] suggested using Kernel Density Estimation (KDE). Suppose

$\{X_1, \ldots, X_n\} \in R^d$ be an independent, identically distributed (iid) random sample from an unknown probability distribution P. KDE aims at estimating P by [38]:

$$\hat{p}(x) = \frac{1}{nh^d} \sum_{i=1}^{n} K\left(\frac{x - X_i}{h}\right) \tag{2}$$

where K is a kernel function and $h \in \mathbb{R}^+$ is called bandwidth. KDE assigns a probability value to each sample. Then, the values are summed to yield a density estimator. The form of the distribution depends on the kernel function used in the equation. Two commonly employed kernels are Gaussian and spherical kernels. A Gaussian kernel is defined as [39]:

$$K(x) = \frac{exp\left(-\frac{\|x\|^2}{2}\right)}{\int exp\left(-\frac{\|x\|^2}{2}\right)dx} \tag{3}$$

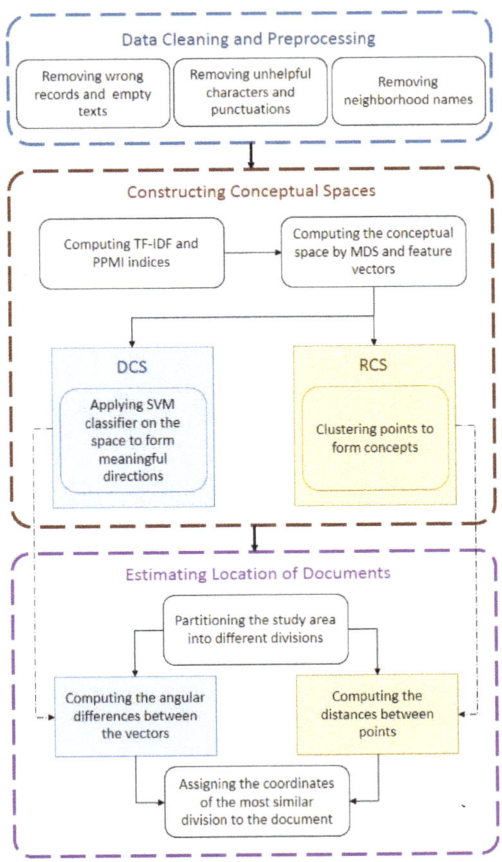

Figure 1. The workflow of the proposed methodology.

3.2. Conceptual Spaces

The theory of conceptual spaces [40] is a knowledge representation method [32] that uses geometrical structures to represent information, allowing for the definition of similarity relations between concepts [41]. This approach is distinct from associationist and symbolic levels of knowledge representation. Associationist models focus on connections among the many parts of a system, while symbolic models manipulate symbols using rules written in First-Order Logic (FOL). Symbolic models typically process information sequentially, one symbol or rule at a time. This is due to the fact that symbolic representations are discrete and well-defined, and each symbol or rule can be processed independently of the others [42]. In contrast, associationist models often process information in parallel, with multiple connections between concepts being activated simultaneously [43]. This is due to the fact that associationist representations are distributed and overlapping, and the activation of one concept can lead to the activation of many others.

The theory of conceptual spaces tries to represent information based on geometrical structures. A significant advantage of this kind of representation is the ability to define similarity relations on information, and to relate the concepts topologically [44]. Moreover, inductive reasoning can be studied more conveniently at the conceptual level [45]. The geometrical forms of concepts can be used in both the explanatory tasks (for example, see Poth [46]), in which a cognitive theory is formulated and verified by means of experiments and observations, and the constructive tasks (for example, see Pol, et al. [47]), in which an artificial agent or system is designed to solve a specific problem. In this article, the focus is on the latter, as we aim to utilize the conceptual spaces to solve the task of inferring spatial information rather than to fathom the underlying complexities of human thinking regarding places.

A conceptual space is a similarity space spanned by a set D of quality dimensions. Each dimension $d \in D$ is an aspect by which an object can be identified. Some quality dimensions that are related to each other are integrated into domains. While the initial works on Gärdenfors' conceptual spaces mainly included explicitly defined quality dimensions, there has been some recent effort to infer the conceptual space in a data-driven approach [48]. In this paper, we construct the conceptual space using a data-driven manner. In what follows, we introduce the elements used in the theory of conceptual spaces. While in the literature there are other elements such as fuzzy concepts [49] or hierarchical conceptual spaces [50], we define only the core elements needed by a conceptual space for the sake of simplicity.

Definition 1. *A concept is defined as* $\mathcal{C} = \langle \mathcal{R}, \mathcal{P} \rangle$, where $\mathcal{R}$ *denotes a set of convex regions in the space.* $\mathcal{P}$ *represents a prototype of the concept in the space. The prototype instance of the concept is assumed to be a point in the space, theoretically located in the center of the convex region representing the concept.*

Definition 2. *A property is defined as the value of a concept in only one quality dimension. Properties can be understood as seeing a concept from only one aspect.*

Definition 3. *An instance* $\imath$ *is defined as a point within a conceptual space that represents an object in the real world. An instance is produced when a value is determined for all quality dimensions in the space.*

In this article, we construct the conceptual space in a data-driven manner and propose a modification in the representation of concepts.

3.3. Data-Driven Conceptual Spaces

As stated in the previous section, the theory of conceptual spaces represents concepts as convex regions in a metric space. In most cases, since there is no prior knowledge about the underlying phenomenon, the quality dimensions cannot be identified explicitly. Therefore, a data-driven approach is essential to building the conceptual space from the bottom up. In this article, we learn the conceptual space of our corpus using some techniques used in machine learning and Natural Language Processing (NLP).

Suppose we are going to represent a corpus C containing documents $\{D_1, \ldots, D_N\}$. Each document is first tokenized into a Bag of Words (BOW) containing all of the words in the documents. Similar to any NLP task, and as some tasks need only specific parts of speech (POS), the BOWs are preprocessed, and the tokens that are not useful for the specific task are removed. Then, each token in each BOW is assigned an index to provide the salience of the words in the documents. TF-IDF is a common index to this aim. Given a set D of documents, TF-IDF scores the word w in the document d as [51]:

$$TF\_IDF(w, D, C) = TF(w, D) \times IDF(w, C) \tag{4}$$

where

$$TF(w, D) = \frac{c(w, D)}{|D|} \tag{5}$$

and

$$IDF(w, C) = log\frac{|C|}{|\{D \in C | w \in D\}|} \tag{6}$$

where $c(w, D)$ is the number of times a term w occurs in the document D. Also, $|D|$ and $|C|$ denote the number of words in the document D and the number of documents in the corpus C, respectively.

Positive point-wise mutual information (PPMI) is similar to TF-IDF in that it also scores the words in a document. Some researchers have shown that it works better than other scoring methods for the purpose of semantic similarity [46]. The index is calculated as:

$$PPMI(w, D) = max\left(0, log\left(\frac{p_{w,D}}{p_{w,*} \times p_{*,D}}\right)\right) \tag{7}$$

where

$$p_{w,D} = \frac{c(w, D)}{\sum_{w'} \sum_{D'} c(w', D')} \tag{8}$$

and

$$p_{w,*} = \sum_{D'} p_{w,D'} \tag{9}$$

and

$$p_{*,D} = \sum_{w'} p_{w',D} \tag{10}$$

By applying the scoring index to each document D, a vector v_D is formed. This vector contains the index values of all words in the corpus. Since the index vectors will be very sparse, they are not suitable to be directly used in constructing the conceptual space. Hence, a Multi-Dimensional Scaling (MDS) technique [52] is applied to the vectors to transform them into a similarity space of documents. While MDS itself is not specific to conceptual spaces theory, it has been denoted by Gärdenfors [40] and others [53] as a method of constructing the conceptual spaces and identifying the quality dimensions. Given the pairwise similarities of a set of objects, MDS estimates their coordinates as points in a similarity space [52]. It has been widely used in cognitive sciences [53] and also in the theory of conceptual spaces [54].

TF-IDF and PPMI offer a clear and easily understandable representation of words based on their significance within a specific document. The assigned weights by TF-IDF and PPMI indicate how relevant a word is to a document relative to a larger corpus, which is particularly useful when analyzing the geo-indicativeness of words in our spatial inference approach. In contrast, more recent methods such as doc2vec employ neural networks for model training, which reduces their interpretability. It is essential to note that one advantage of conceptual spaces over connectionist methods such as neural networks is their interpretability. In addition, TF-IDF and PPMI are relatively simple and computationally efficient methods for representing textual data. In contrast, word embedding methods often involve more complex models and require extensive training on large corpora.

In order to represent concepts as regions in the resulting space, spatial clustering would classify similar documents, represented by points, into single classes. Finally, the convex hull of the points yields the desired regions. In this paper, the conceptual spaces provided by the above procedure are called *regional conceptual spaces* (*RCS*).

Although the original theory of conceptual space revolves around the prototype theory [41], in the case of inferring the conceptual space from data, it might be rare to have a prototypical instance that represents the ideal case of the category in the middle of each class. Therefore, we propose a modification to the way classes are represented. To this end, we partition the resulting similarity space for each word using a Support Vector Machine (SVM) classifier. This leads to two classes of points: a class representing documents that do not contain a given word (negative class), and another class representing documents that contain that word (positive class). Then, we calculate the equation of the hyper-plane classifier. The direction perpendicular to the hyper-plane demonstrates the direction in which the word plays a prominent role. The farther a specific document is placed from the hyper-plane in the direction of positive documents, the more salience the word has in the document. As this classification does not potentially yield perfect accuracy, we consider only those words for which the classification has promising accuracy. This is to ascertain that the directions found have strong meanings. However, even after constructing the hyper-plane only for those words with high accuracy of classification, there may remain a high number of directions depending on the number of words considered. In addition, some vectors may have similar meanings. As we are handling a similarity space, the vectors with similar meanings are pointing towards a close direction. The directions can be clustered to achieve more robust and meaningful directions within the constructed space. We call the result of this procedure *directional conceptual spaces* (*DCS*). Figure 2 compares the two approaches to constructing the data-driven conceptual space mentioned above. Abbasi and Alesheikh [24] presented a method similar to DCS with a focus on utilizing it in recommendation systems. They measure the similarity of items of a recommender system by calculating the difference between the lengths of vectors projected on meaningful directions. The DCS method used in the present study uses the angular differences between the vectors to calculate their closeness. They evaluated the results by real users of a recommender system. The main difference between the present study and [24] is that we propose both DCS and RCS and compare their performance against LDA a random estimator, without focusing on recommendation algorithms. In addition, we partitioned the study area into equally-sized cells to compensate for the problem of the density of points in KDE, which is discussed in the 'Results' section. Furthermore, we suggest utilizing Moran's I, along with the Shannon entropy, as a measure for assessing the geo-indicativeness of terms in textual content.

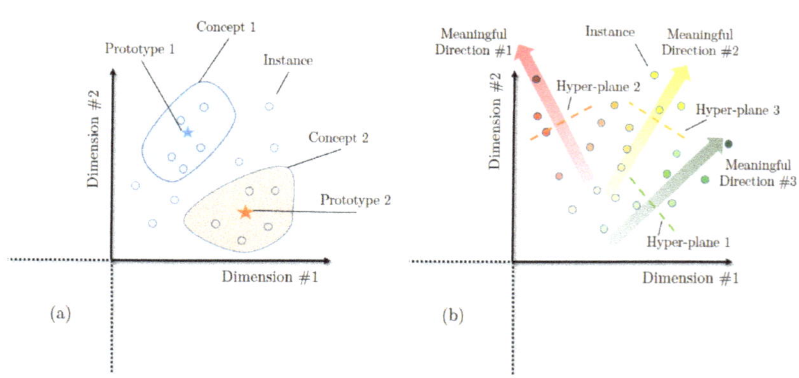

Figure 2. A schematic comparison of the two proposed versions of data-driven conceptual spaces (modified from [24]). (**a**) Regional Conceptual Spaces represent concepts as convex regions in a high-dimensional space; (**b**) Directional Conceptual Spaces represent concepts as directions in a high-dimensional space.

After constructing the conceptual space, the documents need to be located in the geographical space. To this aim, the study area is partitioned into different divisions. Then, for a new document, the most appropriate division is determined based on the similarity of their associated concepts or their closeness to meaningful directions. That is, in the case of RCS, the division with which the document shares the most common regions in the conceptual space is selected as the estimated division, and its geographical centre is assigned to the document. In the case of DCS, the division with which the document has the least angular difference is selected as the estimated division.

4. Datasets

In this paper, we used a dataset of advertisements for rental properties and a dataset of Persian Wikipedia articles, both related to the city of Tehran, Iran. To collect the first dataset, we crawled the Divar website (https://divar.ir, accessed on 21 March 2023), a popular platform used in Iran to advertise properties for sale or rent. The dataset contains about 200,000 records. After removing records that do not contain the coordinates of the property or lack a textual description, 11,393 records of advertisements remained. Each record includes a title and a description of the property in Persian, pricing information, the construction year, the area of the property, the name of the neighborhood where it is located, and locational data including latitude and longitude. In our study, we only processed the descriptions related to each advertisement. In addition, all place names were removed from the descriptions. In order to process the descriptions, we utilized the Stanza library [55], a Python natural language processing package that is equipped with pre-trained models for Persian language. The second dataset was extracted from the Wikimedia Query Service by writing queries in SPARQL. It contains 704 Persian Wikipedia articles that are geolocated in Tehran, Iran. The articles are full-length and cover various topics such as cultural sites, monuments, parks and gardens, and events. In our study, we use 80% of the records for the training purpose, and employ the remaining records for testing. In Table 2, the datasets used in the study are listed, and an example of each dataset is provided.

Table 2. The description of the datasets used in the study.

Dataset	No. of Records	Example [1]
Divar	11,393	"220 m, 3 bedrooms, very well designed, large and spacious living room, fully furnished kitchen, southern exposure, very large meeting hall, the most cozy region, very stylish and luxurious and furnished lobby, the roof garden is very well equipped and wooded, lobby, and resident janitor; it has two parking lots, private pool, sauna, and Jacuzzi."
Persian Wikipedia	704	"The National Museum of Iran is located in Tehran, Iran. It is an institution formed of two complexes; the Museum of Ancient Iran and the Museum of Islamic Archaeology and Art of Iran, which were opened in 1937 and 1972, respectively. The institution hosts historical monuments dating back through preserved ancient and medieval Iranian antiquities, including pottery vessels, metal objects, textile remains, and some rare books and coins. It also includes a number of research departments, categorized by different historical periods and archaeological topics."

[1] The examples are translated into English.

5. Results and Discussion

In this section, we present the results of applying the methods described in Section 3. Specifically, we apply our proposed approaches using conceptual spaces and compare the results with those of the LDA method.

In order to preprocess the textual descriptions, all of those properties for which the textual description is less than 10 words were removed. Considering that the length of feature vectors is equal to the number of all words in the corpus, this would help in compensating for the issue of feature vectors' sparsity. First, the descriptions were segmented into sentences, and then sentences were tokenized into words. To reduce the volume of data in processing, in this paper we only considered nouns and adjectives in sentences. On the other hand, by emphasizing nouns and adjectives, we aim to prioritize words that provide more explicit geo-indicative and descriptive context. We emphasize that the choice of parts of speech is dependent on the use case. In our study, verbs do not seem to be geo-indicative for rental property advertisements and Wikipedia articles about Tehran. However, the use of verbs might be better suited in scenarios where identifying functional regions is important. In addition, we considered only those words that were pointed out in the whole corpus at least 50 times. This choice is based on a heuristic threshold to filter out less frequent and potentially noise-inducing words and to focus on the most informative and discriminative words, capturing the key features and characteristics of the documents. Since data-driven conceptual spaces are very prone to the quality of the data, if the data is noisy or incomplete, then the conceptual space may be inaccurate or misleading. This ensures the strength of the meaning found in future steps and enhances the accuracy of classification. The same pre-processing procedure was used for both LDA and the proposed methods to ensure a fair comparison between them.

We selected the number of LDA topics k through qualitative evaluation of topic interpretability on each dataset. The appropriate k value depends on the semantic domain of the corpus. Since the rental advertisements cover a narrower domain than Wikipedia, their content reflects fewer meaningful topics. After testing different k values, we chose $k = 6$ for the rental advertisements and $k = 10$ for the Wikipedia articles by judging which values produced the most coherent, interpretable topics. In Table 3, example topics from each dataset, both geo-indicative and non-geoindicative, are shown. After training the LDA model, we construct a probability surface using IDW to test whether a routine interpolation technique can suitably represent the probabilities. Figure 3 illustrates the geographical distribution of advertisements belonging to Topic #D1 in Table 3, which contains the words *pool, sauna, jacuzzi, conference_hall,* and *luxurious*. The surfaces are drawn based on the probabilities yielded by the LDA topic modeling.

Table 3. Some topics identified in each dataset along with their entropy values.

Dataset	ID	Topic [1]	Entropy	Moran's I
Divar	#D1	*(pool, sauna, Jacuzzi, conference_hall, luxurious)*	4.52	0.344
	#D2	*(city_center, bazaar, store, accessibility, commercial)*	5.67	0.472
	#D3	*(loan, under_construction, cooperation, phase, vacant)*	10.60	0.022
	#D4	*(owner, room, alley, door, telephone)*	33.02	−0.471
Persian Wikipedia	#W1	*(museum, register, national, the_King, traditional)*	7.51	0.453
	#W2	*(university, faculty, science, institute, research_center)*	10.21	0.320
	#W3	*(station, line, metro, neighborhood, square)*	19.55	0.154
	#W4	*(neighborhood, district, city, municipality, area)*	29.02	−0.745

[1] The words are translated into English.

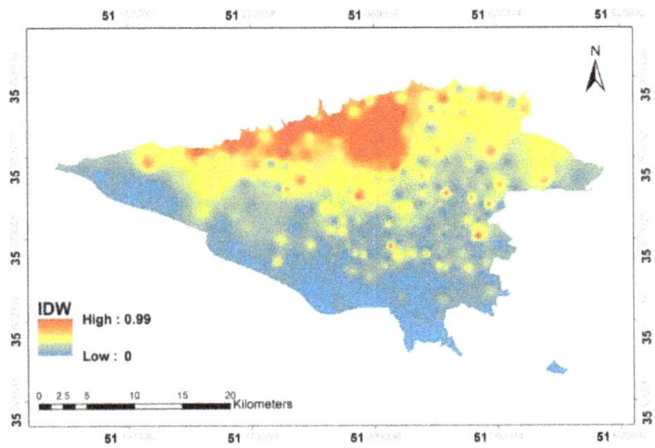

Figure 3. The probability surface calculated by IDW for Topic #D1 (refer to Table 3) found in the corpus.

As seen in the figure, the surface shows a probability distribution for this topic, which is strongly correlated with the northern areas of the city. However, the surface highly respects single points, and the map is too rough. We calculated KDE as another approach to representing the probability surface. As discussed in Section 3.1, the location of documents is involved in the calculations of KDE. Therefore, those areas with more advertisements would have higher probabilities. In order to compensate for this issue, following Adams and Janowicz [23], we partitioned the study area into different divisions. A grid comprising cells of size 1 km in 1 km each divided the city. Then, the averages of probabilities for each topic were calculated and assigned to the centroid of the cells. Figure 4 manifests the distribution of the same topic (Topic #D1) represented by KDE.

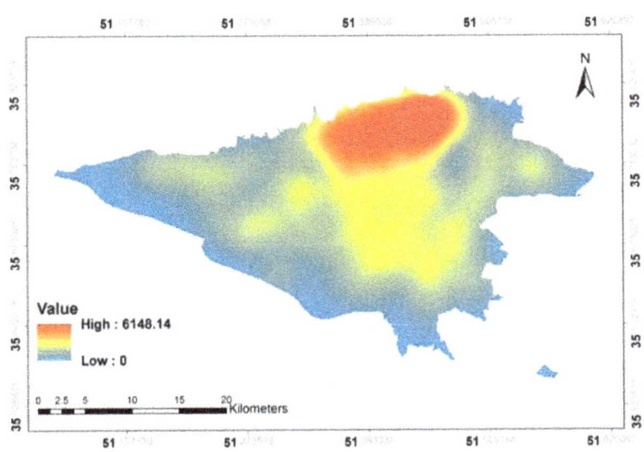

Figure 4. The probability surface calculated by KDE for Topic #D1.

The resulting surface provided by KDE is much smoother than that provided by IDW. This is due to the fact that, according to Equation (2), KDE averages the probability distributions over sample points, compensating the role of single points with high values. However, in IDW, single points can have significant effect on the resulting surface. Following Abbasi and Alesheikh [24], we further analyzed the topics found to measure the degree of geo-indicativeness of each topic. From one point of view, a topic is geo-indicative if it is associated with some neighborhoods while having low association probabilities with other neighborhoods. As a result, a geo-indicative topic would not be uniformly dispersed through the study area. To measure the dispersion, we compute the Shannon entropy of the probabilities by [56]:

$$H(p) = -\sum_{1}^{N} p_i log(p_i) \tag{11}$$

Shannon entropy calculates how much the outcome of a probability distribution would be *surprising* [57]. The lower the entropy, the lower the amount of surprise. This is interestingly in accordance with the concept of geo-indicativeness. The terms of a geo-indicative topic have to, *unsurprisingly*, describe the region to which they belong. Table 3 lists some topics found and their associated entropies.

In addition to Shannon entropy, we propose the Moran's I index to measure the geo-indicativeness of topics. The Moran's I is a measure of spatial autocorrelation of data. It works based on both feature location and attributes. That is, it measures how much near features have similar attributes [58]. It is calculated as:

$$I = \frac{n}{S_0} \frac{\sum_{i=1}^{n} \sum_{j=1}^{n} w_{i,j} z_i z_j}{\sum_{i=1}^{n} z_i^2} \tag{12}$$

where z_i is the deviation of an attribute for feature i from its mean, and $w_{i,j}$ is the spatial weight between feature i and j. S_0 is the aggregate of all of the spatial weights, and is calculated as:

$$S_0 = \sum_{i=1}^{n} \sum_{j=1}^{n} w_{i,j} \tag{13}$$

For a detailed explanation of computing Moran's I, the reader should refer to [59]. The value of Moran's I fluctuates between -1 and 1, where a value of -1 indicates a strong dispersion of similar attributes, a value of 0 indicates a random spatial distribution of similar attributes, and a value of 1 indicates a strong clustering of similar attributes. We expect that geo-indicative topics should have a positive Moran's I. If the value of Moran's I is close to one, it is highly probable that the topic is geo-indicative, as it means the topic is spatially clustered. Conversely, if it is negative, the topic is not geo-indicative. If the Moran's I is close to zero, then the topic is probably not geo-indicative. We suggest that the two criteria, i.e., the Shannon entropy and the Moran's I, be considered simultaneously. This is especially important in the case that the Moran's I shows a borderline value for a topic.

The first three topics in the rental property advertisements listed in Table 3 show high degrees of geo-indicativeness (low entropy). Topic #D1 is the one illustrated in Figures 3 and 4, which refers to properties mostly located in northern neighborhoods of the city. Topic #D2 is vividly related to the central neighborhoods that are the commercial heart of the city and are known for their ease of accessibility. Topic #D3 refers to those properties that are under construction and are mostly located in the vicinity of the city. While the above topics clearly represent specific regions of the city, we could not interpret the rest of the topics as geo-indicative. For instance, Topic #D4 contains general features that can be seen in any neighborhood in the study area. This is in contrast with Topics #D1 and #D2 where their related words are associated with northern and central neighborhoods, respectively. In the case of the Wikipedia articles, the entropy values of topics are generally higher than those of the advertisements dataset. This is due to the fact that the themes of the Wikipedia articles are diverse, and many of them describe features that are not associated with their underlying space. For example, Topic #W2 refers to the articles about universities and research institutes. The universities and research institutions in Tehran are not located in a specific region. The entropy of Topic #W4 is more interesting. The topic contains articles that introduce neighborhoods of the city. Although the documents in this topic inherently describe their underlying space, they are, expectedly, distributed over the study area, leading to high entropy and negative Moran's I. Topic #W1 is related to museums and historical sites, such as palaces, which in turn relate mostly to central neighborhoods and some of the northern neighborhoods. Figure 5 compares a geo-indicative topic (Topic #D2) with a non-geo-indicative topic (Topic #D4).

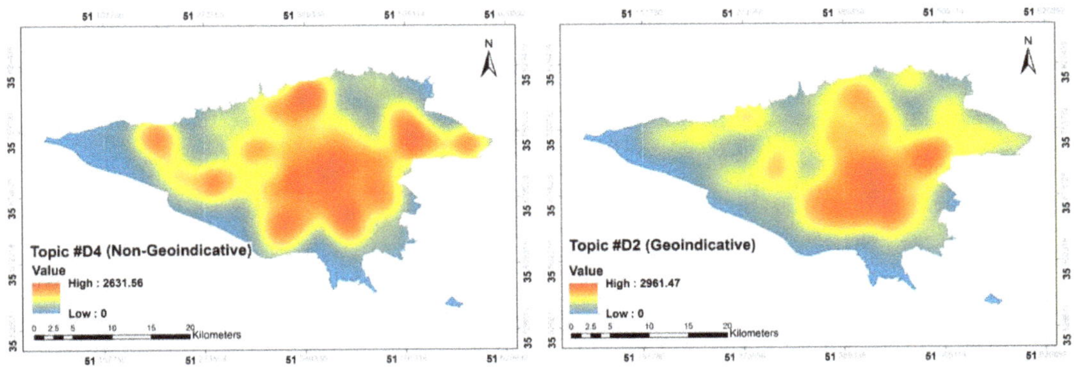

Figure 5. A comparison of the distribution of a geo-indicative (Topic #D2) versus a non-geo-indicative (Topic #D4) topic.

The right panel of Figure 5 verifies the centrality, explained above, associated with Topic #D2. However, for the left panel, the hotspots are almost evenly distributed over the study area, and a specific region cannot be easily adopted for the topic.

We also applied the proposed data-driven conceptual spaces methods (Section 3.3) to the same datasets. For this purpose, we considered only nouns and adjectives to reduce the volume of BOWs and enhance the speed of computations. In addition, all unnecessary tokens, such as pronouns and stop words, were removed from the BOWs. Both the TF-IDF and PPMI indices were calculated to score the salience of each word in the documents. For the purpose of implementing MDS, clustering the points in RCS, and applying SVM to partition the space in DCS, the scikit-learn package was employed. An issue with the DCS approach is that considering all words in the corpus and applying SVM to each is very time-consuming and computationally intensive. In addition, many words do not appear to have signs of being geo-indicative. Therefore, after computing TF-IDF and PPMI, we proceeded with only the top 10 percent of words in terms of these indices. Then, a dissimilarity matrix, as the input for MDS, was computed for the documents. Since we have no information about the dimensions of the resulting conceptual space, we computed spaces with 5, 10, and 20 dimensions. An SVM classifier with a linear kernel was trained to construct the DCS. The accuracy score of the classification was estimated, and those higher than 80 percent were considered as strong directions in the corpus. Table 4 shows the accuracy score of classification for 10 words also found by LDA and listed in Table 3 as Topic #D1 and Topic #W1.

In most cases, the PPMI score outperforms the TF-IDF, which suggests that it is a better scoring scheme for constructing directional conceptual spaces. Nevertheless, the classification for broad terms such as *luxurious* (Persian: شیک) yields the worst results, regardless of the scoring index. The importance of the number of dimensions should not be overlooked. The pattern of convergence can be seen in the accuracies in most cases, and higher dimensions generally provide better classification accuracy. We applied a vector clustering algorithm to group the five directions of the words. Analogous to the prototypes in RCS, the farthest document from the classifier hyper-plane is identified as the most prominent feature for the word under consideration. In order to find the similarities among documents, we first projected the vectors onto the found direction and calculated the distance between those documents and the others in the found direction. Then, we sorted out the documents based on their normalized similarities. Figure 6 illustrates the location of the most prominent document for the words *Jacuzzi* and *museum*, and their corresponding similarity surface.

Table 4. The accuracy score of SVM classification using TF-IDF and PPMI in 5, 10, and 20-dimensional conceptual spaces.

Dataset	Word	TF-IDF			PPMI		
		5D	10D	20D	5D	10D	20D
Divar	*pool* (استخر)	0.36	0.57	0.61	0.52	0.62	0.76
	sauna (سونا)	0.65	0.68	0.69	0.70	0.73	0.76
	Jacuzzi (جکوزی)	0.44	0.52	0.53	0.50	0.55	0.56
	conference_hall (سالن اجتماعات)	0.52	0.54	0.55	0.47	0.50	0.52
	luxurious (شیک)	0.24	0.41	0.54	0.27	0.41	0.55
Persian Wikipedia	*museum* (موزه)	0.65	0.71	0.75	0.65	0.72	0.75
	national (ملّی)	0.43	0.46	0.50	0.40	0.45	0.49
	register (ثبت)	0.49	0.52	0.54	0.51	0.55	0.58
	Shah (شاه)	0.53	0.56	0.56	0.55	0.56	0.56
	traditional (سنتی)	0.39	0.42	0.44	0.41	0.44	0.46

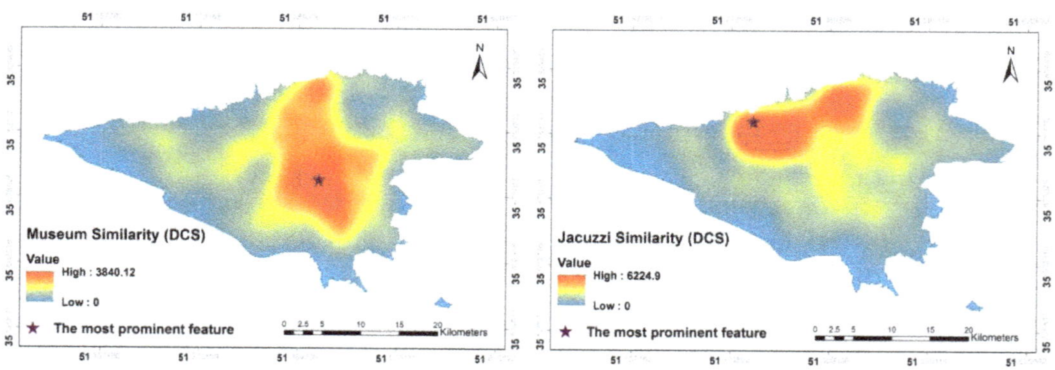

Figure 6. The locations of the most prominent documents containing the word *jacuzzi* in the Divar dataset and the word *museum* in the Wikipedia dataset and their similarity surface.

In the case of RCS, documents with similar words were grouped into one cluster. To group similar documents, we utilized k-means clustering. We tested k in the range of 1–30 clusters, running k-means with each value five times to account for variability. To select the best k, we evaluated each clustering using the Silhouette Coefficient metric, which measures how tightly grouped the data is within each cluster. The Silhouette Coefficient was highest with $k = 20$ clusters. In the case of rental property advertisements, out of 20 clusters, only two seemed geo-indicative. The results of RCS are shown in Figure 7, where the top panels illustrate the two geo-indicative clusters and the bottom panel shows a non-geo-indicative group.

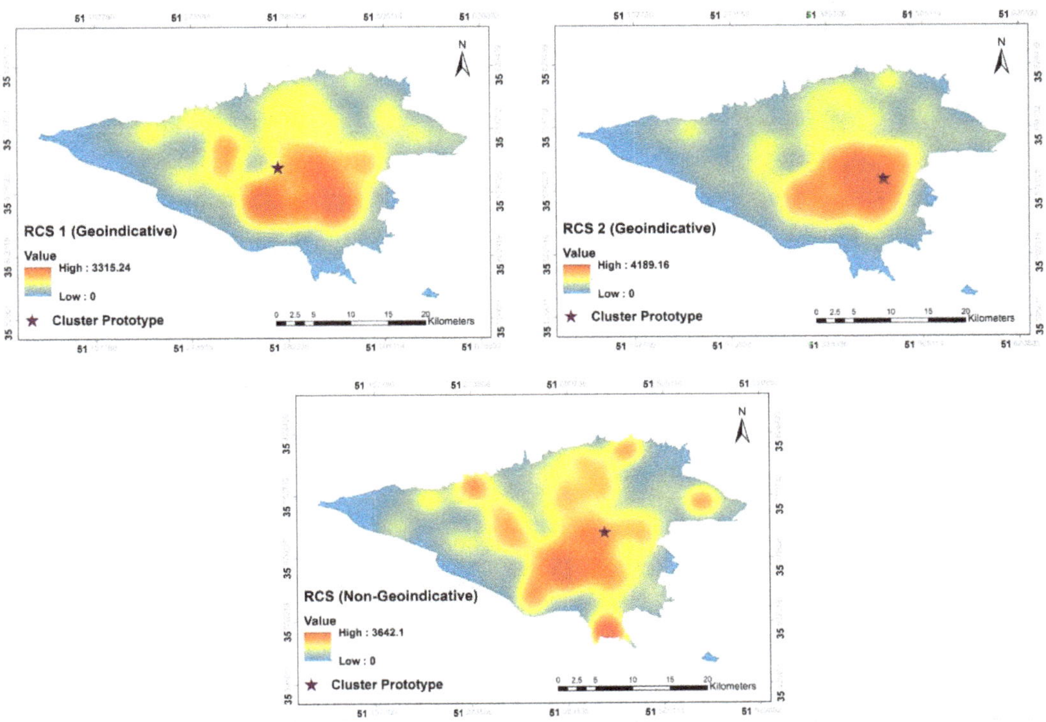

Figure 7. The probability surfaces generated by the RCS method for two geo-indicative and one non-geo-indicative regions in the case of rental property advertisements.

A disadvantage of RCS is that the meaning of clusters is not intuitive, as we have no interpretation of the dimensions or the clusters. Since the classification is done manually in DCS, the issue is more or less compensated in DCS. A suitable measure of the geo-indicativeness of the clusters in RCS is the distance between the prototype and the cluster mapped on the geographical space. As seen in Figure 7, although for the two geo-indicative clusters the prototype is located almost in the middle of the hot spots, the location of the prototype for the non-geo-indicative cluster is far from the hot spots. This means that while the documents in the semantic space have been close to each other, they are dispersed in the geographical space, pertaining to the fact that the semantics for this cluster do not imply a specific region in the geographical space. In order to compare the results of the three approaches, we computed the accuracy of the location prediction in terms of the distance between the predicted locations and the true locations for the test dataset (20 percent). Figure 8 plots the cumulative distance error for all methods and datasets.

The plot also shows the results of a random estimator as a benchmark for the prediction. All of the methods have better accuracies than the random estimator. However, the accuracy of RCS is worse than the other two in both datasets and, in some places, is even equal to the random prediction. DCS is the most accurate approach with an average distance error of 13.3 km for the Divar dataset and 10.3 km for the Wikipedia dataset. Since the size of the city is about 40 km by 20 km, this accuracy is suitable to predict the general region in which a property is located. That is, it can be inferred that a given document is located in the northern parts of Tehran. The worst prediction of DCS has been about 22.5 km, which is approximately equal to half of the length of the city. This value for RCS and LDA is about 31 km and 28 km, which shows that they completely fail in predicting some documents. DCS can promisingly predict the cardinal direction in which the property is located. It is predicting the actual locations better than LDA by 10% and 13% on average, respectively, for the Divar dataset and the Wikipedia articles.

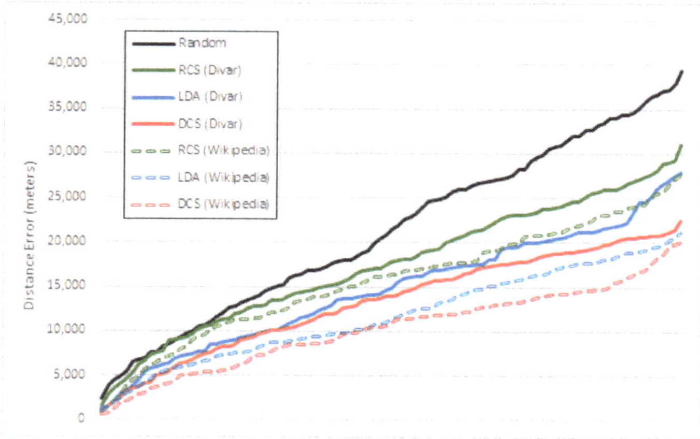

Figure 8. A comparison of the implemented methods in terms of cumulative distance error for both datasets.

6. Conclusions

In this paper, we propose a method to extract spatial information and estimate the location of textual content lacking place names or spatial relations. We argued that some terms within texts do not directly refer to places in geographical space but do imply spatial information. The focus of past studies on inferring spatial information from text has been on extracting spatial relations and place name identification and disambiguation. We proposed two different approaches to infer data-driven conceptual spaces (DCS and RCS) in an unsupervised manner. The proposed method was applied to a dataset of rental properties and a dataset of Persian Wikipedia articles, both for Tehran, Iran. We calculated the degree of geo-indicativeness of terms by their entropy. Then, we compared the results with those of LDA. The results showed that DCS outperformed LDA and RCS in terms of accuracy of prediction. By utilizing DCS, the location of properties can be predicted at 9.6 km, which is suitable for inferring the general region in which a property is located. Considering the volume of textual content lacking spatial information, this method can be a promising approach to inferring spatial information from texts.

The main advantage of our proposed method is that it is grounded in the theory of conceptual spaces, which provides a natural way to represent the meaning of words and their relationships in a geometric space. This means that the properties of the space can be discussed and interpreted in terms of the concepts that they represent. For example, we can visualize the relationships between different concepts in the space and see how they relate to each other. In contrast, LDA relies on statistical models that are not easily interpretable. While it can provide useful insights into the topics and themes that are present in a corpus, it can be challenging to understand how these topics relate to each other and to the underlying concepts that they represent. On the other hand, the proposed method offers a more flexible knowledge representation approach. Since documents are represented by points in conceptual spaces, they can easily be grouped into clusters and form concepts, which provides a multitude of analytical capabilities, such as the topological relationships between different concepts. Furthermore, since similarity computations are more intuitive in conceptual spaces, the proposed method is of great help in recommender systems.

One potential implication of our approach is that it can be used to analyze social media data and identify the geographic locations of social media users. This can be useful for understanding the spatial distribution of social media users and their behavior patterns, which can be valuable information for marketing, social science research, and public policy. While our proposed method infers spatial information of texts, rather users, respecting individuals' privacy is a fundamental principle, and any location inference or data processing should adhere to applicable privacy laws and regulations. Furthermore, our approach is scalable and can handle large volumes of textual data efficiently. This makes it suitable for processing large datasets such as social media posts, news articles, historical documents, and online reviews. In terms of uses, our approach can be applied in various domains such as disaster management, tourism, and urban planning. For example, in tourism, it can be used to identify popular tourist destinations and understand the behavior patterns of tourists. In urban planning, it can be used to analyze social media data and understand the spatial distribution of urban activities, which can inform city planning decisions.

The identification of geo-indicative topics in our paper is based on interpreting the entropy and Moran's I values. This interpretation is manually carried out by comparing the entropies among different topics. As a suggestion, future studies can focus on defining a baseline by which a topic is considered geo-indicative. In addition, various operations can be defined over the conceptual spaces, which makes logical inferences possible. For example, the intersection of two concepts (i.e., regions or directions) may lead to the construction of a new meaningful concept. By studying such operations, the analytical capabilities of the proposed approaches can be improved.

Author Contributions: Conceptualization, O.R.A.; Data curation, O.R.A.; Formal analysis, O.R.A.; Funding acquisition, S.V.R.-T.; Investigation, O.R.A. and A.A.A.; Methodology, O.R.A.; Project administration, A.A.A.; Resources, S.V.R.-T.; Software, O.R.A.; Supervision, A.A.A.; Validation, O.R.A. and S.V.R.-T.; Visualization, O.R.A.; Writing—original draft, O.R.A.; Writing—review & editing, S.V.R.-T. and A.A.A. All authors have read and agreed to the published version of the manuscript.

Funding: This research received no external funding.

Institutional Review Board Statement: Not applicable.

Informed Consent Statement: Not applicable.

Data Availability Statement: The data that support the findings of this study are available at https://figshare.com/s/abe341b40e90c566b061 (accessed on 1 August 2023).

Conflicts of Interest: The authors declare no conflict of interest.

References

1. Wilson, J.P. Local, national, and global applications of GIS in agriculture. In *Geographical Information Systems: Principles, Techniques, Management, and Applications*; John Wiley & Sons: Hoboken, NJ, USA, 1999; pp. 981–998.
2. Qiu, Y.; Duan, H.; Xie, H.; Ding, X.; Jiao, Y. Design and development of a web-based interactive twin platform for watershed management. *Trans. GIS* **2022**, *26*, 1299–1317. [CrossRef]
3. Qiu, Y.; Liu, H.; Liu, F.; Li, D.; Liu, C.; Liu, W.; Huang, J.; Xiao, Q.; Luo, J.; Duan, H. Development of a collaborative framework for quantitative monitoring and accumulation prediction of harmful algal blooms in nearshore areas of lakes. *Ecol. Indic.* **2023**, *156*, 111154. [CrossRef]
4. Choi, Y.; Baek, J.; Park, S. Review of GIS-based applications for mining: Planning, operation, and environmental management. *Appl. Sci.* **2020**, *10*, 2266. [CrossRef]
5. Suh, J. An Overview of GIS-Based Assessment and Mapping of Mining-Induced Subsidence. *Appl. Sci.* **2020**, *10*, 7845. [CrossRef]
6. Wheatley, D.; Gillings, M. *Spatial Technology and Archaeology: The Archaeological Applications of GIS*; CRC Press: Boca Raton, FL, USA, 2013.
7. Kahila-Tani, M.; Kytta, M.; Geertman, S. Does mapping improve public participation? Exploring the pros and cons of using public participation GIS in urban planning practices. *Landsc. Urban Plan.* **2019**, *186*, 45–55. [CrossRef]
8. Masser, I.; Ottens, H. Urban planning and geographic information systems. In *Geographic Information Systems to Spatial Data Infrastructure*; CRC Press: Boca Raton, FL, USA, 2019; pp. 3–28.
9. Wang, F. Why public health needs GIS: A methodological overview. *Ann. GIS* **2019**, *26*, 1–12. [CrossRef]
10. Sharma, A.; Kumar, M.; Hasteer, N. Applications of GIS in Management of Water Resources to Attain Zero Hunger. In *Advances in Water Resources Engineering and Management*; Springer: Berlin/Heidelberg, Germany, 2020; pp. 211–218.

11. Karimi, M.; Mesgari, M.S. Extracting Place Functionality from Crowdsourced Textual Data Using Semantic Space Modeling. *IEEE Access* **2023**, *11*, 129217–129229. [CrossRef]
12. Karimi, M.; Mesgari, M.S.; Purves, R.S. A comparative assessment of machine learning methods in extracting place functionality from textual content. *Trans. GIS* **2022**, *26*, 3225–3252. [CrossRef]
13. Kayed, M.; Dakrory, S.; Ali, A.A. Postal address extraction from the web: A comprehensive survey. *Artif. Intell. Rev.* **2021**, *55*, 1085–1120. [CrossRef]
14. Papadias, E.; Kokla, M.; Tomai, E. Educing knowledge from text: Semantic information extraction of spatial concepts and places. *Agil. GIScience Ser.* **2021**, *2*, 38. [CrossRef]
15. Purves, R.S.; Clough, P.; Jones, C.B.; Hall, M.H.; Murdock, V. Geographic information retrieval: Progress and challenges in spatial search of text. *Found. Trends Inf. Retr.* **2018**, *12*, 164–318. [CrossRef]
16. Wang, W.; Stewart, K. Spatiotemporal and semantic information extraction from Web news reports about natural hazards. *Comput. Environ. Urban Syst.* **2015**, *50*, 30–40. [CrossRef]
17. Qiu, Q.; Xie, Z.; Ma, K.; Chen, Z.; Tao, L. Spatially oriented convolutional neural network for spatial relation extraction from natural language texts. *Trans. GIS* **2021**, *26*, 839–866. [CrossRef]
18. Wang, Y.; Qiao, M.; Liu, H.; Ye, X. Qualitative spatial reasoning on topological relations by combining the semantic web and constraint satisfaction. *Geo-spatial Inf. Sci.* **2017**, *21*, 80–92. [CrossRef]
19. Stock, K.; Jones, C.B.; Russell, S.; Radke, M.; Das, P.; Aflaki, N. Detecting geospatial location descriptions in natural language text. *Int. J. Geogr. Inf. Sci.* **2021**, *36*, 547–584. [CrossRef]
20. Randell, D.A.; Cui, Z.; Cohn, A.G. A spatial logic based on regions and connection. *KR* **1992**, *92*, 165–176.
21. Younis, E.M.G.; Jones, C.B.; Tanasescu, V.; Abdelmoty, A.I. Hybrid Geo-spatial Query Methods on the Semantic Web with a Spatially-Enhanced Index of DBpedia. In Proceedings of the International Conference on Geographic Information Science, Columbus, OH, USA, 18–21 September 2012; pp. 340–353.
22. Stock, K.; Pasley, R.C.; Gardner, Z.; Brindley, P.; Morley, J.; Cialone, C. Creating a corpus of geospatial natural language. In Proceedings of the International Conference on Spatial Information Theory, Scarborough, UK, 2–6 September 2013; Springer: Berlin/Heidelberg, Germany, 2013; pp. 279–298.
23. Adams, B.; Janowicz, K. On the geo-indicativeness of non-georeferenced text. In Proceedings of the International AAAI Conference on Web and Social Media, Dublin, Ireland, 4–8 June 2012; Volume 6, pp. 375–378.
24. Abbasi, O.R.; Alesheikh, A.A. A Place Recommendation Approach Using Word Embeddings in Conceptual Spaces. *IEEE Access* **2023**, *11*, 11871–11879. [CrossRef]
25. Han, B.; Cook, P.; Baldwin, T. Geolocation prediction in social media data by finding location indicative words. In Proceedings of the COLING 2012, Mumbai, India, 8–15 December 2012; pp. 1045–1062.
26. Qiu, Q.; Xie, Z.; Wang, S.; Zhu, Y.; Lv, H.; Sun, K. ChineseTR: A weakly supervised toponym recognition architecture based on automatic training data generator and deep neural network. *Trans. GIS* **2022**, *26*, 1256–1279. [CrossRef]
27. Hollenstein, L.; Purves, R. Exploring place through user-generated content: Using Flickr tags to describe city cores. *J. Spat. Inf. Sci.* **2010**, *1*, 21–48.
28. Wang, Z.; Zhu, Z.; Xu, M.; Qureshi, S. Fine-grained assessment of greenspace satisfaction at regional scale using content analysis of social media and machine learning. *Sci. Total. Environ.* **2021**, *776*, 145908. [CrossRef]
29. Chang, H.-w.; Lee, D.; Eltaher, M.; Lee, J. @ Phillies tweeting from Philly? Predicting Twitter user locations with spatial word usage. In Proceedings of the 2012 IEEE/ACM International Conference on Advances in Social Networks Analysis and Mining, IEEE, Istanbul, Turkey, 26–29 August 2012; pp. 111–118.
30. McKenzie, G.; Janowicz, K. The effect of regional variation and resolution on geosocial thematic signatures for points of interest. In *Societal Geo-Innovation: Selected Papers of the 20th AGILE Conference on Geographic Information Science, Wageningen, The Netherlands, 9–12 May 2017*; Springer: Cham, Switzerland, 2017; pp. 237–256.
31. Adams, B.; Janowicz, K. Thematic signatures for cleansing and enriching place-related linked data. *Int. J. Geogr. Inf. Sci.* **2015**, *29*, 556–579. [CrossRef]
32. Gärdenfors, P. Conceptual spaces as a framework for knowledge representation. *Mind Matter* **2004**, *2*, 9–27.
33. Adams, B.; Raubal, M. A metric conceptual space algebra. In Proceedings of the International Conference on Spatial Information Theory, Aber Wrac'h, France, 21–25 September; Springer: Berlin/Heidelberg, Germany, 2009; pp. 51–68.
34. Blei, D.M.; Ng, A.Y.; Jordan, M.I. Latent dirichlet allocation. In *Advances in Neural Information Processing Systems*; MIT Press: Cambridge, MA, USA, 2001; Volume 14.
35. Blei, D.M.; Ng, A.Y.; Jordan, M.I. Latent dirichlet allocation. *J. Mach. Learn. Res.* **2003**, *3*, 993–1022.
36. Jelodar, H.; Wang, Y.; Yuan, C.; Feng, X.; Jiang, X.; Li, Y.; Zhao, L. Latent Dirichlet allocation (LDA) and topic modeling: Models, applications, a survey. *Multimed. Tools Appl.* **2019**, *78*, 15169–15211. [CrossRef]
37. Chen, C.; Ren, J. Forum latent Dirichlet allocation for user interest discovery. *Knowl.-Based Syst.* **2017**, *126*, 1–7. [CrossRef]
38. Chen, Y.-C. A tutorial on kernel density estimation and recent advances. *Biostat. Epidemiol.* **2017**, *1*, 161–187. [CrossRef]
39. Wang, S.; Li, A.; Wen, K.; Wu, X. Robust kernels for kernel density estimation. *Econ. Lett.* **2020**, *191*, 109138. [CrossRef]
40. Gärdenfors, P. *Conceptual Spaces: The Geometry of Thought*; MIT Press: Cambridge, MA, USA, 2004.
41. Douven, I.; Elqayam, S.; Gärdenfors, P.; Mirabile, P. Conceptual spaces and the strength of similarity-based arguments. *Cognition* **2021**, *218*, 104951. [CrossRef]

42. Janowicz, K.; Wilkes, M. SIM-DL A: A novel semantic similarity measure for description logics reducing inter-concept to inter-instance similarity. In Proceedings of the European Semantic Web Conference, Heraklion, Greece, 31 May–4 June 2009; Springer: Berlin/Heidelberg, Germany, 2009; pp. 353–367.
43. Shanks, D.R. Associationism and cognition: Human contingency learning at 25. *Q. J. Exp. Psychol.* **2007**, *60*, 291–309. [CrossRef]
44. Mormann, T. Prototypes, poles, and tessellations: Towards a topological theory of conceptual spaces. *Synthese* **2021**, *199*, 3675–3710. [CrossRef]
45. Sznajder, M. What conceptual spaces can do for Carnap's late inductive logic. *Stud. Hist. Philos. Sci. Part A* **2016**, *56*, 62–71. [CrossRef] [PubMed]
46. Poth, N.L. Conceptual spaces, generalisation probabilities and perceptual categorization. In *Conceptual Spaces: Elaborations and Applications*; Springer: Berlin/Heidelberg, Germany, 2019; pp. 7–28.
47. Pol, M.; Dessalles, J.-L.; Diaconescu, A. Explanatory AI for Pertinent Communication in Autonomic Systems. In Proceedings of the SAI Intelligent Systems Conference, London, UK, 6–9 September 2019; pp. 212–227.
48. Banaee, H.; Schaffernicht, E.; Loutfi, A. Data-driven Conceptual Spaces: Creating Semantic Representations for Linguistic Descriptions Of Numerical Data. *J. Artif. Intell. Res.* **2018**, *63*, 691–742. [CrossRef]
49. Ahlqvist, O. A Parameterized Representation of Uncertain Conceptual Spaces. *Trans. GIS* **2004**, *8*, 493–514. [CrossRef]
50. Lewis, M.; Lawry, J. Hierarchical conceptual spaces for concept combination. *Artif. Intell.* **2016**, *237*, 204–227. [CrossRef]
51. Alammary, A.S. Arabic Questions Classification Using Modified TF-IDF. *IEEE Access* **2021**, *9*, 95109–95122. [CrossRef]
52. Hout, M.C.; Papesh, M.H.; Goldinger, S.D. Multidimensional scaling. *Wiley Interdiscip. Rev. Cogn. Sci.* **2013**, *4*, 93–103. [CrossRef]
53. Jaworska, N.; Chupetlovska-Anastasova, A. A Review of Multidimensional Scaling (MDS) and its Utility in Various Psychological Domains. *Tutorials Quant. Methods Psychol.* **2009**, *5*, 1–10. [CrossRef]
54. Augustin, D.; Leder, H. Art expertise: A study of concepts and conceptual spaces. *Psychol. Sci.* **2006**, *48*, 135.
55. Qi, P.; Zhang, Y.; Zhang, Y.; Bolton, J.; Manning, C.D. Stanza: A Python Natural Language Processing Toolkit for Many Human Languages. *arXiv* **2020**, arXiv:2003.07082.
56. Lesne, A. Shannon entropy: A rigorous notion at the crossroads between probability, information theory, dynamical systems and statistical physics. *Math. Struct. Comput. Sci.* **2014**, *24*, e240311. [CrossRef]
57. Baldi, P. A computational theory of surprise. In *Information, Coding and Mathematics*; Springer: Boston, MA, USA, 2002; pp. 1–25.
58. Getis, A. Spatial autocorrelation. In *Handbook of Applied Spatial Analysis: Software Tools, Methods and Applications*; Springer: Berlin/Heidelberg, Germany, 2009; pp. 255–278.
59. Myint, S.W. Spatial autocorrelation. In *Encyclopedia of Geography*; Sage: London, UK; Thousand Oaks, CA, USA, 2010; pp. 2607–2608.

mathematics

MDPI

Article

Enhancing Data Security in IoT Networks with Blockchain-Based Management and Adaptive Clustering Techniques

Ajmeera Kiran [1], Prasad Mathivanan [2], Miroslav Mahdal [3,*], Kanduri Sairam [4], Deepak Chauhan [5,6] and Vamsidhar Talasila [7]

1 Department of Computer Science and Engineering, MLR Institute of Technology, Hyderabad 500043, India; ajmeerakiran@mlrinstitutions.ac.in
2 School of Computer Science and Engineering, Vellore Institute of Technology, Chennai 600127, India; prasad.m@vit.ac.in
3 Department of Control Systems and Instrumentation, Faculty of Mechanical Engineering, VSB-Technical University of Ostrava, 17. Listopadu 2172/15, 70800 Ostrava, Czech Republic
4 Department of Electronics and Communication Engineering, NITTE Deemed to Be University, Udupi 574110, India; drsairam@nitte.edu.in
5 School of Computing, Graphic Era Hill University, Dehradun 248002, India; dchauhan@gehu.ac.in
6 Graphic Era Deemed to Be University, Dehradun 248002, India
7 Department of Computer Science and Engineering, Koneru Lakshmaiah Education Foundation, Guntur 522502, India; talasila.vamsi@kluniversity.in
* Correspondence: miroslav.mahdal@vsb.cz

Abstract: The rapid proliferation of smart devices in Internet of Things (IoT) networks has amplified the security challenges associated with device communications. To address these challenges in 5G-enabled IoT networks, this paper proposes a multi-level blockchain security architecture that simplifies implementation while bolstering network security. The architecture leverages an adaptive clustering approach based on Evolutionary Adaptive Swarm Intelligent Sparrow Search (EASISS) for efficient organization of heterogeneous IoT networks. Cluster heads (CH) are selected to manage local authentication and permissions, reducing overhead and latency by minimizing communication distances between CHs and IoT devices. To implement network changes such as node addition, relocation, and deletion, the Network Efficient Whale Optimization (NEWO) algorithm is employed. A localized private blockchain structure facilitates communication between CHs and base stations, providing an authentication mechanism that enhances security and trustworthiness. Simulation results demonstrate the effectiveness of the proposed clustering algorithm compared to existing methodologies. Overall, the lightweight blockchain approach presented in this study strikes a superior balance between network latency and throughput when compared to conventional global blockchain systems. Further analysis of system under test (SUT) behavior was accomplished by running many benchmark rounds at varying transaction sending speeds. Maximum, median, and lowest transaction delays and throughput were measured by generating 1000 transactions for each benchmark. Transactions per second (TPS) rates varied between 20 and 500. Maximum delay rose when throughput reached 100 TPS, while minimum latency maintained a value below 1 s.

Keywords: blockchain; Internet of Things (IoT); Evolutionary Adaptive Swarm Intelligent Sparrow Search (EASISS); Network Efficient Whale Optimization (NEWO); data security; clustering techniques

MSC: 90C27; 68T07

1. Introduction

The Internet of Things (IoT) architecture can be employed to install pervasive, interconnected devices utilizing cloud platforms in a centralized network. The predicted heterogeneity of IoT network devices is yet constrained by factors including scalability, scarce resources, throughput, centralized control, overhead, and latency [1]. The devices are controlled and managed by the server using a centralized network architecture. Centralized systems, however, have several shortcomings [2]. When several smart devices are connected to a network, it is not uncommon for a great deal of data to be produced. A large amount of network bandwidth is required, as well as excellent effectiveness and memory, for a cloud software network operator. Additionally, there is always a chance that the failure of the critical components of the centralized network would result in a catastrophic (or total) failure of the system [3]. A third party is frequently needed to perform additional manipulation on the data that the central cloud storage has acquired. The privacy of the end user could be jeopardized as a result of data leaks. Coordination of external computer resources presents an additional difficulty in demonstrating IoT performance and security in centralized systems [4]. Because of this, the majority of centralized systems in use today are unable to provide organizations with a guarantee of confidentiality and security of data [5]. Most IoT devices can only transmit over limited distances due to their low-power wireless receivers and transmitters [6]. The Multihop Cellular Network (MCN) idea, which enables a large reduction in signal coverage, can be used to benefit IoT networks [7]. Creating a large-scale network with diverse nodes is more challenging since typical blockchain implementations require high-performance nodes [8]. The dispersed nature of IoT networks and the sheer volume of connected things mean that some kind of autonomous defense mechanism is also required [9].

The primary challenges in the new IoT era powered by 5G include an ever-increasing variety of devices that are connected, the heterogeneous nature of gadgets and automakers, interconnection, an enormous amount of gathered data and network traffic, the need for a great deal of bandwidth, communication latency, and trust. In most cases, their technical worth has been acknowledged. Unfortunately, at the moment, they simply provide the ability to store transactional data. To address difficulties with complexity and expand the approach for securing IoT connections, this research analyzes a multi-level structure according to a special clustering method appropriate for the blockchain system. To accommodate blockchain integration in IoT systems, a modified clustering technique was developed by factoring in network performance measurements in predetermined cost functions [10,11]. This means that hacked nodes cannot insert invalid blocks onto the public ledger without being discovered. Hence, the multi-level blockchain prevents compromised entities from accessing it. Hence, the validity of data stored on a blockchain cannot be disputed [12]. Furthermore, crucial is the fact that the new multi-level architecture makes it possible to update the existing primary cloud server. This allows for widespread rollouts to become a reality. In addition, smart contracts guarantee that only authorized users have access to network resources by implementing a lightweight authorization and authentication method that operates locally inside each cluster [13].

Several systems for maintaining security and privacy have been created using tools like public key infrastructure, blockchains, smart cards, and passwords. Although some of these protocols are too sophisticated to be used with IoT devices, many of them have severe privacy and security flaws. This study suggests numerous contributions to overcome these problems.

- First, we present a novel method for direct, non-centralized authentication of communicating entities using biometrics and elliptic curve cryptography. By doing this, any possible single-point of failure problems are resolved;
- In order to prevent user cooperation and privileged insider assaults, we also design an authorization framework employing security tokens for access and membership validation as well as admittance rights groups;
- Additionally, we do formal security analysis utilizing the well-known BAN logic, which verifies the existence of a session key between interacting entities. To show that the suggested system is secure under all of the threat model assumptions, informal security analysis is also carried out;

- To show how effective the suggested protocol is, we conducted a comparative performance study. The results reveal that it has the fewest computing and communication difficulties when compared to other protocols.

The rest of the paper is organized as follows: Section 2 describes the literature review; Section 3 provides the proposed methodology; Section 4 details the result and discussion; and Section 5 illustrates the conclusion.

2. Literature Review

The study [14] suggests a different strategy for building trust in supply chains that contain a variety of IoT components. By employing simulations, we assess the proposed model and demonstrate its applicability. The goal of the research [15] was to investigate the use of blockchain technology to protect a loT-enabled WBMS's communication and data against harmful cyberattacks [16]. Experimental experiments provide proof of concept for the proposed blockchain-based IoT network for WBMSs. The cloud storage option for sensors is taken into consideration in the paper [17]. A unique group signature system that has minimal computational and communication costs is created to first establish anonymous authentication and then provide secure and effective data storage and sharing. Then, based on the suggested group signature scheme, a unique blockchain-based cloud storage protocol for sensors in the IIoT is built. This protocol makes use of smart contracts and proxy re-encryption to enable safe data transfer with minimal computational overhead [18]. The intent of this research [19] was to suggest Edge Share, a blockchain-based data-sharing structure for edge data-sharing offerings across heterogeneous network contexts. This is achieved by implementing a two-level overlay network architecture and an edge computing concept, which can substantially alleviate scheme stress and decrease transmission delay to improve information effectiveness. The suggested framework [20] has been carefully evaluated against several security criteria in comparison to conventional mechanisms. With the use of a multi-level authorization hub and a hierarchical attribute structure, the paper [21] suggests a novel attribute-based encryption method. By spreading different user attributes to distinct authorization centers, the technique enables flexible and fine-grained access control. It was then integrated with the Fabric blockchain technology to address the issue of high decryption costs for Internet of Things consumers [22]. To reduce consumers' decryption overhead, smart contracts on blockchain use high-complexity partial decryption algorithms. These methods use a blockchain-based dynamic secret-sharing mechanism. The power blockchain sharing approach, which can additionally share power trading books, is used to create a dependable trading center. The power data consensus technique and dynamically linked memory are intended to enable secure power data transport matching [23]. Secure SVM, a privacy-preserving SVM training scheme over blockchain-based encrypted IoT data, was used in this study [24]. Blockchain technology creates a secure and dependable data-sharing platform for numerous data sources, in which IoT data is encrypted and then stored on a distributed ledger [25]. The study [26] offers a unique Decentralized Blockchain-based Security (DeBlock-Sec) solution to address the security challenge in resource-constrained IoT environments. The limitation of the proposed model is the decentralized nature of blockchain can make it difficult to manage and update security protocols across a large number of devices. Finally, the integration of blockchain with existing IoT networks and protocols may require significant changes to the underlying infrastructure, which can be costly and time-consuming.

Motivation of the Research

In order to improve message relay between devices, many are turning to 5G networks as the demand for higher quality of service (QoS) and security in IoT networks develops. However, there are difficulties specific to this adoption. High-capacity 5G networks also enable extensive private data exchanges between 5G-IoT devices. In the event that malevolent actors compromised this data, serious consequences could result.

Furthermore, devices typically need to authenticate themselves during handovers in order to stop assaults due to the ultra-densification of base stations in 5G networks. Conventional security mechanisms, however, with their high processing, communication, and storage needs, are not appropriate for this context due to the resource-constrained nature of IoT devices.

In 5G-IoT networks, device heterogeneity poses additional difficulties because different device architectures for communication and security may exist. More effective key exchange and authentication techniques are required in order to guarantee the highest levels of security and privacy protection while preserving the best QoS. Overall, it is important but challenging to strike a balance between security, privacy, and QoS in 5G-IoT networks.

3. Proposed Methodology

The proposed network architecture is meant to make use of the cellular systems' capabilities and performance while providing an efficient and reliable security feature for IoT networks. To encrypt the multi-level structure, Evolutionary Adaptive Swarm Intelligent Sparrow Search (EASISS) is used. This research offers a framework to make it easier for IoT networked devices (objects and nodes) to be authenticated and authorized in a lightweight manner using blockchain technology. The entire cellular-enabled IoT network is divided into several levels by the suggested multi-level network paradigm. IoT nodes and diverse clusters make up Level 1. Sink nodes and controlling elements like cluster heads are included in level 2. Cellular network base stations are located at Level 3. The decentralized blockchain mechanism can be implemented at Level 3 by the BSs with the necessary servers and CPUs. The total system model is depicted in Figure 1.

The introduction of blockchain technology may result in more work and scalability problems. Figure 2 depicts the multi-level network concept that may be used to cut down on redundancy, speed up responses, better organize data, keep conversations private, and accommodate future growth.

Several modules and nodes with varying computational and resource needs are found on the first level. For the IoT system's regional authorization program, the cluster heads offer authentication and authorization functions for regionally embedded systems. In a blockchain setting, CH nodes may safely connect because of a thin consensus. On this level, the permission-based local HLF blockchain is implemented. Cellular network BSs make up the top level. At this level, reliable asymmetric cryptography method deployment is possible. The use of the global system and cutting-edge security measures at the highest levels ensures privacy and security (Level 3).

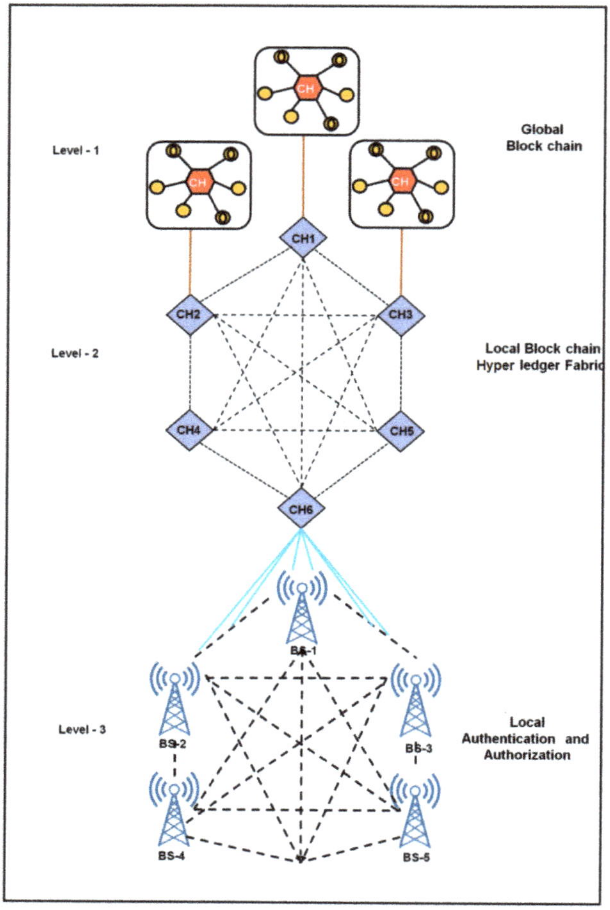

Figure 1. IoT network multi-level model.

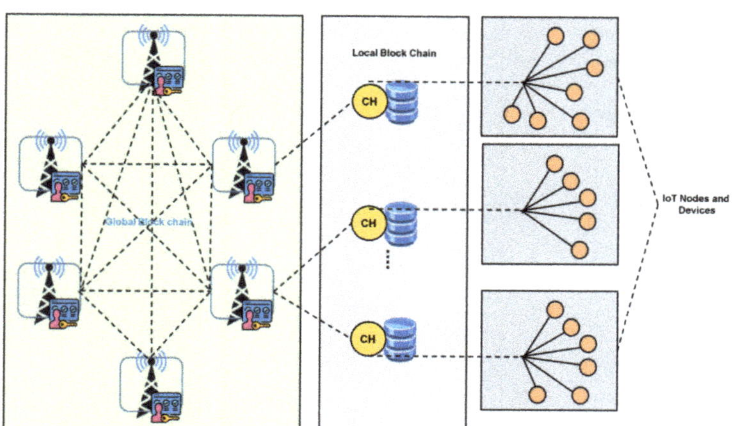

Figure 2. The network concept has three levels: an infrastructure level with a local authorization service, a local blockchain, and a public chain.

3.1. Implementation of the Framework

3.1.1. Network Self-Clustering

The foundation of metaheuristic algorithms is the intimate relationship between computational techniques and optimization. These techniques' main benefit is their freedom from local optimum spots. As a result, these methods search everywhere in the search space. Additionally, the metaheuristic algorithms' entire distribution of individual control (network nodes and participants). Localized communication is how these people interact with one another. The system responds strongly, and the application is quick to adapt to environmental changes. The IoT network's clustering strategy is seen in Figure 3.

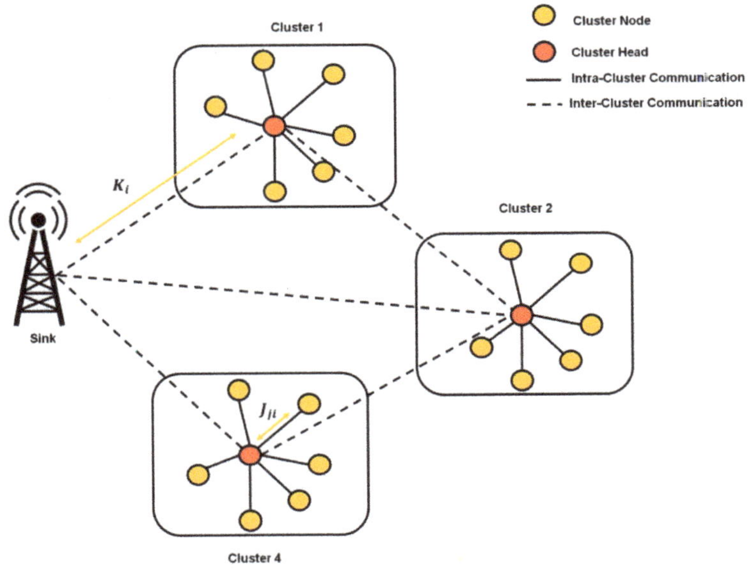

Figure 3. Scheme for cellular IoT network clustering.

Therefore, Evolutionary Adaptive Swarm Intelligent Sparrow Search (EASISS) is presented herein for the clustering of heterogeneous IoT networks. In this study, NEWO is used to improve the clustering of the proposed IoT network since it shows extremely strong performance in this regard. By reducing the average distance that data must travel between IoT devices and the selected cluster heads, the clustering method helps keep IoT systems as efficient as possible. By grouping, just a fraction of the nodes will need to make far-reaching connections to the Base Station (BS). As a result, the system's overall energy usage drops while the network's reach expands. By streamlining the deployment process, the clustering-based method improves the effectiveness of blockchain applications. The CH nodes are in charge of organizing the whole network into functionally independent clusters. To exchange information, other nodes in the cluster have to talk to the CH nodes. Clustering is the goal, and the idea is to get there by using evolutionary computing techniques to run on the network itself. This is done to simulate the disparate components that make up the IoT infrastructure. There is no hard-and-fast rule for how many clusters there should be or how many nodes there should be in each cluster. Moreover, the suggested clustering improves the adaptability of the IoT network's node placement. Taking into account the fact that nodes aren't evenly dispersed within clusters is a major factor.

3.1.2. CH Selection with Evolutionary Adaptive Swarm Intelligent Sparrow Search (EASISS)

Satisfying energy restrictions is the most important aspect of the IoT network architecture. To extend network operation, communications, and transmission links can be made shorter, and power consumption can be decreased. By arranging nodes into autonomous clusters, shorter communication linkages can be achieved. Due to the necessity of each cluster member exchanging information with the associated CH, such an approach makes data aggregation and forwarding easier. EASISS is used in our suggested paradigm to represent vector data. The placements of the sparrow are determined at random inside the limits and are shown in a linear system in the form of Equation (1). Rand [0, 1], a randomly generated distribution number, makes up the beginning vector. For the optimization problem, each row represents a fitness function solution.

$$f(x) = \begin{bmatrix} f(SP_{1,1}, & SP_{1,2} & \cdots & SP_{1,d}) \\ f([SP_{2,1} & SP_{2,2} & \cdots & SP_{2,d}]) \\ \vdots & \vdots & \vdots & \vdots \\ f([SP_{n,1} & SP_{n,2} & \cdots & SP_{n(k),d}\big]) \end{bmatrix} \begin{bmatrix} f_{obj_1} \\ f_{obj_2} \\ \vdots \\ f_{obj_n} \end{bmatrix} \quad (1)$$

The mutation operator uses the Mutation Rate (MR) to help choose the best vector TaV from the population. The new progeny is displayed in Equation (2). The likelihood that $SP_{(g+1)}$ would produce two unique vectors is random, and it is also modified according to three random values. The vectors $\in [0, 1]$ differences in the mutation operation $r_1, r_2,$ and r_3 may be the conscious cause of the resultant's unfavorable outcome. Because of this, the method has to generate the difference in the vector's components in a manner that satisfies the range. Every single sparrow goes through a population-wide evolution of every single process until the optimal solution is identified.

$$SP_{g+1} = SP_{r1} + MR * (SP_{r_2} - SP_{r_3}) \quad (2)$$

Equation (3) calculates the Euclidean distance between the best-chosen sparrow and each sparrow.

$$dist = \sqrt{\sum_{i=1}^{N} (SP_{i.j} - SPbest_{i,j})^2} \quad (3)$$

Similar to Equation (4) and undergoes recombination to create a new generation of children is known as $DV\ TaV$ a trial vector. In this case, a preset Crossover Probability Rate (CPR) is employed with binomial cross-over. If ($r \leq CPR\ BestDV$), the donor vector is chosen; if not, the target vector is.

$$SP_{i,j,g+1} = \begin{Bmatrix} BestDV_j\ if\ (r \leq CPR) \\ BestTaV_j\ if\ (r > CPR) \end{Bmatrix} \quad (4)$$

When $AV < ST$ indicates that there are no predators, producers move into a wider search mode, as shown in Equation (5). When the $AV \geq ST$ predator is present, which is rare, sparrows must defend themselves by fleeing to safer areas. Only infrequent scavengers follow producers significantly.

$$SP_{i,j}^{(t+1)} = \begin{cases} SP_{i,j}^{t}.\exp\left(\frac{-1}{a.I_{max}}\right) & if\ AV < ST \\ SPTV_{i,j}^{t} + Q.L & if\ AV \geq ST \end{cases} \quad (5)$$

For a fixed number of iterations, the location of the scavenger may be found in Equation (6), where I max β mean is the variable step controlling the parameter's average value and variation is 1. Algorithm 1 is EASISS. The equation combines the current position of the particle with the difference between the global best and personal best positions, scaled by β. The goal is to update the particle's position towards the global best position, while taking into account its own personal best position. This helps the particle swarm algorithm to explore the search space efficiently and converge to a good solution.

$$SP_{i,j}^{(t+1)} = SP_{GBest}^{t} + \beta \cdot \left[SP_{i,j}^{t} - SP_{GBest}^{t} \right]^{(t+1)} \quad (6)$$

Algorithm 1: EASISS

Create nodes at random in 2D space,
Initialize population G at random
Evaluate each (node) sparrow's objective function
While (not utilizing the most iteration)
  Use every node i in a cluster k
  Do with $i = 1$ to N nodes
    Choose three nodes at random from the population
    Use the Euclidean distance to calculate the distance between nodes
    Observe the mutant sparrow population
    Produce a new population
    Perform the recombination process
    If ($r < CPR$) then
      Do a selection operation
    Else
      Optimal fitness should replace the sparrow population G.
      For each sparrow, reevaluate the objective function.
      Update the population's SP_{best} position and chose as CH
    End
    $t = t + 1$
  End Do
Stop while
Return Best solution

3.1.3. Network Changes Optimization Using Network Efficient Whale Optimization (NEWO)

The network modifications, such as node creation, movement into, and removal from the IoT network, are carried out using the meta-heuristic algorithm NEWO. NEWO is mathematically designed to encircle prey, assault with a bubble net, and seek prey, which is quite similar to the Whale Optimization Algorithm in terms of feeding behavior. The following steps are taken to adopt NEWO, as well as initializing the whale's population:

$$X = \begin{bmatrix} X_1 \\ X_2 \\ \cdots \\ X_3 \end{bmatrix} = \begin{bmatrix} x_{1,1} & \cdots & x_{1,d} \\ x_{2,1} & \cdots & x_{2,d} \\ \vdots & \cdots & \vdots \\ x_{n,1} & \cdots & x_{n,d} \end{bmatrix} \quad (7)$$

where n is the number of whales and d is the dimension. Calculating and assessing the whale population's fitness value. There will be at least two fitness values for each whale to consider depending on the issue. This evaluation is formulated as follows for problems with two objectives, such as F_1 and F_2:

$$FX = \begin{bmatrix} F_1(X_1) & F_2(X_1) \\ F_1(X_2) & F_2(X_2) \\ \vdots & \vdots \\ F_1(X_n) & F_2(X_n) \end{bmatrix} \quad (8)$$

The search agent's initial spawn location is chosen at random, and when it does, the nearby node's information is cloned to that node's position. Calculating each search agent's fitness value and selecting the best for reference are the following steps. Afterward,

NEWO's parameters are modified so that further search agents are generated based on the position of the best agent. A fitness function is used to determine which cluster head should be used. This function is essential to the exploring mechanism of NEWO (search for prey). The characteristics of the node, such as its energy level (*Er*) and the density of nodes nearby are the function's determining factors.

$$f(CH_i) = p_1[N(CH_i)] + p_2\sum(CH_E) \tag{9}$$

In Equation (9), the variables p_1 and p_2 are chosen at random between 0 and 1, and the variable (CH_i) is a list of all the nodes that are close to the CH. CH_i and CH_E are the remaining energies of the neighboring nodes. In this case, a node needs to have a lot of neighbors nearby in addition to enough remaining energy to become the CH; hence, the higher the fitness value, the better the solution. The base station transmits the cluster heads' IDs throughout the network after determining the best arrangement of cluster heads and the nodes they belong to. Each of these cluster heads transforms into a command center for its area, in charge of gathering and transmitting information. Algorithm 2 provides the pseudocode for the proposed approach.

Algorithm 2: NEWO

```
while (r < r_max)
  while (t < t_max)
    For every search agent's CH_i
      Clone the CH_i node closest to it
      Calculate fitness
      Compute the coefficient vectors
      Select the whale's location
      Adjust the X* best position if X is greater than X*
      For all search agents, update the fitness function.
    End for
  End while
CH = nearest node to the X* position
End while
```

4. Result and Discussion

An IoT network simulation was carried out to test the effectiveness of the proposed clustering techniques. In a 2-D network, there are 100 randomly chosen nodes. Since MATLAB 2018a provided a trustworthy environment for clustering algorithms and made it simple to simulate techniques, it was chosen so that the final results could be compared.

4.1. Clustering Results

To test the efficiency of the suggested clustering approach, it was compared to four other algorithms already reported in the literature: the cascading model [27], Low Energy Adaptive Clustering Hierarchy (LEACH) [28], Gateway Clustering Energy-Efficient Centroid (GCEEC) [29], and Black Hole and Ant Colony Optimization (BH-ACA) [30]. The same simulated network architecture was used to assess each protocol. The number of clusters and CHs produced by each method after its optimization varied. The simulation outcomes show that the proposed clustering model is effective and that the algorithm for minimizing distances and overall network energy is also effective. By reducing network load, decreasing distance, and increasing network coverage, as shown in Figures 4–6, the suggested EASISS-NEWO outperforms the existing algorithms.

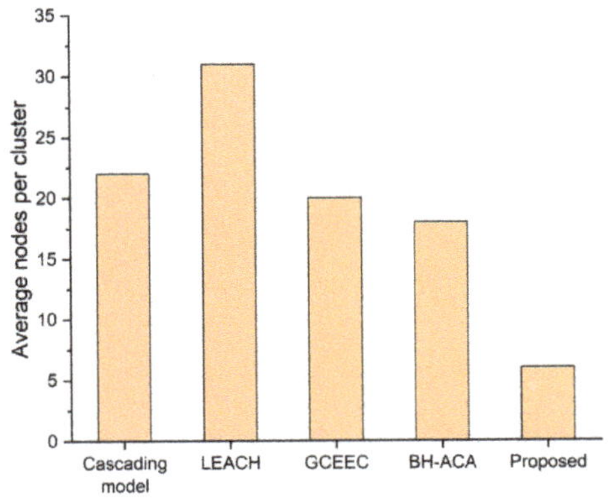

Figure 4. Comparison of average nodes per cluster of the proposed model with cascading model [27], LEACH [28], GCEEC [29], and BH-ACA [30].

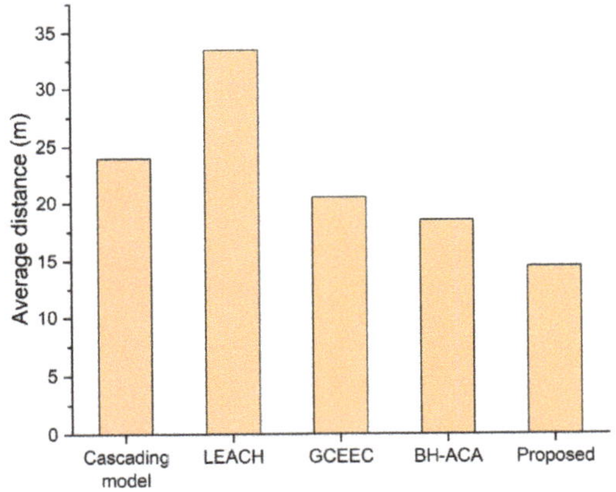

Figure 5. Comparison of average distance of the proposed model with cascading model [27], LEACH [28], GCEEC [29], and BH-ACA [30].

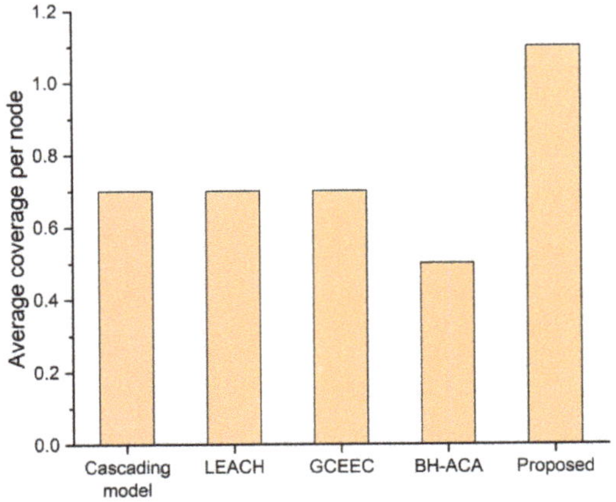

Figure 6. Comparison of average coverage per node of the proposed model with cascading model [27], LEACH [28], GCEEC [29], and BH-ACA [30].

As the blockchain applications' BS server, a workstation was used to build the Level-3 simulation model. Ethereum and Hyperledger private networks' throughput and latency characteristics might be easily assessed in this environment. All of the networks had the same physical infrastructure and were given the same virtual workload. For the experiment, it was planned to use a distributed ecosystem that would incorporate the aforementioned two blockchain networks. The basis for the simulation models was a workstation with an Intel Core i7-3770 processor running at 3.4 GHz and 16 GB of RAM. For simplicity, just one mining node was used to start the Ethereum network.

4.2. Performance Evaluation

A blockchain application's principal goal is to keep track of the many transactions that its users have submitted. A block is generated, and the transaction outcome is stored on the blockchain ledger as a result of the submitted transactions' subsequent processing through the verification and ordering procedures. The latency and throughput of the system were measured to evaluate the effectiveness of the proposed model. To demonstrate the value of the proposed framework, the results were compared to the parameters established in the literature. Hyperledger Caliper was used throughout the study to allow for the administrator's unique blockchain setup. Both node and network latency are significantly influenced by block size. The system's latency is determined by how long it takes to gain consensus when a node begins to notice fresh block validations. Several transactions, including Open, Transfer, and Query, were used to analyze the system. Results were reported for Ethereum and Hyperledger Fabric (level 2 suggested blockchain). Table 1 shows the performance of Hyperledger and Ethereum.

Table 1. Performance of Hyperledger and Ethereum.

Name	Send Rate (TPS)		Max Latency (s)		Min Latency (s)		Avg Latency (s)		Throughput (TPS)	
Blockchain	E	H	E	H	E	H	E	H	E	H
Transfer	10.7	10	7.13	0.38	2.07	0.06	4.63	0.19	6.7	10
Open	22.7	20.2	7.05	0.38	2.12	0.04	4.58	0.18	10	20.1
Query	10.2	10	0.02	0.07	0.01	0.01	0.01	0.01	10.2	10

E—Ethereum; H—Hyperledger Fabric.

By using a multi-level architecture, the average latency is reduced. Only some of the nodes (i.e., CHs) in this architecture contribute to the validation of new blocks. The outcomes of the Ethereum implementation are also shown in Table 1. As a worldwide blockchain technology, it is clear that the proposed lightweight HLF blockchain outperforms Ethereum.

When choosing a suitable blockchain platform for IoT applications, latency and throughput are crucial performance parameters, along with security and privacy. The latency goals can only be met if the blockchain network's distribution of resources is optimized for a given input load. Further analysis of System under Test (SUT) behavior was accomplished by running many benchmark rounds at varying transaction sending speeds. Maximum, median, and lowest transaction delays and throughput were measured by generating 1000 transactions for each benchmark. Transactions per second (TPS) rates varied between 20 and 500. In Figure 7, we can see the highest, median, and lowest transaction latencies that occurred throughout our experiment's many runs. Maximum delay rose when throughput reached 100 TPS, while minimum latency maintained a value below 1 s. The transaction throughput results for different transaction sending rates are shown in Figure 8.

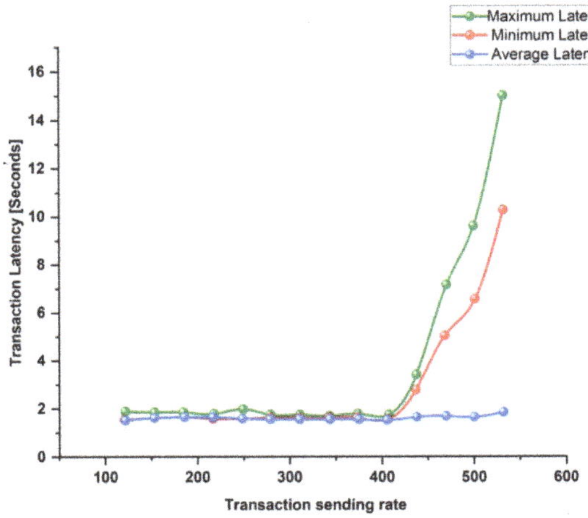

Figure 7. Latency vs. transaction sending rate.

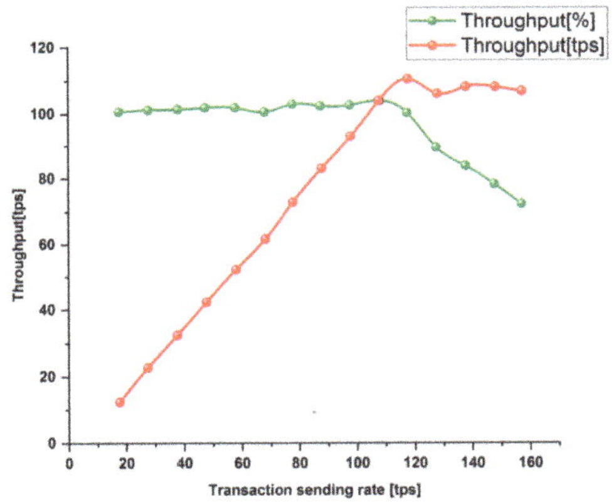

Figure 8. Throughput vs. transaction sending rate.

The throughput was consistent at roughly 100%, even though the sending rate peaked at 110 transactions per second. When the sending rate was raised to 110 TPS, the maximum acceptable sending rate for the SUT, a drastic decrease in throughput was noticed. For the sake of this experiment, every blockchain transaction was produced by a separate client. The HLF has a new three-step design called execute-order-validate, where each stage is dependent on the results of previous stages. As the real throughput is very close to 100%, a 100 TPS transmission rate is doable. Yet, there is only a marginal reduction in throughput when sending at 100 and 200 TPS. We calculated that this means our infrastructure can handle sending at a pace of roughly 110 TPS. Hence, our proposed architecture might allow for the real-time provisioning of a wide variety of 5G-enabled IoT applications with little additional delay. The maximum delay approaches 15 s as the number of input transactions rises. Because of limitations imposed by the containers' allocation to peer nodes, this is the scenario. Since the peer nodes are not initially subjected to heavy loads, the minimum latency almost always remains constant. The block size, channel count, ordering service, users, and endorsing nodes are just a few examples of how blockchain configuration can affect latency. It is evident that, in each case, every transaction is completed, i.e., no transaction loss takes place.

5. Conclusions

This research proposes a distributed security model for IoT devices communicating through cellular networks with multiple hops, using a blockchain-based approach with a multi-level architecture. The IoT network is divided into clusters using the self-clustering EC method, designed with the EASISS-NEWO technique to increase network lifespan and security and reduce processing burden, network load, and latency. The proposed solution addresses issues related to IoT security, such as framework privacy, authentication, heterogeneity, flexibility, and scalability. Four existing methods were compared with the suggested clustering method through simulation studies, and the proposed technique outperformed the alternatives in terms of network load, coverage, and distance. The proposed multi-level system was evaluated, and the results indicated the lightweight blockchain was preferable over Ethereum's global network. Future studies will focus on building a realistic, scalable testbed to investigate, assess, and compare the performance of IoT devices in a real-world environment. Blockchain technology is known to be energy-intensive, which can be a significant issue for IoT systems that rely on battery-powered devices. Future research could explore ways to optimize blockchain solutions for energy efficiency, such as by using lightweight consensus algorithms or off-chain processing.

Author Contributions: A.K.—Conceptualization, methodology, formal analysis, investigation, writing—original draft; P.M.—Conceptualization, methodology, formal analysis, investigation, writing—original draft; M.M.—Conceptualization, methodology, resources, writing—review and editing; K.S.—Conceptualization, formal analysis, investigation, writing—original draft; D.C.—methodology, validation, writing—original draft; V.T.—methodology, validation, writing—review and editing. All authors have read and agreed to the published version of the manuscript.

Funding: This work was supported by the project SP2023/074 Application of Machine and Process Control Advanced Methods supported by the Ministry of Education, Youth and Sports, Czech Republic.

Institutional Review Board Statement: Not applicable.

Informed Consent Statement: Not applicable.

Data Availability Statement: The data presented in this study are available through email upon request to the corresponding author.

Conflicts of Interest: The authors declare no conflict of interest.

References

1. Pal, K. Privacy, security, and policies: A review of problems and solutions with blockchain-based internet of things applications in manufacturing industry. *Procedia Comput. Sci.* **2021**, *191*, 176–183. [CrossRef]
2. Yazdinejad, A.; Parizi, R.M.; Dehghantanha, A.; Karimipour, H.; Srivastava, G.; Aledhari, M. Enabling drones in the internet of things with decentralized blockchain-based security. *IEEE Internet Things J.* **2020**, *8*, 6406–6415. [CrossRef]
3. Gong, S.; Tcydenova, E.; Jo, J.; Lee, Y.; Park, J.H. Blockchain-based secure device management framework for an internet of things network in a smart city. *Sustainability* **2019**, *11*, 3889. [CrossRef]
4. Firoozjaei, M.D.; Lu, R.; Ghorbani, A.A. An evaluation framework for privacy-preserving solutions applicable for blockchain-based internet-of-things platforms. *Secur. Priv.* **2020**, *3*, e131. [CrossRef]
5. Lv, Z.; Qiao, L.; Li, J.; Song, H. Deep-learning-enabled security issues in the internet of things. *IEEE Internet Things J.* **2021**, *8*, 9531–9538. [CrossRef]
6. Yu, K.; Tan, L.; Yang, C.; Choo, K.-K.R.; Bashir, A.K.; Rodrigues, J.J.P.C.; Sato, T. A blockchain-based shamir's threshold cryptography scheme for data protection in industrial internet of things settings. *IEEE Internet Things J.* **2021**, *9*, 8154–8167. [CrossRef]
7. Ali, G.; Ahmad, N.; Cao, Y.; Asif, M.; Cruickshank, H.; Ali, Q.E. Blockchain based permission delegation and access control in Internet of Things (BACI). *Comput. Secur.* **2019**, *86*, 318–334. [CrossRef]
8. Fu, Y.; Li, C.; Yu, F.R.; Luan, T.H.; Zhao, P.; Liu, S. A survey of blockchain and intelligent networking for the metaverse. *IEEE Internet Things J.* **2023**, *10*, 3587–3610. [CrossRef]
9. Yang, X.; Yang, X.; Yi, X.; Khalil, I.; Zhou, X.; He, D.; Huang, X.; Nepal, S. Blockchain-based secure and lightweight authentication for Internet of Things. *IEEE Internet Things J.* **2021**, *9*, 3321–3332. [CrossRef]
10. Lv, Z.; Qiao, L.; Hossain, M.S.; Choi, B.J. Analysis of using blockchain to protect the privacy of drone big data. *IEEE Netw.* **2021**, *35*, 44–49. [CrossRef]

11. Wang, S.; Sheng, H.; Zhang, Y.; Yang, D.; Shen, J.; Chen, R. Blockchain-empowered distributed multi-camera multi-target tracking in edge computing. In *IEEE Transactions on Industrial Informatics*; IEEE: Piscataway, NJ, USA, 2023; pp. 1–10.
12. Yan, L.; Yin-He, S.; Qian, Y.; Zhi-Yu, S.; Chun-Zi, W.; Zi-Yun, L. Method of reaching consensus on probability of food safety based on the integration of finite credible data on block chain. *IEEE Access* **2021**, *9*, 123764–123776. [CrossRef]
13. Almakhour, M.; Sliman, L.; Samhat, A.E.; Mellouk, A. Verification of smart contracts: A survey. *Pervasive Mob. Comput.* **2020**, *67*, 101227. [CrossRef]
14. Al-Rakhami, M.S.; Al-Mashari, M. A blockchain-based trust model for the internet of things supply chain management. *Sensors* **2021**, *21*, 1759. [CrossRef]
15. Faika, T.; Kim, T.; Ochoa, J.; Khan, M.M.; Park, S.-W.; Leung, V.C.M. A blockchain-based Internet of Things (IoT) network for security-enhanced wireless battery management systems. In Proceedings of the 2019 IEEE Industry Applications Society Annual Meeting, Baltimore, MD, USA, 29 September–3 October 2019.
16. Li, B.; Zhou, X.; Ning, Z.; Guan, X.; Yiu, K.-F.C. Dynamic event-triggered security control for networked control systems with cyber-attacks: A model predictive control approach. *Inf. Sci.* **2022**, *612*, 384–398. [CrossRef]
17. Lu, J.; Shen, J.; Vijayakumar, P.; Gupta, B.B. Blockchain-based secure data storage protocol for sensors in the industrial internet of things. *IEEE Trans. Ind. Inform.* **2021**, *18*, 5422–5431. [CrossRef]
18. Krichen, M.; Lahami, M.; Al-Haija, Q.A. Formal methods for the verification of smart contracts: A review. In Proceedings of the 2022 15th International Conference on Security of Information and Networks (SIN), Sousse, Tunisia, 11–13 November 2022.
19. Yang, L.; Zou, W.; Wang, J.; Tang, Z. EdgeShare: A blockchain-based edge data-sharing framework for Industrial Internet of Things. *Neurocomputing* **2022**, *485*, 219–232. [CrossRef]
20. Rathee, G.; Ahmad, F.; Sandhu, R.; Kerrache, C.A.; Azad, M.A. On the design and implementation of a secure blockchain-based hybrid framework for Industrial Internet-of-Things. *Inf. Process. Manag.* **2021**, *58*, 102526. [CrossRef]
21. Xu, H.; He, Q.; Li, X.; Jiang, B.; Qin, K. BDSS-FA: A blockchain-based data security sharing platform with fine-grained access control. *IEEE Access* **2020**, *8*, 87552–87561. [CrossRef]
22. Cao, K.; Ding, H.; Wang, B.; Lv, L.; Tian, J.; Wei, Q.; Gong, F. Enhancing physical-layer security for IoT with nonorthogonal multiple access assisted semi-grant-free transmission. *IEEE Internet Things J.* **2022**, *9*, 24669–24681. [CrossRef]
23. Liang, W.; Tang, M.; Long, J.; Peng, X.; Xu, J.; Li, K.C. A secure fabric blockchain-based data transmission technique for industrial Internet-of-Things. *IEEE Trans. Ind. Inform.* **2019**, *15*, 3582–3592. [CrossRef]
24. Shen, M.; Tang, X.; Zhu, L.; Du, X.; Guizani, M. Privacy-preserving support vector machine training over blockchain-based encrypted IoT data in smart cities. *IEEE Internet Things J.* **2019**, *6*, 7702–7712. [CrossRef]
25. Lekssays, A.; Landa, L.; Carminati, B.; Ferrari, E. PAutoBotCatcher: A blockchain-based privacy-preserving botnet detector for Internet of Things. *Comput. Netw.* **2021**, *200*, 108512. [CrossRef]
26. Narayanan, U.; Paul, V.; Joseph, S. Decentralized blockchain based authentication for secure data sharing in Cloud-IoT: DeBlock-Sec. *J. Ambient. Intell. Humaniz. Comput.* **2021**, *13*, 769–787. [CrossRef]
27. Fu, X.; Yao, H.; Yang, Y. Modeling and analyzing cascading dynamics of the clustered wireless sensor network. *Reliab. Eng. Syst. Saf.* **2019**, *186*, 1–10. [CrossRef]
28. Safaa, S.S.; Mabrouk, T.F.; Tarabishi, R.A. An improved energy-efficient head election protocol for clustering techniques of wireless sensor network (June 2020). *Egypt. Inform. J.* **2021**, *22*, 439–445.
29. Qureshi, K.N.; Bashir, M.U.; Lloret, J.; Leon, A. Optimized cluster-based dynamic energy-aware routing protocol for wireless sensor networks in agriculture precision. *J. Sens.* **2020**, *2020*, 1–19. [CrossRef]
30. Sefati, S.; Abdi, M.; Ghaffari, A. Cluster-based data transmission scheme in wireless sensor networks using black hole and ant colony algorithms. *Int. J. Commun. Syst.* **2021**, *34*, e4768. [CrossRef]

mathematics

MDPI

Article

Local Property of Depth Information in 3D Images and Its Application in Feature Matching

Erbing Yang [1], Fei Chen [2,*], Meiqing Wang [1], Hang Cheng [1] and Rong Liu [1]

[1] College of Mathematics and Statistics, Fuzhou University, Fuzhou 350108, China
[2] College of Computer and Data Science, Fuzhou University, Fuzhou 350108, China
* Correspondence: chenfei314@fzu.edu.cn

Abstract: In image registration or image matching, the feature extracted by using the traditional methods does not include the depth information which may lead to a mismatch of keypoints. In this paper, we prove that when the camera moves, the ratio of the depth difference of a keypoint and its neighbor pixel before and after the camera movement approximates a constant. That means the depth difference of a keypoint and its neighbor pixel after normalization is invariant to the camera movement. Based on this property, all the depth differences of a keypoint and its neighbor pixels constitute a local depth-based feature, which can be used as a supplement of the traditional feature. We combine the local depth-based feature with the SIFT feature descriptor to form a new feature descriptor, and the experimental results show the feasibility and effectiveness of the new feature descriptor.

Keywords: image registration; keypoint match; depth map; SIFT

MSC: 68T45

1. Introduction

In the era of global automation and artificial intelligence, image and vision processing technologies play an important role in many areas, such as autonomous driving, 3D reconstruction, and positioning and navigation. Image matching, also known as image registration or correspondence, is a key and fundamental problem in these complex tasks [1]. Image matching refers to establishing a corresponding relationship between the two images (image pair) before and after the camera moves (or the camera does not move, the object moves), in which two pixels from the two images corresponding to the same object point consist of a pair. Image matching roughly includes three steps: image key feature extraction and positioning, feature description, and feature matching. The key features of the image, also called keypoints, refer to points with particularly prominent properties in a certain aspect, such as corner points and points that are invariant to affine transformation. Feature matching refers to identifying and matching pixels with the same or similar features in two images [2].

Regarding the extraction of image keypoints, Lowe proposed a method to extract unique invariant features from images (SIFT, scale-invariant feature transform) [3]. These features are invariant to the scale and rotation of the image and are shown to provide robust matching of affine distortions, changes in 3D viewpoint, the addition of noise, and changes in illumination. Subsequently, Bay et al. proposed a new scale and rotation invariant interest point detector and descriptor SURF (speeded up robust features) [4], which improved efficiency while maintaining repeatability, specificity, and robustness. Alahi et al. proposed a new keypoint descriptor [5], which was inspired by the human visual system, and more precisely, from the retina, called fast retinal keypoints (FREAK). In terms of key feature matching, Bellavia designed a general matching framework with a new matching strategy and a new local spatial filter [6], combining multiple strategies, including pre-screening as well as many-to-many and symmetric matching, to find and adjust keypoint neighborhood consistency, enabling global improvements to each individual strategy. Another approach for robust feature matching between two images captured from different viewpoints is view synthesis [7–10]. This method generates multiple affine or projective deformations of each of the two images, then extracts and matches the features of each deformation. Toft proposed that CNN-based depth inferred from a single RGB image is very helpful, which can be used to pre-distort images and correct perspective distortion, significantly enhancing SIFT and BRISK capabilities [11]. However, the method is mainly for situations where the camera is looking at the same scene but in the opposite direction.

Most of the current image matching methods work on two images with small changes. When the image pair parallax is large, there is a possibility that similar objects both far and nearby are considered to be the same object. The reason is that the existing matching methods do not consider the depth information of the image, that is, the distance of the object from the camera. Recently, Chen et al. proposed a new de-mismatching algorithm that combines depth prediction and feature matching [12], but they simply used the key depth information as an indicator for threshold judgment and did not use image depth information as a part of the feature.

In 3D computer graphics and computer vision, the depth map records the distance from the shooting device to each point in the scene, where each pixel value of the depth map represents the distance from the object plane to the camera center [13]. Since the depth map contains the depth information of the scene, it can be combined with the traditional RGB image to form the RGBD image which extends the image information from two-dimensional to three-dimensional and can be calculated and used for 3D shape simulation and 3D reconstruction. There are many methods to obtain image depth information; 3D cameras (depth cameras) [14–16], for examples, LiDAR or Kinect, can be used to obtain depth maps. Binocular stereo vision [17,18] uses the principle of parallax to obtain two images of the measured object from different positions and calculates the positional deviation between the corresponding points of the images to obtain the depth information of the object. Monocular depth estimation [19–22] trains a neural network to predict the depth map of a single RGB image. There are also many commonly used public RGBD datasets, for examples, SUN RGBD, TUM, SCAN NET, and NYU depth dataset V2. Among them, the NYU dataset consists of video sequences of various indoor scenes recorded by Microsoft Kinect's RGB and depth cameras [23]. It consists of pairs of RGB and depth frames that are synchronized and densely labeled for each image.

In this paper, we study the property of the depth map and construct a new feature descriptor with depth information on the basis of SIFT. We prove that, when the camera moves, the ratio of the depth difference of a keypoint and its neighbor pixel before and after the camera movement approximates a constant. That means the depth difference of a keypoint and its neighbor pixel after normalization is invariant to the camera movement. Based on this property, all the depth differences of a keypoint and its neighbor pixels constitute a local depth-based feature, which can be used as a supplement of the traditional feature. We combine the local depth-based feature with the SIFT feature descriptor to form a new feature descriptor. To validate the effectiveness of the new feature descriptor, 20 images in the NYU dataset are extracted to form 10 pairs for comparative experiments. The experimental results show that the proposed method can eliminate the wrong matching pairs and improve the accuracy.

In summary, the contributions of this work are as follows:

- We explore the local property of the depth information of image pairs taken before and after the camera movements.
- We prove the local property of the depth information: the ratio of the depth difference of a keypoint and its neighbor pixels before and after the camera movement approximates to a constant.

- Based on the local property, a local depth-based feature descriptor is proposed, which can be used as a supplement of the traditional feature.

The rest of this paper is structured as follows: Related work is discussed in the second section. The third and fourth sections introduce the key features of SIFT and its mismatching problem. Then, the local property of depth information is proved and used for constructing a supplement of the SIFT descriptor. Finally, the experiment results and analysis are given.

2. Related Work

This section gives a brief introduction of existing methods on image matching and depth information.

2.1. Image Matching

In 2004, Lowe et al. proposed the famous scale-invariant feature transform (SIFT) algorithm, with the features extracted being invariant to the scale and rotation of the image. The details of the SIFT algorithm will be given in the next section. However, SIFT has some limits. For example, it has a mismatch problem produced by the movements of cameras or lack of colors, low search efficiency, etc.

To improve the SIFT algorithm, Abdel constructed a colored SIFT (CSIFT) descriptor in color invariant space [24], extending the SIFT descriptor to the color space. Bay et al. proposed a new scale and rotation invariant interest point detector and descriptor SURF (speeded up robust features) [4], which used a blob detector based on a Hessian matrix to find the interest points and improved efficiency while maintaining repeatability, specificity, and robustness. However, it relied too much on the gradient direction of the local area pixel in the stage of finding the host direction, which enlarged the error in the subsequent feature matching even if the deviation angle was not large. Rublee et al. proposed directional FAST and Rotating BRIEF algorithms ORB [25], which combined FAST [26] and BRIEF [27] algorithms to form new feature detectors and feature descriptors. The ORB method used a fast and accurate corner orientation component with intensity centroid and the efficient computation of BRIEF characteristics which enabled real-time computation. However, it is not scale-invariant and is sensitive to brightness. Balammal et al. proposed an image local feature extraction method based on SIFT and KAZE fusion [28], which preserves the unique property of an image representation. In this method, the bag of visual word model [29] was introduced to enhance the scalability and the relevance feedback system was included to reduce the semantic gap. Tang et al. proposed an improved SIFT algorithm [30] in which a stability factor was added in scale space for accurate matching, and the feature descriptor was simplified to shorten the matching time. Feng et al. proposed the concept of inter-feature relative azimuth and distance (IFRAD) [31] and constructed the corresponding feature descriptor to improve the scale invariance and matching accuracy. In this method, the FAST method [26] is used to detect a series of features, and the criteria based on IFRAD is used to select the stable features. Finally a special feature-similarity evaluator was designed to match features in two images. Chung et al. proposed a new cooperative RANSAC (COOSAC) [32] method using a geometry histogram-based (GH-based) constructed to reduce the correspondence set for remote sensing matching [33].

Besides the SIFT descriptor and its variants, some other descriptors, especially for 3D images, were proposed. Tombari proposed a local 3D descriptor for surface matching [34], which was located at the intersection between signatures and histograms so that a better balance between descriptive and robust was possible. Later, they proposed the SHOT method aimed at a more favorable balance between descriptive power and robustness [35]. Based on the SHOT descriptor, Prakhya et al. proposed the B-SHOT descriptor, a binary 3D feature descriptor for keypoint matching on 3D point clouds [36]. In this method, a binary quantization method converting a real valued vector to a binary vector is proposed which reduces the memory requirements in keypoint matching. Steder et al. proposed the normal-aligned radial features (NARF) for feature descriptor calculation in 3D range data [37], which considered object boundary information and the surface structure and extracted the keypoints located on the stable areas for normal estimation or descriptor calculation.

2.2. Depth Information Estimation

Deep neural network is the state-of-the-art technology for obtaining depth information or a depth map. The current mainstream method of depth information estimation is monocular depth estimation [38–40], which alleviates the limitation of binocular estimation due to its high cost, large size, and fixed location.

In 2017, Ummenhofer et al. proposed DeMoN, which is the first depth network that learns to estimate depth and camera motion from two unconstrained images [20]. Unlike networks that estimate depth from a single image, DeMoN can take advantage of motion parallax, a powerful clue that can be generalized to new types of scenes. However, it does not have the flexibility of the classic approach when it comes to dealing with cameras with different intrinsic parameters. So, their next challenge was to lift this restriction and extend this work to more than two images. In 2019, they proposed a new convolution CAM-Convs [21], which could take camera parameters into account, thus allowing neural networks to learn to calibrate perceptual patterns, which greatly improved the generalization ability of depth prediction networks. In 2020, they proposed DeepTAM, a fully learned camera tracking and depth mapping estimation system based on dense keyframes [22]. For tracking, the small attitude increment between the current camera image and the composite viewpoint is estimated.

Reza Mahjourian et al. proposed a method for unsupervised learning of depth and ego-motion from monocular videos [41]. This method explicitly considered the inferred 3D geometry of the whole scene and enforced consistency of the estimated 3D point clouds and ego-motion across consecutive frames. Jun Wang et al. proposed a self-supervised framework for RGB-guided depth enhancement [42]. In this method, the dependency between RGB and depth was exploited and a multi-scale edge-guided network model was designed for the learning of depth information enhancement.

3. The SIFT Key Features

Scale-invariant feature transform (SIFT) [3] is an algorithm used to detect and describe local features of images, effectively solving the problems of object rotation, scaling, translation, image affine/projection transformation, lighting effects, object occlusion, fragmentation scenes, etc. The Gaussian Laplacian of the image reflects the second-order variation of the color or brightness of the image, and its maximum and minimum points can be used as the keypoints of the image [3]. The SIFT algorithm introduces scale variables σ, approximates the Laplacian operator with the difference of Gaussian (DoG, difference of Gaussian) in the scale space [43], and finds three-dimensional (space + scale) extreme points in the Gaussian difference images as keypoints. Finally, the gradient distribution information of the surrounding neighborhood of each keypoint is expressed as a 128-dimensional vector used as the feature descriptor.

- Scale space and Gaussian difference space

The scale space $L(x,y,\sigma)$ of the image $I(x,y)$ is defined as the convolution of the variable-scale Gaussian function $G(x,y,\sigma)$ with the original image $I(x,y)$ [43]:

$$L(x,y,\sigma) = G(x,y,\sigma) * I(x,y) \tag{1}$$

$$G(x,y,\sigma) = \frac{1}{2\pi\sigma^2} e^{-\frac{(x-m/2)^2+(y-n/2)^2}{2\sigma^2}} \tag{2}$$

where (x,y) represents the pixel location of the image, and $m \times n$ are the size of the two-dimensional template of the Gaussian function. In addition, σ is the scale space factor, the smaller the σ value, the smaller the local area used for image smoothing. The large scale reflects the contour characteristics of the image, while the small scale reflects the detailed characteristics of the image.

The difference of two nearby scale spaces separated by a constant multiplicative factor k is defined as the difference of Gaussian (DoG) space:

$$D(x,y,\sigma) = L(x,y,k\sigma) - L(x,y,\sigma) = (G(x,y,k\sigma) - G(x,y,\sigma)) * I(x,y) \tag{3}$$

- Gaussian difference pyramid

For the same physical object in the imaging system, the farther from the center of the camera, the smaller and blurrier the object in the image is. The details of the objects close to the camera can be seen clearly, but the contour profile of the farther object may not be seen completely; on the other hand, the contour profile of the object far from the camera can be exhibited, but the details may be blurred. The SIFT method uses the scale space and the pyramid method to describe this phenomenon and reduce the computational cost [44].

Suppose the original image $I_1(x,y)$ is of size $M \times N$. With different scales σ_1, $k\sigma_1$, $k^2\sigma_1$, $k^3\sigma_1$, and $k^4\sigma_1$ where k is a constant, a group of Gaussian difference spaces can be obtained:

$$\begin{aligned} D_1(x,y,\sigma_1) &= (G(x,y,k\sigma_1) - G(x,y,\sigma_1)) * I_1(x,y) \\ D_1(x,y,k\sigma_1) &= (G(x,y,k^2\sigma_1) - G(x,y,k\sigma_1)) * I_1(x,y) \\ D_1(x,y,k^2\sigma_1) &= (G(x,y,k^3\sigma_1) - G(x,y,k^2\sigma_1)) * I_1(x,y) \\ D_1(x,y,k^3\sigma_1) &= (G(x,y,k^4\sigma_1) - G(x,y,k^3\sigma_1)) * I_1(x,y) \end{aligned} \tag{4}$$

Denote

$$P_1(I_1,\sigma_1) = \{D_1(x,y,\sigma_1), D_1(x,y,k\sigma_1), D_1(x,y,k^2\sigma_1), D_1(x,y,k^3\sigma_1)\} \tag{5}$$

Downsample the image $I_1(x,y)$ by a factor of two to obtain an image $I_2(x,y)$ with size $\frac{M}{2} \times \frac{N}{2}$. Taking $\sigma_2 = 2\sigma_1$ as the initial scale, a new set of Gaussian difference spaces can be obtained by a calculation similar to Formula (4):

$$P_2(I_2,\sigma_2) = \{D_2(x,y,\sigma_2), D_2(x,y,k\sigma_2), D_2(x,y,k^2\sigma_2), D_2(x,y,k^3\sigma_2)\} \tag{6}$$

Two-factor continuous downsampling will produce images $I_3(x,y)$ and $I_4(x,y)$. Let $\sigma_3 = 2\sigma_2, \sigma_4 = 2\sigma_3$, respectively. Then, the other two Gaussian difference groups $P_3(I_3,\sigma_3)$ and $P_4(I_4,\sigma_4)$ are obtained similarly. The four groups of Gaussian difference spaces $P_1(I_1,\sigma_1)$, $P_2(I_2,\sigma_2)$, $P_3(I_3,\sigma_3)$, and $P_4(I_4,\sigma_4)$ consist of a pyramid.

- Keypoint positioning

The SIFT algorithm selects the three-dimensional (coordinates x, y, and the scale σ) extreme points in the Gaussian difference pyramid as the keypoints. For each point (x,y,σ) in the pyramid, it needs to compare its eight neighbors of the same scale σ space, and nine neighbors of adjacent scales $k\sigma$ and σ/k. In a group of Gaussian difference spaces $P_i(I_i,\sigma_i), i = 1,\ldots,4$, since the first scale space and the last scale space each have only one neighbor in the scale direction, there are no three-dimensional extreme points. Only the middle two scale spaces of a group of Gaussian difference spaces may exist as extreme points. All Gaussian difference spaces that may have extreme points are listed following:

$$\begin{aligned} &D_1(x,y,k\sigma_1), D_1(x,y,k^2\sigma_1), D_2(x,y,k\sigma_2), D_2(x,y,k^2\sigma_2), \\ &D_3(x,y,k\sigma_3), D_3(x,y,k^2\sigma_3), D_4(x,y,k\sigma_4), D_4(x,y,k^2\sigma_4) \end{aligned} \tag{7}$$

Change scales $\sigma_i, i = 1,2,3,4$ to the initial scale σ_1 :

$$\begin{aligned} &D_1(x,y,k\sigma_1), D_1(x,y,k^2\sigma_1), D_2(x,y,2k\sigma_1), D_2(x,y,2k^2\sigma_1), \\ &D_3(x,y,4k\sigma_1), D_3(x,y,4k^2\sigma_1), D_4(x,y,8k\sigma_1), D_4(x,y,8k^2\sigma_1) \end{aligned} \tag{8}$$

In the SIFT algorithm, taking $k = \sqrt{2}$, then the above list is rewritten as :

$$\begin{aligned} &D_1(x, y, \sqrt{2}\sigma_1), D_1(x, y, 2\sigma_1), D_2(x, y, 2\sqrt{2}\sigma_1), D_2(x, y, 4\sigma_1), \\ &D_3(x, y, 4\sqrt{2}\sigma_1), D_3(x, y, 8\sigma_1), D_4(x, y, 8\sqrt{2}\sigma_1), D_4(x, y, 16\sigma_1) \end{aligned} \tag{9}$$

Now we can see that the scale factor uniquely determines the scale space where the extreme point is located.

- Keypoint offset correction

The Gaussian difference pyramid is discrete, and the above calculation of the extreme points can only determine the approximate position of keypoints [45], which may not be a pixel point when transformed to the original images. The offset to the precise position can be calculated by the Taylor expansion.

Let $X = (x, y, \sigma)^T$ be the approximate position of a keypoint and $\Delta X = (\Delta x, \Delta y, \Delta\sigma)^T$ be the offset. The Taylor expansion of the DoG function (3) in the precise point is:

$$D(X + \Delta X) \approx D(X) + \frac{\partial D^T}{\partial X} \cdot \Delta X + \frac{1}{2}\Delta X^T \frac{\partial^2 D}{\partial X^2} \Delta X \tag{10}$$

Since the keypoint is an extreme point, the derivative of this function with respect to ΔX is zero:

$$\frac{\partial D(X + \Delta X)}{\partial \Delta X} = 0 \tag{11}$$

and the offset of the extreme point can be obtained:

$$\Delta X \approx -\left(\frac{\partial^2 D}{\partial X^2}\right)^{-1} \frac{\partial D^T}{\partial X} \tag{12}$$

When the offset of ΔX in any dimension (i.e., Δx, Δy, or $\Delta\sigma$) is greater than 0.5, the keypoint has been offset to its adjacent point, and the position of the current keypoint must be changed.

- Key feature descriptors

The SIFT algorithm uses gradient information of neighboring pixels around keypoints in the same scale space as key features. The gradient direction is discretized into eight directions as described in Figure 1: up, down, left, right, and the diagonals [46]:

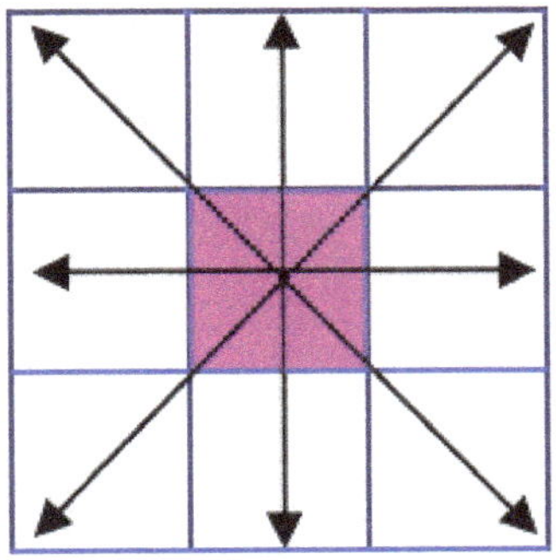

Figure 1. The discretization of the gradient direction.

For the detected keypoint (x, y, σ) in the DoG pyramid, the 16×16 neighborhood of (x, y, σ) in the scale space image $L(x, y, \sigma)$ is divided into 16 small blocks of 4×4. In every small block, the gradient magnitudes with the same direction are weighted accumulated and form an eight-dimensional vector. All the 8-dimensional vectors of 16 small blocks are collaged to a 128-dimensonal vector, which is the SIFT key feature descriptor.

In the actual calculation, in order to obtain better scale invariance and rotation invariance, the SIFT algorithm also performs operations, such as edge response removal and direction normalization. One may find the detailed description in reference [3].

4. False matching Based on SIFT Feature Descriptors

When the parallax of the image pair (I_1, I_2) is not small, the use of basic SIFT feature descriptors may lead to false matching results in subsequent feature matching. For example, Figure 2 shows an image pair taken from different shooting angles in which the SIFT algorithm gives false matching pairs 1, 2, and 3 marked by green boxes. The reason is that the keypoints corresponding to the different object points have similar SIFT features.

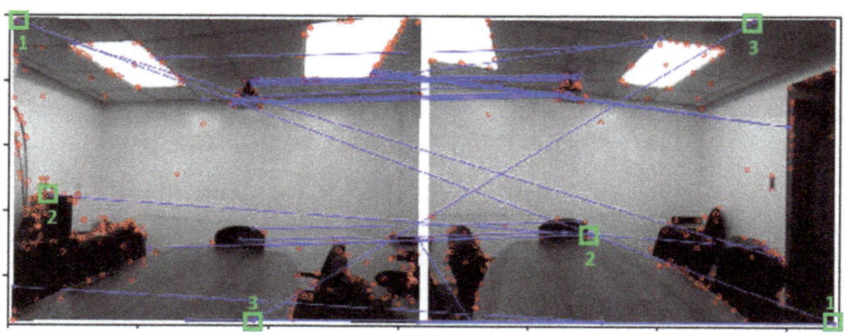

Figure 2. Mismatching results of the traditional SIFT algorithm (keypoint pairs 1, 2, and 3 marked in green boxes).

In order to alleviate this problem, in this paper, we introduce the depth information of the image, that is, the distance between the object plane and the camera, as the supplement of the SIFT feature. We prove that when the camera moves, the ratio of the depth difference of the keypoint and its neighbor pixels before and after the camera movement approximates a constant. Based on this property, the depth differences of all pixels in the keypoint neighborhood with the keypoint constitute a local depth-based feature used as a supplement of the SIFT feature.

5. Local Property of Depth Information under the Camera Movement

In image registration, an image pair to be matched can be regarded as two image planes generated by the same natural scene before and after the camera moves. The camera movement can be decomposed as the movement of the camera center along the optic axis and the rotation of the optic axis around the camera center.

Figure 3 depicts the relationship among the physical coordinate system $(O_W - X_W Y_W Z_W)$, the camera coordinate system $(O_1 - X_C Y_C Z_C)$,and the image coordinate system $(O_{I_1} - XY)$. O_1 is the initial camera center, and the ray $O_1 Z_C$ is the optical axis. I_1 is the image plane, and $f = (f_x, f_y)$ is the focal length. M is an object point, and the object plane P_1 in which M is located is vertical to the optic axis with N_1 the intersection point.

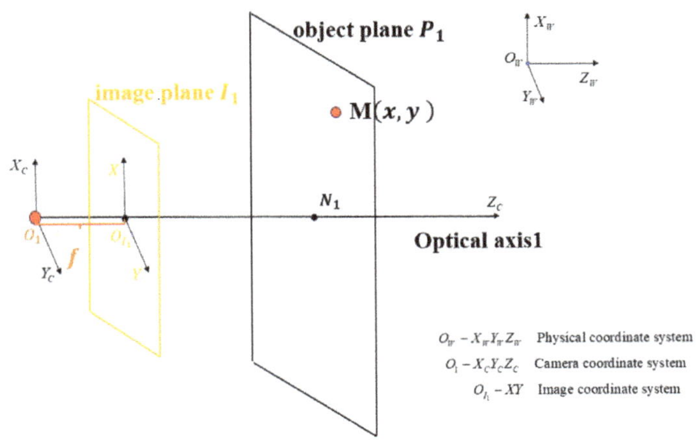

Figure 3. The relationship among three coordinates.

5.1. The Movement of the Camera Center along the Optic Axis

Figure 4 depicts the relationship among the camera, the image planes, the point in the scene, and its projections in image planes when the camera moves along the optic axis. $M = (x, y)$ is the point in the scene with the distance d_1 to the camera center O_1. $Z_1 = (u_1, v_1)$ is the projection of M on the image plane I_1. The three-dimensional coordinate of point M relative to the image plane I_1 is (u_1, v_1, d_1).

When the camera center moves along the optic axis from point O_1 to point O_2 with distance T and the direction of the optic axis does not change, the new image plane I_2 is produced. The projection of M on the image plane I_2 is $Z_2 = (u_2, v_2)$, and the depth of M to the camera center changes to d_2. The three-dimensional coordinate of the object point M relative to the image plane I_2 is changed as (u_2, v_2, d_2).

The relationship of $Z_1 = (u_1, v_1)$ and $Z_2 = (u_2, v_2)$ is described in the following lemma.

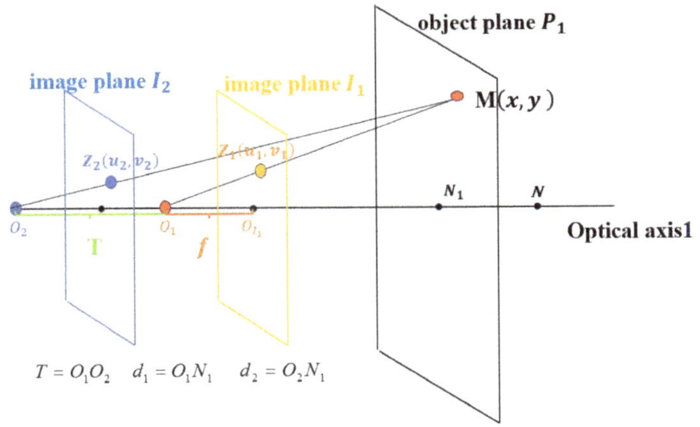

Figure 4. The projection coordinates and related physical quantities corresponding to the object point M when the camera moves along the optical axis with distance T.

Lemma 1. *When the camera moves along the optical axis with distance T and the direction of the optic axis does not change, the two projections $Z_1 = (u_1, v_1)$ and $Z_2 = (u_2, v_2)$ of the physical object point $M(x, y)$ before and after the camera movement satisfy the following relationships:*

$$\begin{aligned} u_2 &= u_1 \cdot \frac{d_1}{d_2} = u_1 \cdot \frac{d_1}{d_1 + T} \\ v_2 &= v_1 \cdot \frac{d_1}{d_2} = v_1 \cdot \frac{d_1}{d_1 + T} \end{aligned} \tag{13}$$

Proof of Lemma 1. According to the principle of pinhole optical imaging,

$$\begin{aligned} u_1 &= \frac{f_x}{d_1} x \\ u_2 &= \frac{f_x}{d_2} x \end{aligned} \tag{14}$$

It is easy to find

$$u_2 = u_1 \cdot \frac{d_1}{d_2} \tag{15}$$

According to the definition of depth, it is obvious that $d_2 = d_1 + T$; thus

$$u_2 = u_1 \cdot \frac{d_1}{d_2} = u_1 \cdot \frac{d_1}{d_1 + T}. \tag{16}$$

Similarly, the following relation can be obtained:

$$v_2 = v_1 \cdot \frac{d_1}{d_2} = v_1 \cdot \frac{d_1}{d_1 + T} \tag{17}$$

□

5.2. The Rotation of the Camera's Optical Axis

The relationship of two image planes produced before and after the rotation of the camera's optical axis is shown in Figure 5. The camera center position O_1 remains unchanged, and the camera optical axis rotates around O_1 with an angle $\varphi = (\varphi^x, \varphi^y)$ and produces a new object plane P_2 and new image plane I_3. The intersection point of the new object plane P_2, and the new optical axis is N_2.The projection of M on the image plane I_3 is $Z_3 = (u_3, v_3)$, and the depth is d_3. The three-dimensional coordinate of the object point M relative to the image plane I_3 is (u_3, v_3, d_3).

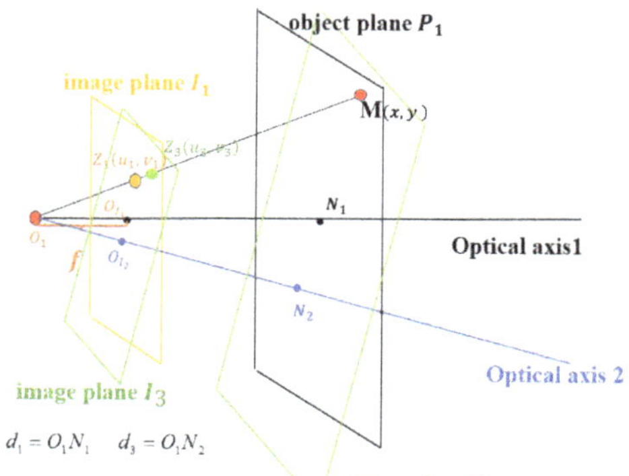

Figure 5. Projection coordinates and related physical quantities corresponding to the object point M when the optical axis rotates with angle φ.

Lemma 2 gives the relationships among the distance d_1, d_3, and the rotation angle φ.

Lemma 2. *If the camera optical axis rotates around the camera center with angle φ while the position of the camera center keeps unchanged, then the two projections $Z_1 = (u_1, v_1)$ and $Z_3 = (u_3, v_3)$ of the physical object point $M(x, y)$ before and after the rotation satisfy the following relationships:*

$$d_3 = d_1 \cdot \frac{\cos\left(\arctan\frac{u_1}{f_x} - \varphi^x\right) \cdot \cos\varphi^x}{\cos\left(\arctan\frac{u_1}{f_x}\right) \cdot \cos\varphi} = d_1 \cdot \frac{\cos\left(\arctan\frac{v_1}{f_y} - \varphi^y\right) \cdot \cos\varphi^y}{\cos\left(\arctan\frac{v_1}{f_y}\right) \cdot \cos\varphi} \tag{18}$$

Proof of Lemma 2. Figure 6 depicts the spatial relationships among the object plane P_1, image plane I_1, and the optical axis O_1N_1. In the object plane P_1, the points M^x and M^y are the projections of M to the x-axis and y-axis, respectively. Suppose the angle between the initial optical axis O_1N_1 and the line connecting the object point M and the camera center P_1 is $\theta = (\theta^x, \theta^y)$, where $\theta^x = \angle N_1O_1M^x$, $\theta^y = \angle N_1O_1M^y$.

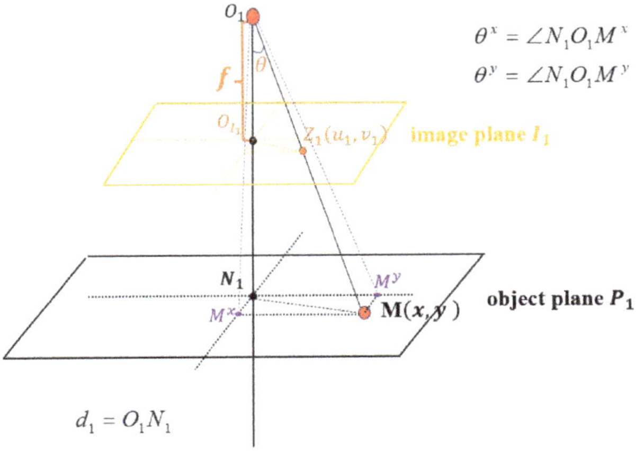

Figure 6. The spatial relationships among the object plane P_1, image plane I_1, and the optical axis O_1N_1.

It is easy to find that,

$$\begin{aligned} \tan\theta^x &= \frac{|N_1M^x|}{|O_1N_1|} = \frac{x}{d_1}, \\ \tan\theta^y &= \frac{|N_1M^y|}{|O_1N_1|} = \frac{y}{d_1} \end{aligned} \tag{19}$$

Combine with Formula (14), and we have

$$\begin{aligned} \theta^x &= \arctan\frac{x}{d_1} = \arctan\frac{u_1}{f_x} \\ \theta^y &= \arctan\frac{y}{d_1} = \arctan\frac{v_1}{f_y} \end{aligned} \tag{20}$$

When the optical axis rotates around the camera center O_1 with angle $\varphi = \varphi^x, \varphi^y)$, a new object plane P_2 and new image plane I_3 are produced. Figure 7 depicts the spatial relationships among them. The intersection point of the new object plane P_2 and the new optical axis is N_2. So, we call the new optical axis O_1N_2 which passes through the first object plane P_1 in the point N_3. Figure 8 gives some details of Figure 7. Three points, M, N_1, and N_3, all locate in the object plane P_1, which is depicted in Figure 8a. N_3^x and N_3^y are the projections of N_3 to the x- and y-axis in the plane P_1; $\varphi^x = \angle N_1O_1N_3^x$, $\varphi^y = \angle N_1O_1N_3^y$. $\eta = \angle MO_1N_3 = \theta - \varphi$.

At the same time, let N_2^y be the projection of N_2 to the plane $O_1N_1M^y$. So, the points O_1, N_1, M^y, N_2^y, and N_3^y, all fall in the same plane which is depicted in Figure 8b.

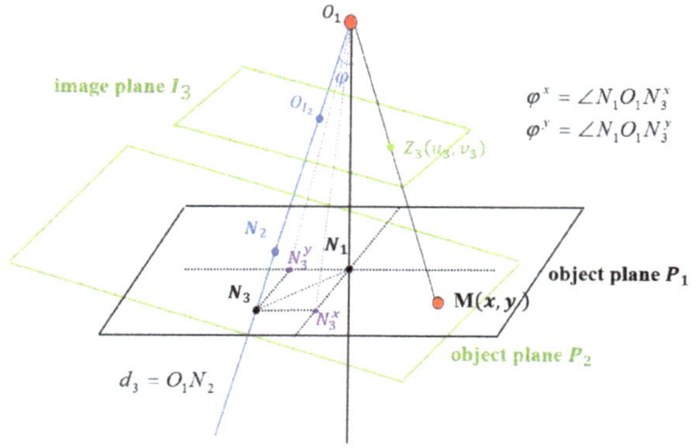

Figure 7. The spatial relationships before and after the optical axis rotation with angle $\varphi = (\varphi^x, \varphi^y)$.

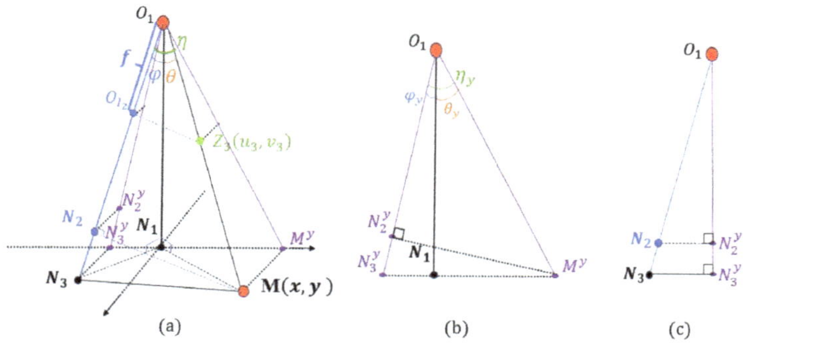

Figure 8. Some details of Figure 7: (**a**) the object plane P_1 contains points, M, N_1, and N_3; (**b**) the plane contains points O_1, N_1, M^y, N_2^y, and N_3^y; (**c**) the plane $O_1N_3N_3^y$ of (**a**).

Considering $\theta^y = \angle N_1O_1M^y$, $\varphi^y = \angle N_1O_1N_3^y$ and $\eta^y = \angle N_2^yO_1M^y$ in Figure 8b, respectively, we have

$$\begin{aligned} &\cos\theta^y = \frac{|O_1N_1|}{|O_1M^y|} = \frac{d_1}{|O_1M^y|}, \cos\varphi^y = \frac{|O_1N_1|}{|O_1N_3^y|} = \frac{d_1}{|O_1N_3^y|}, \cos\eta^y = \frac{|O_1N_2^y|}{|O_1M^y|} \\ &\Rightarrow |O_1N_2^y| = d_1 \cdot \frac{\cos\eta^y}{\cos\theta^y}, \ |O_1N_3^y| = d_1 \cdot \frac{1}{\cos\varphi^y} \end{aligned} \quad (21)$$

On the other hand, the triangle $O_1N_1N_3$ in Figure 8a gives the following:

$$\cos\varphi = \frac{|O_1N_1|}{|O_1N_3|} = \frac{d_1}{|O_1N_3|} \Rightarrow |O_1N_3| = d_1 \cdot \cos\varphi \quad (22)$$

Figure 8c details the plane $O_1N_3N_3^y$ of Figure 8a, in which we may find that $\Delta O_1N_2N_2^y \sim \Delta O_1N_3N_3^y$, giving the following:

$$\begin{aligned} &\frac{|O_1N_2^y|}{|O_1N_2|} = \frac{|O_1N_3^y|}{|O_1N_3|} \\ &\Rightarrow \frac{|O_1N_2^y|}{d_3} = \frac{|O_1N_3^y|}{|O_1N_3|} \\ &\Rightarrow d_3 = \frac{|O_1N_3| \cdot |O_1N_2^y|}{|O_1N_3^y|} \end{aligned} \tag{23}$$

Combining with Equations (21) and (22), we have

$$d_3 = d_1 \cdot \frac{\cos\eta^y \cdot \cos\varphi^y}{\cos\theta^y \cdot \cos\varphi} \tag{24}$$

Similarly, we have

$$d_3 = d_1 \cdot \frac{\cos\eta^x \cdot \cos\varphi^x}{\cos\theta^x \cdot \cos\varphi} \tag{25}$$

Combining with Equation (20), we have the following results:

$$\Rightarrow \begin{cases} d_3 = d_1 \cdot \dfrac{\cos\eta^x \cdot \cos\varphi^x}{\cos\theta^x \cdot \cos\varphi} = d_1 \cdot \dfrac{\cos(\theta^x - \varphi^x) \cdot \cos\varphi^x}{\cos\theta^x \cdot \cos\varphi} = d_1 \cdot \dfrac{\cos\left(\arctan\frac{u_1}{f_x} - \varphi^x\right) \cdot \cos\varphi^x}{\cos\left(\arctan\frac{u_1}{f_x}\right) \cdot \cos\varphi} \\ d_3 = d_1 \cdot \dfrac{\cos\eta^y \cdot \cos\varphi^y}{\cos\theta^y \cdot \cos\varphi} = d_1 \cdot \dfrac{\cos(\theta^y - \varphi^y) \cdot \cos\varphi^y}{\cos\theta^y \cdot \cos\varphi} = d_1 \cdot \dfrac{\cos\left(\arctan\frac{v_1}{f_y} - \varphi^y\right) \cdot \cos\varphi^y}{\cos\left(\arctan\frac{v_1}{f_y}\right) \cdot \cos\varphi} \end{cases} \tag{26}$$

□

5.3. The Local Property of Depth Information Irrelevant to Camera Movements

Theorem 1. *Assume that the image planes before and after the camera movement are I and $\bar{I}$, respectively, and the corresponding depth functions are d and bard, respectively. $Z = (u, v)$ is a keypoint on the image plane I, and $Z' = (u', v')$ is a neighbor pixel of Z in I. The corresponding pixels of Z and Z'on the image plane $\bar{I}$ are $\bar{Z} = (\bar{u}, \bar{v})$ and $\bar{Z}' = (\bar{u}', \bar{v}')$, respectively. Then, the ratio of the depth difference between $\bar{Z}'$ and $\bar{Z}$ to the depth difference between Z' and Z approximates a constant which is irrelevant to Z'. That is:*

$$\frac{\bar{d}(\bar{Z}') - \bar{d}(\bar{Z})}{d(Z') - d(Z)} \approx \alpha \tag{27}$$

Proof of Theorem 1. As mentioned before, the camera movement can be decomposed as the movement of the camera center along the optic axis and the rotation of the optic axis around the camera center.

(1) In the first case, Let $I = I_1, \bar{I} = I_2$ according to Lemma 1. We have

$$\begin{aligned} \bar{d}(\bar{Z}') - \bar{d}(\bar{Z}) &= d_2(\bar{Z}') - d_2(\bar{Z}) \\ &= \left[d_1(Z') + T\right] - \left[d_1(Z) + T\right] \\ &= d_1(Z') - d_1(Z) = d(Z') - d(Z) \end{aligned} \tag{28}$$

(2) In the second case, Let $I = I_1, \bar{I} = I_3$, and according to Lemma 2,

$$\begin{aligned} &\bar{d}(\bar{Z}') - \bar{d}(\bar{Z}) = d_3(\bar{Z}') - d_3(\bar{Z}) \\ &\quad = d_1(Z') \cdot \frac{\cos\left(\arctan\frac{u'}{f_x} - \varphi^x\right) \cdot \cos\varphi^x}{\cos\left(\arctan\frac{u'}{f_x}\right) \cdot \cos\varphi} - d_1(Z) \cdot \frac{\cos\left(\arctan\frac{u}{f_x} - \varphi^x\right) \cdot \cos\varphi^x}{\cos\left(\arctan\frac{u}{f_x}\right) \cdot \cos\varphi} \end{aligned} \tag{29}$$

Define

$$\beta_1 \triangleq \frac{\cos\left(\arctan\frac{u}{f_x} - \varphi^x\right)\cdot\cos\varphi^x}{\cos\left(\arctan\frac{u}{f_x}\right)\cdot\cos\varphi} \beta_2 \triangleq \frac{\cos\left(\arctan\frac{u'}{f_x} - \varphi^x\right)\cdot\cos\varphi^x}{\cos\left(\arctan\frac{u'}{f_x}\right)\cdot\cos\varphi} \tag{30}$$

then

$$\frac{\beta_2}{\beta_1} = \frac{\frac{\cos\left(\arctan\frac{u'}{f_x} - \varphi^x\right)\cdot\cos\varphi^x}{\cos\left(\arctan\frac{u'}{f_x}\right)\cdot\cos\varphi}}{\frac{\cos\left(\arctan\frac{u}{f_x} - \varphi^x\right)\cdot\cos\varphi^x}{\cos\left(\arctan\frac{u}{f_x}\right)\cdot\cos\varphi}} = \frac{\cos\left(\arctan\frac{u'}{f_x} - \varphi^x\right)}{\cos\left(\arctan\frac{u'}{f_x}\right)}\cdot\frac{\cos\left(\arctan\frac{u}{f_x}\right)}{\cos\left(\arctan\frac{u}{f_x} - \varphi^x\right)} \tag{31}$$

By using the Taylor expansion, it can be seen that:

$$\begin{aligned}\frac{\beta_2}{\beta_1} &= \frac{2-\left(\frac{u'}{f_x}-\varphi^x\right)^2}{2-\left(\frac{u'}{f_x}\right)^2}\cdot\frac{2-\left(\frac{u}{f_x}\right)^2}{2-\left(\frac{u}{f_x}-\varphi^x\right)^2} = \frac{2-\left(\frac{u'}{f_x}-\varphi^x\right)^2}{2-\left(\frac{u}{f_x}-\varphi^x\right)^2}\cdot\frac{2-\left(\frac{u}{f_x}\right)^2}{2-\left(\frac{u'}{f_x}\right)^2} \\ &= \frac{2-\left(\frac{u}{f_x}-\varphi^x+\frac{u'-u}{f_x}\right)^2}{2-\left(\frac{u}{f_x}-\varphi^x\right)^2}\cdot\frac{2-\left(\frac{u}{f_x}\right)^2}{2-\left(\frac{u}{f_x}+\frac{u'-u}{f_x}\right)^2}\end{aligned} \tag{32}$$

Because $Z' = (u', v')$ is a neighbor pixel of $Z = (u, v)$, the difference $u' - u$ is a small value compared to the optical focal length, that is, $\frac{u'-u}{f_x} \ll 1$. So,

$$\frac{2-\left(\frac{u}{f_x}-\varphi^x+\frac{u'-u}{f_x}\right)^2}{2-\left(\frac{u}{f_x}-\varphi^x\right)^2} \to 1 \ and \ \frac{2-\left(\frac{u}{f_x}\right)^2}{2-\left(\frac{u}{f_x}+\frac{u'-u}{f_x}\right)^2} \to 1 \tag{33}$$

thus,

$$\frac{\beta_2}{\beta_1} \to 1 \tag{34}$$

Then, we have

$$\frac{d_3(\bar{Z}') - d_3(\bar{Z})}{d_1(Z') - d_1(Z)} \approx \frac{\cos\left(\arctan\frac{u}{f_x} - \varphi^x\right)\cdot\cos\varphi^x}{\cos\left(\arctan\frac{u}{f_x}\right)\cdot\cos\varphi} \tag{35}$$

It can be seen that the right-hand-side term is relevant to the rotation angle φ and the position of the keypoint Z but irrelevant to the position of the neighbor pixel Z'. □

6. Depth-Based Supplemental Vector of the SIFT Feature Descriptor

The image pair I and $\bar{I}$ that needs to be matched can be regarded as the image plane before and after the camera movement. From Theorem 1, the ratio of the depth difference of the keypoint and its neighbor pixels before and after the camera movement is almost a constant. By normalization with the constant, the depth difference is an invariant and can be used as a local feature of the key points.

Select the $n \times n$ neighborhood of the keypoint Z on the image I, and the neighboring pixels are listed as $Z_i^{'}$, $i = 1, \ldots, n^2 - 1$. Define

$$\Delta d_i = \left|d\left(Z_i^{'}\right) - d(Z)\right|,\ i = 1, 2, \ldots, n^2 - 1 \tag{36}$$

$$d^* = \min_i \{\Delta d_i | \Delta d_i \neq 0\} \tag{37}$$

$$\Delta D = \begin{cases} 0, & if\ d* = 0 \\ (\frac{\Delta d_1}{d^*}, \frac{\Delta d_2}{d^*}, \ldots, \frac{\Delta d_{n^2-1}}{d^*}), & else \end{cases} \quad (38)$$

The vector ΔD is the depth-based local feature information, which can be used as a supplement to the SIFT feature descriptor. The properties of the theorem can be applied not only to SIFT, but also to other methods, such as SURF.

7. Experiments

7.1. Experimental Data

The experimental images are taken from the NYU dataset [47] which has a total of 1449 indoor scene RGB images and corresponding depth images with size of 640 × 480. In this paper, 20 images were selected to form 10 pairs for experimental comparison. We show the results of the second pair (numbered 25 and 26 in the database) and the sixth pair (numbered 191 and 192 in the database).

The experiments were performed under a Windows 10 operating system, using the Python-Open CV computer vision library for image processing.

Figures 9 and 10 show the original RGB image pairs and the corresponding depth images. Figure 9a and 9b correspond to the RGB images numbered 25 and 26 in the database, respectively. Figure 9c and 9d are depth images of Figure 9a and 9b, respectively. Figure 10a and 10b correspond to the RGB images of numbered 191 and 192 in the database, respectively. Figure 10c and 10d are depth images of Figure 10a and 10b, respectively.

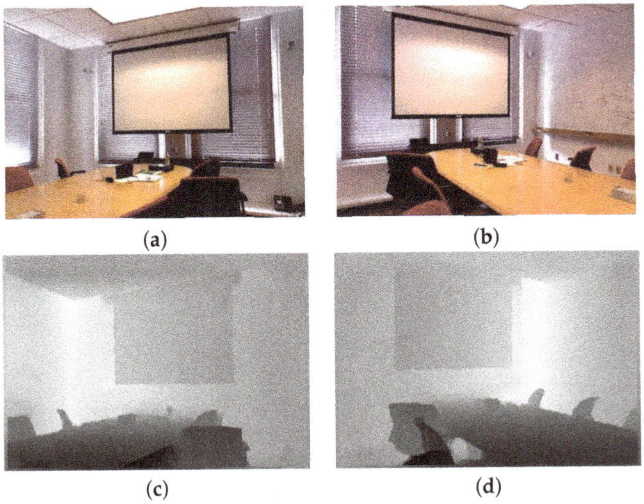

Figure 9. The RGB image of the second pair and their depth images: (**a**) original RGB image numbered 25; (**b**) original RGB image numbered 26; (**c**) depth image of (**a**); (**d**) depth image of (**b**).

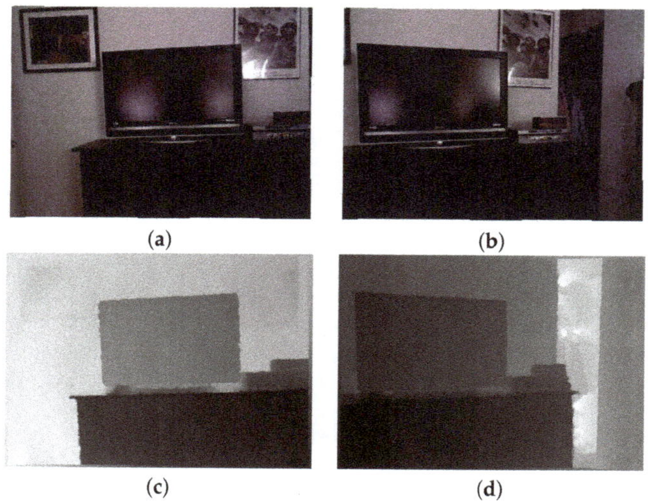

Figure 10. The RGB image of the sixth pair and their depth images: (**a**) original RGB image numbered 191; (**b**) original RGB image numbered 192; (**c**) depth image of (**a**); (**d**) depth image of (**b**).

7.2. Pixel Pre-Classification

Figure 11 presents the depth information histograms of Figures 9c,d and 10c,d. It is observed that each histogram can be roughly considered to have three peaks, that is, the objects in the image roughly are classified as falling in the foreground, middle ground, or background. Accordingly, the pixels of the RGB image are divided into three categories corresponding to the depth information. This paper simply uses 30% and 70% of the cumulative frequency as the threshold.

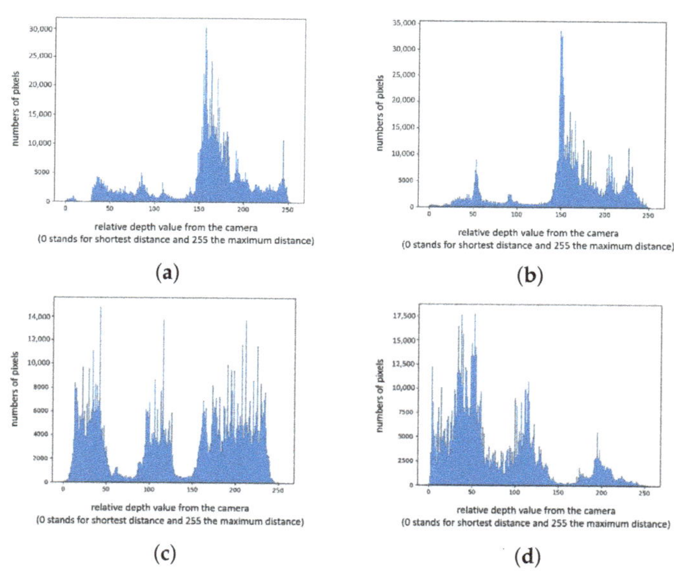

Figure 11. Image pair depth information histogram: (**a**) depth information histogram of image No. 25; (**b**) depth information histogram of image No. 26; (**c**) depth information histogram of image No. 191; (**d**) depth information histogram of image No. 192.

7.3. Experimental Results

We use the SIFT algorithm to extract image keypoint features and use the Kd-tree algorithm [24] to perform keypoint feature matching. The results are shown in Figure 12. Figure 12a is the matching result of the SIFT algorithm in Figure 9a,b, and Figure 12b is the matching result of the SIFT algorithm in Figure 10a,b.

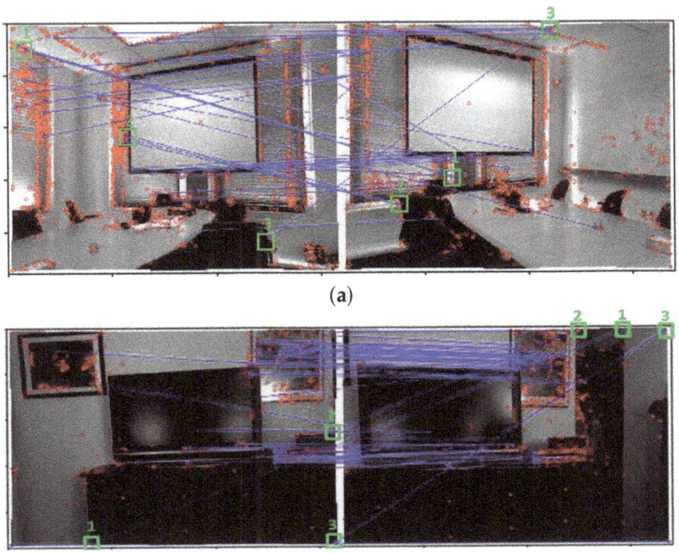

(**a**)

(**b**)

Figure 12. SIFT algorithm matching results: (**a**) Figure 9a,b matching result of SIFT algorithm; (**b**) Figure 10a,b matching result of SIFT algorithm.

It can be seen from Figure 12 that there are many false matching pairs in the results of SIFT matching. For example, in Figure 12a, the keypoint pair 1 marked by the green box is located at the coordinates (46, 54) in the left image which is on the ceiling; however, in the right image, the corresponding coordinates (212, 297) are located at the bottom of the screen. Obviously, this is not the same object point. Similarly, the other green box matching point pairs are false matches.

In this paper, the SIFT algorithm is used to obtain the keypoints first, the SIFT descriptor combining with a depth-information supplemental vector is constructed as the new descriptor for matching, and the Kd-tree algorithm is used to match the similar keypoints in the image pair.

Figure 13 presents the improved matching results for the second and sixth image pairs. It can be seen that the keypoint pairs that were originally incorrectly matched have been removed.

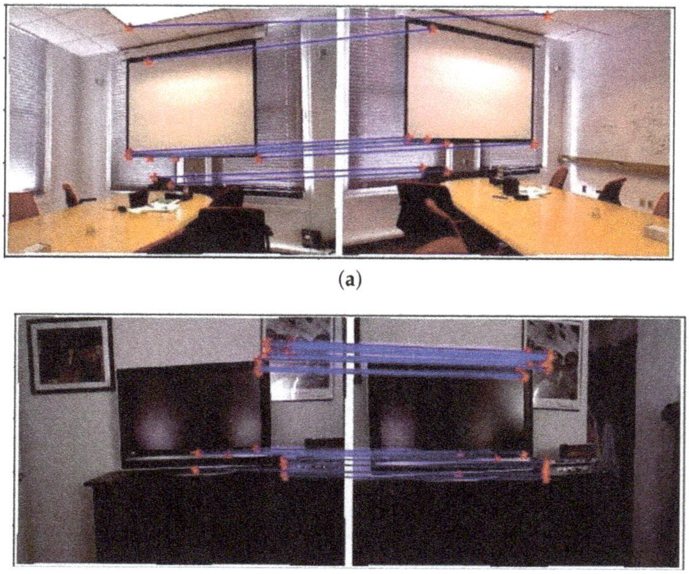

Figure 13. Matching results of the improved algorithm: (**a**) Figure 8a,b matching results of the improved algorithm; (**b**) Figure 9a,b matching results of the improved algorithm.

Table 1 gives the comparison of the matching results of all ten pairs of images.

Table 1. Comparison of image pair matching before and after improvement.

Image Pair	Total Number of Keypoints	Number of SIFT Matching Pairs	Accuracy Rate	Number of Matching Pairs after Improvement	Accuracy Rate
1(18,19)	629	28	67.86%	14	78.57%
2(25,26)	2756	41	36.59%	9	88.89%
3(69,70)	2038	28	96.43%	13	100.00%
4(124,125)	2800	514	77.43%	375	90.13%
5(131,132)	2424	378	89.15%	280	96.43%
6(191,192)	1101	66	80.30%	22	100.00%
7(411,412)	1282	46	89.13%	19	94.74%
8(510,511)	1065	39	84.62%	16	93.75%
9(568,569)	2238	84	83.33%	43	93.02%
10(975,976)	1681	52	82.69%	24	91.67%

From Table 1, we can see that the method proposed in this paper can effectively filter out the wrong matching point pairs, and the accuracy rate of image pair matching has been significantly improved. The accuracy rate has been increased by an average of 13.967 percentage points. Since the relative depth information is used, there is no specific requirement about the accuracy of the depth information. In practice, an estimated depth without a scale factor or an approximate dense depth map can be used.

8. Conclusions

In this paper, we analyzed the local property of the depth information of image pairs taken before and after camera movements and proved that the ratio of the depth difference of a keypoint and its neighbor pixels before and after the camera movement approximates a constant. Based on this property, a depth-based feature vector was constructed as a supplement of SIFT feature descriptor. All the keypoints were classified as foreground, middle ground, and background and the feature matching was performed within the same class.

Experiments were performed to validate the effectiveness of the proposed method. The experimental images were taken from the NYU dataset. total of 20 images were selected to form 10 pairs for experimental comparison. The experimental results show that the accuracy rate of image pair matching increased by an average of 13.967 percentage points, which means the method proposed in this paper can effectively filter out mismatch point pairs.

The local property described in the theorem proposed in this paper is a common feature which may be used for other situations, not only for a supplement of the SIFT feature descriptor and can be widely used in all aspects of image matching. The relative depth differences used for the local property do not require accurate depth information. In addition, we intend to further optimize the related model of image stitching according to the proposed method.

Author Contributions: Conceptualization and methodology, M.W., F.C. and H.C.; software and validation, E.Y., H.C. and R.L.; writing—original draft preparation, E.Y.; writing—review and editing, M.W. and F.C.; project administration, F.C. All authors have read and agreed to the published version of the manuscript.

Funding: This research was funded in part by the National Natural Science Foundation of China (62172098, 61771141) and the Natural Science Foundation of Fujian Province (2021J01620, 2020J01497).

Data Availability Statement: Not applicable.

Conflicts of Interest: The authors declare no conflict of interest.

References

1. Ma, J.; Jiang, X.; Fan, A.; Jiang, J.; Yan, J. Image Matching from Handcrafted to Deep Features: A Survey. *Int. J. Comput. Vis.* **2020**, *1*, 23–79. [CrossRef]
2. Zitová, B.; Flusser, J. Image Registration Methods: A Survey. *Image Vis. Comput.* **2003**, *21*, 977–1000. [CrossRef]
3. Lowe, D.G. Distinctive Image Features from Scale-Invariant Keypoints. *Int. J. Comput. Vis.* **2004**, *60*, 91–110. [CrossRef]
4. Bay, H.; Tuytelaars, T.; Gool, L.V. SURF: Speeded up robust features. In Proceedings of the 9th European Conference on Computer Vision, Graz, Austria, 7–13 May 2006.
5. Alahi, A.; Ortiz, R.; Vandergheynst, P. FREAK: Fast Retina Keypoint. In Proceedings of the 2012 IEEE Conference on Computer Vision and Pattern Recognition, Providence, RI, USA, 16–21 June 2012.
6. Bellavia, F. SIFT Matching by Context Exposed. *IEEE Trans. Pattern Anal. Mach. Intell.* **2022**, *45*, 2445–2457. [CrossRef] [PubMed]
7. Liu, W.; Wang, Y.; Chen, J.; Guo, J.; Lu, Y. A completely affine invariant image-matching method based on perspective projection. *Mach. Vis. Appl.* **2012**, *23*, 231–242. [CrossRef]
8. Mishkin, D.; Matas, J.; Perdoch, M. MODS: Fast and robust method for two-view matching. *Comput. Vis. Image Underst.* **2015**, *141*, 81–93. [CrossRef]
9. Morel, J.M.; Yu, G. Asift: A new framework for fully affine invariant image comparison. *SIAM J. Imaging Sci.* **2009**, *2*, 438–469. [CrossRef]
10. Pang, Y.; Li, W.; Yuan, Y.; Pan, J. Fully affine invariant surf for image matching. *Neurocomputing* **2012**, *85*, 6–10. [CrossRef]
11. Toft, C.; Turmukhambetov, D.; Sattler, T.; Kahl, F.; Brostow, G.J. Single-Image Depth Prediction Makes Feature Matching Easier. In Proceedings of the Computer Vision—ECCV 2020: 16th European Conference, Glasgow, UK, 23–28 August 2020.
12. Chen, Y.; Wang, G.; Wu, L. Research on feature point matching algorithm improvement using depth prediction. *J. Eng.* **2019**, *2019*, 8905–8909. [CrossRef]
13. Schuon, S.; Theobalt, C.; Davis, J.; Thrun, S. LidarBoost: Depth superresolution for ToF 3D shape scanning. In Proceedings of the 2009 IEEE Conference on Computer Vision and Pattern Recognition, Miami, FL, USA, 20–25 June 2009; pp. 343–350.
14. Liu, C.; Kim, K.; Gu, J.; Furukawa, Y.; Kautz, J. Planercnn: 3d plane detection and reconstruction from a single image. In Proceedings of the IEEE Conference on Computer Vision and Pattern Recognition, Long Beach, CA, USA, 15–20 June 2019; pp. 4450–4459.
15. Liu, C.; Yang, J.; Ceylan, D.; Yumer, E.; Furukawa, Y. Planenet: Piece-wise planar reconstruction from a single rgb image. In Proceedings of the IEEE Conference on Computer Vision and Pattern Recognition, Salt Lake City, UT, USA, 18–22 June 2018; pp. 2579–2588.
16. Li, Z.; Snavely, N. Megadepth: Learning single-view depth prediction from internet photos. In Proceedings of the IEEE Conference on Computer Vision and Pattern Recognition, Salt Lake City, UT, USA, 18–22 June 2018.

17. Li, J.; He, L.; Ren, G. A feature point matching method for binocular stereo vision images based on deep learning. *Autom. Instrum.* **2022**, *2*, 57–60.
18. Zhang, Z.; Huo, W.; Lian, M.; Yang, L. Research on fast binocular stereo vision ranging based on Yolov5. *J. Qingdao Univ. Eng. Technol. Ed.* **2021**, *36*, 20–27.
19. Chen, W.; Fu, Z.; Yang, D.; Deng, J. Single-image depth perception in the wild. In Proceedings of the 30th International Conference on Neural Information Processing Systems, Barcelona, Spain, 5–10 December 2016; pp. 730–738.
20. Ummenhofer, B.; Zhou, H.; Uhrig, J.; Mayer, N.; Ilg, E.; Dosovitskiy, A.; Brox, T. DeMoN: Depth and Motion Network for Learning Monocular Stereo. In Proceedings of the IEEE Conference on Computer Vision and Pattern Recognition (CVPR), Honolulu, HI, USA, 21–26 July 2017.
21. Facil, J.M.; Ummenhofer, B.; Zhou, H.; Montesano, L.; Brox, T.; Civera, J. CAM-Convs: Camera-Aware Multi-Scale Convolutions for Single-View Depth. In Proceedings of the 2019 IEEE/CVF Conference on Computer Vision and Pattern Recognition (CVPR), Long Beach, CA, USA, 15–20 June 2019.
22. Zhou, H.; Ummenhofer, B.; Brox, T. DeepTAM: Deep Tracking and Mapping with Convolutional Neural Networks. *Int. J. Comput. Vis.* **2020**, *128*, 756–769. [CrossRef]
23. Silberman, N.; Hoiem, D.; Kohli, P.; Fergus, R. *Indoor Segmentation and Support Inference from RGBD Images;* Springer: Berlin/Heidelberg, Germany, 2012.
24. Abdel-Hakim, A.E.; Farag, A. CSIFT: A SIFT Descriptor with Color Invariant Characteristics. In Proceedings of the IEEE Computer Society Conference on Computer Vision Pattern Recognition, New York, NY, USA, 17–22 June 2006.
25. Rublee, E.; Rabaud, V.; Konolige, K.; Bradski, G. ORB: An efficient alternative to SIFT or SURF. In Proceedings of the 2011 International Conference on Computer Vision, Barcelona, Spain, 6–13 November 2011; pp. 2564–2571.
26. Rosten, E.; Porter, R.; Drummond, T. Faster and better: A machine learning approach to corner detection. *IEEE Trans. Pattern Anal. Mach. Intell.* **2010**, *32*, 105–119. [CrossRef]
27. Calonder, M.; Lepetit, V.; Strecha, C.; Fua, P. Brief: Binary robust independent elementary features. In Proceedings of the European Conference on Computer Vision, Crete, Greece, 5–11 September 2010; pp. 778–792.
28. Balammal Geetha, S.; Muthukkumar, R.; Seenivasagam, V. Enhancing Scalability of Image Retrieval Using Visual Fusion of Feature Descriptors. *Intell. Autom. Soft Comput.* **2022**, *31*, 1737–1752. [CrossRef]
29. Csurka, G.; Dance, R.; Fan, L.; Willamowski, J.; Bray, C. Visual categorization with bags of keypoints. In Workshop on Statistical Learning in Computer Vision, Prague, Czech Republic, May 2004; pp. 1–22.
30. Tang, L.; Ma, S.; Ma, X.; You, H. Research on Image Matching of Improved SIFT Algorithm Based on Stability Factor and Feature Descriptor Simplification. *Appl. Sci.* **2022**, *12*, 8448. [CrossRef]
31. Feng, Q.; Tao, S.; Liu, C.; Qu, H.; Xu, W. IFRAD: A Fast Feature Descriptor for Remote Sensing Images. *Remote Sens.* **2021**, *13*, 3774. [CrossRef]
32. Fischler, M.A.; Bolles, R.C. Random Sample Consensus: A Paradigm for Model Fitting with Applications to Image Analysis and Automated Cartography. *Commun. ACM* **1981**, *24*, 381–395. [CrossRef]
33. Chung, K.L.; Tseng, Y.C.; Chen, H.Y. A Novel and Effective Cooperative RANSAC Image Matching Method Using Geometry Histogram-Based Constructed Reduced Correspondence Set. *Remote Sens.* **2022**, *14*, 3256. [CrossRef]
34. Tombari, F.; Salti, S.; Stefano, L.D. Unique signatures of histograms for local surface description. In Proceedings of the European Conference on Computer Vision, Crete, Greece, 5–11 September 2010; pp. 356–369.
35. Salti, S.; Tombari, F.; Di Stefano, L. SHOT: Unique Signatures of Histograms for Surface and Texture Description. *Comput. Vis. Image Underst.* **2014**, *125*, 251–264. [CrossRef]
36. Prakhya, S.M.; Liu, B.; Lin, W. B-SHOT: A binary feature descriptor for fast and efficient keypoint matching on 3D point clouds. In Proceedings of the 2015 IEEE/RSJ International Conference on Intelligent Robots and Systems (IROS), Hamburg, Germany, 28 September–2 October 2015; pp. 1929–1934.
37. Steder, B.; Rusu, R.B.; Konolige, K.; Burgard, W. NARF: 3D range image features for object recognition. In Proceedings of the International Conference on Intelligent Robots and Systems (IROS), Taipei, Taiwan, 18–22 October 2010.
38. Johnston, A.; Carneiro, G. Self-supervised monocular trained depth estimation using self-attention and discrete disparity volume. In Proceedings of the IEEE/CVF Conference on Computer Vision and Pattern Recognition (CVPR), Seattle, WA, USA, 13–19 June 2020.
39. Zhou, T.; Brown, M.; Snavely, N.; Lowe, D.G. Unsupervised learning of depth and ego-motion from video. In Proceedings of the IEEE/CVF Conference on Computer Vision and Pattern Recognition (CVPR), Honolulu, HI, USA, 21–26 July 2017; pp. 21–26.
40. Ranjan, A.; Jampani, V.; Balles, L.; Kim, K.; Sun, D.; Wulff, J.; Black, M.J. Competitive collaboration: Joint unsupervised learning of depth, camera motion, optical flow and motion segmentation. In Proceedings of the IEEE/CVF Conference on Computer Vision and Pattern Recognition (CVPR), Long Beach, CA, USA, 15–20 June 2019; pp. 15–20.
41. Mahjourian, R.; Wicke, M.; Angelova, A. Unsupervised Learning of Depth and Ego-Motion from Monocular Video Using 3D Geometric Constraints. In Proceedings of the 2018 IEEE/CVF Conference on Computer Vision and Pattern Recognition (CVPR), Salt Lake City, UT, USA, 18–22 June 2018; pp. 5667–5675.
42. Wang, J.; Liu, P.; Wen, F. Self-supervised learning for RGB-guided depth enhancement by exploiting the dependency between RGB and depth. *IEEE Trans. Image Process.* **2023**, *32*, 159–174. [CrossRef]
43. Marr, D.; Hildreth, E. Theory of Edge Detection. *Proc. R. Soc. Biol. Sci.* **1980**, *207*, 187–217.

44. Olkkonen, H.; Pesola, P. Gaussian Pyramid Wavelet Transform for Multiresolution Analysis of Images. *Graph. Model. Image Process.* **1996**, *58*, 394–398. [CrossRef]
45. Lindeberg, T. Feature Detection with Automatic Scale Selection. *Int. J. Comput. Vis.* **1998**, *30*, 79–116. [CrossRef]
46. Wang, M.; Lai, C.H. *A Concise Introduction to Image Processing using C++*; Chapman and Hall/CRC: Boca Raton, FL, USA, 2009.
47. Ram, P.; Sinha, K. Revisiting kd-tree for nearest neighbor search. In Proceedings of the 25th Acm Sigkdd International Conference on Knowledge Discovery & Data Mining, Anchorage, AK, USA, 4–8 August 2019; pp. 1378–1388.

MDPI AG
Grosspeteranlage 5
4052 Basel
Switzerland
Tel.: +41 61 683 77 34

Mathematics Editorial Office
E-mail: mathematics@mdpi.com
www.mdpi.com/journal/mathematics

www.ingramcontent.com/pod-product-compliance
Lightning Source LLC
Chambersburg PA
CBHW040749250926
51850CB00007BB/1446
9783725868742